C$_1$ Chemistry

C$_1$ Chemistry

Principles and Processes

Saeed Sahebdelfar
Maryam Takht Ravanchi
Ashok Kumar Nadda

CRC Press
Taylor & Francis Group
Boca Raton London New York

CRC Press is an imprint of the
Taylor & Francis Group, an **informa** business

First edition published 2022
by CRC Press
6000 Broken Sound Parkway NW, Suite 300, Boca Raton, FL 33487-2742

and by CRC Press
2 Park Square, Milton Park, Abingdon, Oxon, OX14 4RN

ISBN: 978-1-032-24562-1 (hbk)
ISBN: 978-1-032-24563-8 (pbk)
ISBN: 978-1-003-27928-0 (ebk)

DOI: 10.1201/9781003279280

Typeset in Times
by KnowledgeWorks Global Ltd.

This book is dedicated to the memory of our late teacher, Professor Morteza Sohrabi, Amirkabir University of Technology, who will not actually have the chance to read it but whose teachings were our incentives for this publication.

Contents

With them the Seed of Wisdom did I sow,
And with my own Hand wrought to make it grow;
And this was all the Harvest that I reap'd—
"I came like Water, and like Wind I go."

Ruhdiydt of Omar Khayyam*, XXXI
Fitzgerald translation
*Omar Khayyam (1048–1131), Persian mathematician,
philosopher and poet

Preface

The rapid rise of crude oil price with its depleting reservoirs has stimulated worldwide research for alternative and sustainable sources of raw materials for chemicals and fuels. Many raw materials such as methane, carbon dioxide and biomass have been, mostly independently, investigated. However, for a global solution, more integrated research is necessary. The aim of this work is unifying the approach for a better understanding of interrelationships of these activities and potential cooperations.

The idea of using single-carbon atom molecules as chemical building blocks is not a new one, and many of such compounds have been techno-economically studied as raw materials. Nevertheless, unifying such scientific and technical issues under the topic of C_1 chemistry is not as easy as it may appear. It needs a deep and comprehensive understanding of the chemical transformation from molecular to commercial plant scales in different fields such as chemistry, thermodynamics, chemical engineering and environmental engineering, among others.

C_1 *Chemistry: Principles and Processes* tries to address these issues for academic and industrial researches. It could also be useful to both undergraduate and graduate students in chemistry, chemical engineering and environmental science and technology. Different disciplines from feedstock to final products were covered including raw materials, chemistry, thermodynamics, catalysis, reaction engineering and processing.

In preparation of the book, the authors did their best to present the topics in a logical order such that each chapter could also stand independently for the interested reader.

In *Chapter 1*, an overview about C_1 chemistry is provided. C_1 chemistry provides alternative and sustainable non-petroleum-based routes for valuable chemicals from single-carbon compounds. C_1 feedstocks can be obtained from various sources and are of unlimited supply. By C_1 chemistry processes, organic wastes and biomass could be exploited in synthesis of valuable chemicals. Despite high potential in commercial application of these processes many of them are still in research and development stage especially direct conversion routes which suffer from low selectivities and operational problems (e.g., heat supply or removal). The main challenge is the high stability of the most abundant C_1 molecules. It is expected that C_1 chemistry become a major source of fuels and chemicals in near future.

In *Chapter 2*, important sources of naturally occurring C_1 compounds, and also other carbon sources that can be directly converted to C_1 compounds are presented and the processing and treatments required before their utilization as a feedstock are also reviewed. Because of the dwindling petroleum resources, C_1 compounds have gained attention as carbon source for sustainable chemical synthesis and fuel production.

In *Chapter 3*, C_1 interconversions are presented and examples of processes that have been used to convert methane and carbon dioxide (as the most abundant and less reactive C_1 compounds) to more versatile and active compounds are provided.

Currently, these conversions are the main routes for valorization of natural gas to chemicals and fuels and also chemical fixation of CO_2.

In *Chapter 4*, direct conversions of methane to higher hydrocarbons are provided. Due to its large resources, methane is an attractive feedstock in C_1 chemistry. Its other advantages as feedstock are source-independent composition and ease of purification from heteroatoms. It has a high hydrogen content which can be used in product treatment. Another incentive for methane conversion is its low density which impact its transportation cost. As most methane natural gas fields are located in remote areas, the direct conversion of methane to transportable liquid fuels and chemicals would be highly desirable. Oligocondensation of methane, high-temperature self-coupling of methane, two-step methane homologation, methane dehydroaromatization and oxidative coupling of methane are examples of its direct conversions that are investigated in detail in this chapter.

In *Chapter 5*, reactions of carbon monoxide and hydrogen of syngas for the production of bulk chemicals being relevant to C_1 chemistry are provided. Fischer–Tropsch synthesis and its modification, synthesis of higher alcohols and hydroformylation are examples of the processes reviewed in detail in this chapter.

In *Chapter 6*, hydrogenation and polymerization of carbon dioxide are discussed as potentially large-scale processes for using CO_2 as a chemical building block. For effective utilization of CO_2 relevant to reduced global emissions, only conversion into products with large or even unlimited demands such as polymers and fuels should be targeted.

In *Chapter 7*, methanol conversions to hydrocarbons are focused. These reactions are of importance for the supply of non-petroleum petrochemical feedstocks (such as olefins and aromatics) and fuels. The carbonylation of methanol to acetic acid is also discussed.

In *Chapter 8*, methane derivative routes are discussed in detail (specifically, hydrocarbons from methyl halides and hydrocarbons from sulfurated methanes). The use of halide and sulfur routes is far less developed than the methanol route despite inherent similarities in their catalysts and processes. This chapter deals with the former routes for the production of higher hydrocarbons from methane.

In *Chapter 9*, the challenges and research needs for the development of commercial C_1-based processes are reviewed. The role of C_1 chemistry in realizing a sustainable circular economy by which the wastes are reused and recycled using renewable energy is addressed.

The authors would like to thank many colleagues and researchers in NPC-RT who helped us and also CRC for giving the opportunity to publish this work.

Saeed Sahebdelfar, Maryam Takht Ravanchi, Ashok Kumar Nadda

Authors

Dr. Saeed Sahebdelfar earned his PhD in chemical engineering from Sharif University of Technology, Tehran, Iran. He is now the manager of Catalysis Research Group in Petrochemical Research and Technology Co. Dr. Sahebdelfar main research interests include catalytic conversions of carbon dioxide and natural gas to synfuels and chemicals, petroleum refining, biomass conversions and environmental engineering. He has published over 80 original articles (h index 27) and written 3 book chapters and 5 books.

Dr. Maryam Takht Ravanchi holds a PhD in Chemical Engineering. She is a researcher in catalyst research group in Petrochemical Research and Technology Company. Her research interests are in the fields of natural gas conversion, C_1 chemistry, heterogeneous catalyst synthesis, membrane separation technologies and environmental researches. Her recent research activities focus on carbon dioxide utilization, paraffin dehydrogenation, hydrogenation processes, and methanol synthesis. She has about twenty years of experience in petrochemical industry as a process engineer and researcher. She has published about 70 journal papers, 70 conference papers, 7 national patents and 14 books (h index 21 and i-10 index 28).

Dr. Ashok Kumar Nadda is working as an Assistant Professor in the Department of Biotechnology and Bioinformatics, Jaypee University of Information Technology, Waknaghat, Solan, Himachal Pradesh, India. He holds extensive Research and Teaching experience of more than 10 years in the field of microbial biotechnology, with research expertise focusing on various issues pertaining to nano-biocatalysis, microbial enzymes, biomass, bioenergy and climate change. He holds international work experiences in South Korea, India, Malaysia, and People's Republic of China. He worked as a post-doctoral fellow in the State Key Laboratory of Agricultural Microbiology, Huazhong Agricultural University, Wuhan China. He also worked as a Brain Pool researcher/Assistant Professor at Konkuk University, Seoul, South Korea. Dr. Ashok has published more than 140 scientific contributions in the form of research, review, books, book chapters and others at several platforms in various journals of international repute. The research output includes 91 research articles, 32 book chapters and 16 books.

1 C$_1$ Chemistry: An Overview

1.1 INTRODUCTION

Traditionally, the source of organic chemicals and automotive fuels has been petroleum. The related technologies have been developed and optimized in the past century and most of them are demonstrated and approached to the theoretical yields specially in refining. There are, however, concerns regarding the sustainability of these processes.

The first factor is the limited petroleum sources despite discovering new oil reservoirs and improved recovering technologies and economic availability of shale oil. The oil price is highly fluctuating. To attain a profit margin, integration of refining-petrochemical plants has been considered as a solution.

The available crude oils are also becoming increasingly heavier with more heteroatoms (sulfur, nitrogen, metals and oxygen). Their inclusion (or their fractions) in the refinery diet brings about deep conversion challenges (Ramírez-Corredores, 2000). The emission regulations are becoming more stringent for the emission of sulfur and nitrogen compounds. Furthermore, treating of low-quality oil necessitates more advanced and sophisticated technologies in refining.

Another problem is global warming as a result of increased use of fossil fuels in energy production in household and transportation sectors. International efforts (e.g., Kyoto Protocol and Paris Agreement) are underway to reduce greenhouse gas emissions (e.g., CO_2 and CH_4) worldwide. Paris Agreement (end of 2015) goal is limiting global temperature rise well below 2°C above the pre-industrial level. The carbon footprint and lifecycle analysis are, therefore, becoming additional criteria for evaluating the processes, and carbon taxes may impact the economic viability of many conventional processes in the future.

As a consequence, the use of alternative carbon sources for sustainable chemical industry has received much attention.

Biomass-based chemicals and fuels are an alternative. The ultimate source of carbon in biomass is the nature carbon cycle from CO_2 via photosynthesis. A main criterion here is that it should not compete with human food chain directly. Thus, non-edible lignocellulosic (e.g., second-generation biofuels) and wastes should be considered. This will avoid deforestation and use of fresh water for supplying biomass. The main problems with biomass conversion are limited supply, low energy content and difficulties in treatment of the primary products. The development of active catalysts and effective processes is an active research field.

Another approach could be using small carbon-containing molecules as building block for constructing larger molecules. Compared to conventional methods from petroleum which are cracking-based (less selective), this approach provides the opportunity of higher selectivities. The increasing share of natural gas (methane) as a feedstock has been already underway in refining and petrochemical industries.

DOI: 10.1201/9781003279280-1

These small molecules are potentially much more abundant than petroleum and can be produced from a wide variety of sources. Therefore, they offer higher potentials for being used as chemical feedstocks in the future.

1.2   DEFINITION

C_1 chemistry is the chemistry of single-carbon-bearing molecules and their conversion to more valuable intermediate feedstocks, chemicals and clean fuels. This definition could be extended to include certain other small molecules such as dimethyl ether and methyl formate despite having two carbon atoms per molecule (Lee et al., 1990). These compounds lack a C–C bond in their molecules and could be viewed as "condensed" C_1 molecules.

The main source of C_1 molecules is currently natural gas but they can be obtained also from a variety of sources including biomass, organic wastes and coal or even captured carbon dioxide (Roberts and Elbashir, 2003).

An inherent advantage of C_1 feedstocks is that they could be easily purified by conventional methods and that their conversion is source-independent. Methane contains a high ratio of hydrogen; however, carbon oxides (especially CO_2) need a source of (renewable) hydrogen which is a drawback for their use as chemical feedstock.

Successful and sustainable utilization of C_1 molecules as feedstock is a great task in green and sustainable chemistry (Anastas and Zimmerman, 2019).

1.3 C_1 CHEMISTRY DEVELOPMENTS AND DRIVERS

Research in C_1 chemistry is a very active and rapidly growing field as illustrated by number of publications for individual processes. Within an interval of about 20 years, the number of publications with subject of chemistry and catalysis of methane, CO and CO_2 shows a fourfold increase from about 3,000 in 1996 to about 12,000 in 2017 according to a Sci-Finder search (Cui et al., 2019).

The quest for national energy independence and the need for carbon-neutral renewable fuels free of S and N have resulted in increased attention to C_1-chemistry-based processes. The development of technologies for the capture and storage of anthropogenic CO_2 and the advent of shale gas renewed attention to C_1 conversions (Galadima and Muraza, 2016). Other non-conventional gas reserves such as vast methane hydrates increase the potential of these processes.

The catalytic conversion of C_1 molecules to useful products is still a challenging task. Many potentially attractive conversion processes including direct conversions of methane and CO_2 to chemicals are still far from scale-up and need further research and development efforts. The principal research areas in C_1 chemistry can be divided into new reactions in C_1 chemistry, development of novel and efficient catalysts and scale-up and process development (Fierro, 1993).

An important step in C_1 chemistry is the formation of carbon backbone of the target molecule. Thus, the mechanism of the formation of the first C–C bond in the process is often an important research topic and still is not well understood in many transformations (Olah et al., 2018). Nevertheless, development of advanced and *in-situ* characterization techniques (e.g., low-energy electron diffraction (LEED),

Auger electron spectroscopy (AES) and extended X-ray absorption fine structure (EXAFS) spectroscopy) and theoretical studies (e.g., density functional theory (DFT) calculations) helped to gain better understanding on reaction mechanism and reaction pathways.

Because the formation of a complex molecule from a simple molecule may involve several transformations, many of the catalysts examined for single-step conversion of C_1 molecules are bifunctional or multifunctional catalysts, providing the specific active sites for different individual reactions. This approach promotes the tandem reactions when the initial steps are equilibrium limited. The main challenge here is how to integrate different functionalities within a single catalyst during catalyst preparation. Furthermore, the reactions should need similar operating conditions (e.g., temperature) to achieve acceptable synergies.

Thermodynamic and kinetic coupling of the tandem reactions and achieving good proximity between different active sites are key factors for the design and preparation of multifunctional catalysts (Bao et al., 2019). Physical mixing of the catalysts containing the desired active sites typically results in poor catalyst performances because close proximity between the phases cannot be achieved.

1.4 FEEDSTOCKS

The main sources of C_1 molecules are natural gas, coal and biomass and potentially the captured carbon dioxide. In fact, they can be obtained virtually from any carbon-bearing compound.

Table 1.1 shows C_1 molecules relevant to large-scale C_1-based processes and their gas enthalpy and Gibbs free energy of formations at two temperatures. The first three molecules are the main raw materials and the followings are typically intermediate molecules in synthesis paths from the former. The table implies that the former compounds are very stable which impose an energy barrier in their conversion. The high bound energies of elements attached to carbon (439 kJ/mol for C–H bond in

---

TABLE 1.1
Enthalpy and Gibbs Free Energy of Formation of Gas-Phase Important C_1 Compounds (Yaws, 1999)

Compound	Chemical Formula	ΔH^0_f (kJ/mol)		ΔG^0_f (kJ/mol)	
		25°C	227°C	25°C	227°C
Carbon dioxide	CO_2	−394	−394	−394	−395
Methane	CH_4	−74.9	−80.8	−50.8	−32.6
Carbon monoxide	CO	−111	−110	−137	−156
Methanol	CH_3OH	−201	−208	−163	−134
Methyl bromide	CH_3Br	−37.7	−57.9	−28.2	−12.3
Carbon disulfide	CS_2	117	107	66.9	35.1
Phosgene	$COCl_2$	−219	−219	−206	−197

CH_4, 1076 kJ/mol for C-O bond in CO and 532 kJ/mol in CO_2; Mesters, 2016) make catalytic activation of the bonds as a challenge in catalyst development. Therefore, many of reaction routes rely on converting the C_1 feedstocks into more reactive C_1 molecules (e.g., methanol and methyl halides, Table 1.1) (Lorkovic et al., 2004).

1.5   OVERVIEW OF CONVERSION TECHNOLOGIES

Figure 1.1 shows the inter-relation of main C_1 feedstocks and the main potentially obtainable commodity products. It illustrates that carbon monoxide (syngas) is currently the platform chemical in C_1 chemistry, but its position could be shared by methanol and even by CO_2 in the future. Among the reaction routes, syngas production/conversion has been commercialized and in operation for several decades (Keim, 1986).

Methane is converted into synthesis gas or syngas (a mixture of mostly CO and H_2 of various proportions) using an oxidant such as H_2O, CO_2 and O_2. Accordingly, the process is called steam reforming, dry (or carbon dioxide) reforming and partial oxidation (PO), respectively. The processes differ in energy demand and syngas composition (Jafarbegloo et al., 2015). The trend is toward using combined reforming to adjust the CO/H_2 ratio for the specified application and reducing energy consumption. This process can also be used to utilize biomass and/or its derivatives and also coal as C_1 feedstock sources (Sahebdelfar, 2017).

Syngas has been used in large scale for hydrogen production, methanol synthesis and hydrocarbons via Fischer–Tropsch (FT) synthesis. It has been the main route for valorization of methane to fuels and chemicals.

Despite commercial demonstration of syngas-based methane conversions, direct conversion routes are subject of extensive academic research to eliminate the highly energy intensive syngas production step. Both non-oxidative and oxidative conversions have been under consideration. Non-oxidative routes such as methane aromatization are high-temperature processes and are highly endothermic and equilibrium limited (Moghimpour Bijani et al., 2014).

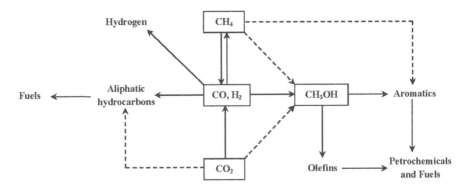

FIGURE 1.1   Main reaction paths in C_1 chemistry (solid lines: commercialized paths, dash lines: research stage); C_1 molecules are shown in boxes.

The use of an oxidant (i.e., oxidative conversion) removes the thermodynamic limitation. Examples include partial oxidation of methane to methanol and formaldehyde and oxidative coupling of methane (OCM). The main problem here is overoxidation and thus low selectivity of the reaction achieved under high conversions with the state-of-the-art catalysts and processes (Sahebdelfar et al., 2012).

The chemical conversion of carbon dioxide is much less developed compared to carbon monoxide due to high stability of the molecule and high oxidation state of carbon. Only dry reforming and recently methanation have been commercialized. In fact, many proposed catalysts and processes are tandem conversions of carbon dioxide to carbon monoxide via reverse water gas shift reaction and subsequent conversion of CO to the desired product. Thus, in CAMERE (carbon dioxide to methanol via reverse water gas shift reaction) process for methanol synthesis, CO_2 is partially reformed to CO and their mixture is subsequently converted in methanol synthesis reactor (Jadhav et al., 2014).

Methanol (and dimethyl ether, DME) is a key intermediate in C_1 chemistry. In methanol economy proposed by George Olah et al. (2006), methanol plays the key role as chemical feedstock and energy carrier and in chemical fixation of the captured carbon dioxide (Figure 1.1). Methanol can be transformed to DME (a potential diesel substitute), olefins (methanol to olefins, MTO) and aromatics which are the building blocks in petrochemical industry.

1.6 CONCLUSIONS

C_1 chemistry provides alternative and sustainable non-petroleum-based routes for valuable chemicals from single-carbon compounds. C_1 feedstocks can be obtained from various sources and are of unlimited supply. By these processes, organic wastes and biomass could be exploited in synthesis of valuable chemicals. Despite high potential in commercial application of these processes, many of them are still in research and development stage especially direct conversion routes which suffer from low selectivities and operational problems (e.g., heat supply or removal). The main challenge is the high stability of the most abundant C_1 molecules. It is expected that C_1 chemistry will become a major source of fuels and chemicals in near future.

REFERENCES

Anastas, P. T., Zimmerman, J. B., The periodic table of the elements of green and sustainable chemistry, Green Chem., 21 (2019) 6545–6566.

Bao, J., Yang, G., Yoneyama, Y., Tsubaki, N., Significant advances in C_1 catalysis: highly efficient catalysts and catalytic reactions, ACS Catal., 9 (2019) 3026–3053.

Cui, W. G., Zhang, G. Y., Hu, T. L., Bu, X. H., Metal-organic framework-based heterogeneous catalysts for the conversion of C_1 chemistry: CO, CO_2 and CH_4, Coord. Chem. Rev., 387 (2019) 79–120.

Fierro, J. L. G., Catalysis in C_1 chemistry: future and prospect, Catal. Let., 22 (1993) 67–91.

Galadima, A., Muraza, O., Revisiting the oxidative coupling of methane to ethylene in the golden period of shale gas: a review, J. Ind. Eng. Chem., 37 (2016) 1–13.

Jadhav, S. G., Vaidya, P. D., Bhanage, B. M., Joshi, J. B., Catalytic carbon dioxide hydrogenation to methanol: a review of recent studies, Chem. Eng. Res. Des., 92 (2014) 2557–2567.

Jafarbegloo, M., Tarlani, A., Wahid Mesbah, A., Sahebdelfar, S., Thermodynamic analysis of carbon dioxide reforming of methane and its practical relevance, Int. J. Hyd. Energy, 40 (2015) 2445–2451.

Keim, W., C₁ Chemistry: potential and developments, Pure Appl. Chem., 58 (1986) 825–832.

Lee, J. S., Kim, J. C., Kim, Y. G., Methyl formate as a new building block in C₁ chemistry, Appl. Catal., 57 (1990) 1–30.

Lorkovic, I., Noy, M., Weiss, M., Sherman, J., McFarland, E., Stuckya, G. D., Forda, P. C., C₁ Coupling via bromine activation and tandem catalytic condensation and neutralization over CaO/zeolite composites, Chem. Commun., (2004) 566–567.

Mesters, C., A selection of recent advances in C₁ chemistry, Ann. Rev. Chem. Biomol. Eng., 7 (2016) 223–238.

Moghimpour Bijani, P., Sohrabi, M., Sahebdelfar, S., Nonoxidative aromatization of CH₄ using C₃H₈ as a co-reactant: thermodynamic and experimental analysis, Ind. Eng. Chem. Res., 53 (2014) 572–581.

Olah, G. A., Goeppert, A., Surya Prakash, G. K., Beyond Oil and Gas: The Methanol Economy, Wiley-VCH, Weinheim (2006).

Olah, G. A., Molnar, A., Surya Prakash, G. K., Hydrocarbon Chemistry, 3rd Ed. Wiley, New Delhi (2018).

Ramírez-Corredores, M. M., Catalysis: New Concepts and New Materials, Proceeding of the 16th World Petroleum Congress, Calgary (2000).

Roberts, C. B., Elbashir, N. O., An overview to 'Advances in C₁ chemistry in the year 2002', Fuel Proc. Tech., 83 (2003) 1–9.

Sahebdelfar, S., Takht Ravanchi, M., Gharibi, M., Hamidzadeh, M., Rule of 100: an inherent limitation or performance measure in oxidative coupling of methane? J. Nat. Gas Chem., 21 (2012) 308–313.

Sahebdelfar, S., Steam reforming of propionic acid: thermodynamic analysis of a model compound for hydrogen production from bio-oil, Int. J. Hydrogen Energy, 42 (2017) 16386–16395.

Yaws, C. L., Chemical Properties Handbook, Physical, Thermodynamic, Environmental, Transport, Safety, and Health Related Properties for Organic and Inorganic Chemicals, McGraw-Hill, New York (1999).

2 C₁ Sources

2.1 INTRODUCTION

Many of C_1 compounds occur widely in nature. They are found in atmosphere, volcanic gases, biological metabolites and large reservoirs. However, only methane and carbon dioxide are sufficiently abundant for being used as a chemical feedstock. Even these compounds are found either in low concentrations or contain undesirable impurities. Others such as carbon monoxide can be only obtained from fossil fuels and renewable carbon sources such as biomass by gasification under partial oxidation conditions.

Thermochemical and biological degradation of organic compounds are two common techniques used for producing C_1 compounds. The product is usually contaminated with heteroatoms (N and S) originated from the raw material. Nevertheless, the removal of these elements from product is much easier than that from the raw material. Therefore, according to the final use, appropriate post-treatment might be necessary.

The conversion of coal and biomass to C_1 intermediates (especially carbon monoxide by gasification) are the initial steps in indirect coal to liquids (CTL) and biomass to liquids (BTL) processes as alternative routes for the production of liquid fuels.

Because of the dwindling petroleum resources, C_1 compounds have gained attention as carbon source for sustainable chemical synthesis and fuel production.

This chapter deals with important sources of naturally occurring C_1 compounds, and also other carbon sources that can be directly converted into C_1 compounds. The processing and treatments required before utilization as a feedstock are also reviewed.

2.2 NATURAL GAS

2.2.1 SOURCES

Methane (CH_4) is the lowest alkane. It occurs abundantly in nature being formed by both geological and biological processes. It is emitted, for example, from volcanos. However, the natural methane is essentially of biological origin. It is a product of aerobic decomposition of plant materials in wet environment (and thus the common name swamp or marsh gas).

Methane is found as a main constituent of many organic-based natural or synthetic gases such as natural, landfill and fermentation gases. As a hydrocarbon, methane can be used as energy source for household and industry or as a raw material for chemical industry.

Natural gas is an important fossil fuel with vast resources. Its main constituent is methane (typically 70–90 vol.%, depending on source) with variable proportions

DOI: 10.1201/9781003279280-2

of higher hydrocarbons (ethane, propane, natural gas liquids), N_2, CO_2, H_2S, He and water. Natural gas with high acid gases (such as CO_2 and H_2S) is termed "sour gas" while that with high heavy hydrocarbons is termed "wet gas". Thus, before being used as fuel or chemical feedstock, it should be purified by removing heteroatoms (N, S and O) (Faramawy et al., 2016).

The uncaptured natural gas in earth crust is estimated as 200×10^{12} cubic meters, which could be as high as 15×10^{15} when methane hydrates are also included (Latimer et al., 2017).

Natural gas may occur in association with coal (coalbed methane) or oil (associated gas) or as nonassociated gas. Many of natural gas sources are considered as unconventional (coalbed methane, tight sandstone gas and methane hydrates), meaning that their extraction is accompanied with economic and/or operational challenges, not found in conventional sources. Shale gas extraction is now considered as a conventional gas by development of combination of directional (horizontal) drilling and hydraulic fracturing technologies (Speight, 2013).

Methane hydrates (clathrates) are crystalline solids in which methane molecules entrapped in water cages formed by hydrogen bonding without any bond between the host and guest molecules. The empirical formula is $CH_4 \cdot 5.75H_2O$. Large reservoirs exist in Siberia, Antarctica and in continental shelves of oceans mostly hundreds of meters below the sea floor. They were formed below 30°C and above 5 kPa. Currently, they are difficult to access but are considered as the main hydrocarbon sources in the long term (Chong et al., 2016).

Many of natural gas reserves are located in off-shore or remote parts. Because of the low density of methane, its transportation to the market in long distances is costly. Thus, its chemical conversion to transportable liquids (gas to liquid (GTL) technologies) are received much attention.

Compared to other feedstocks, natural gas (methane) has the advantage of a high H/C ratio and low concentration of heteroatoms. These make methane as an attractive C_1 feedstock. Never-the-less only a small portion of methane (about 5% by 2002, according to Cornils et al., 2013) is used as chemical feedstock. The main challenges in effective catalytic activation of methane are its chemical stability and symmetrical structure of the molecule.

The commercial conversion of methane has also environmental implications as well. During its production, processing, transmission, distribution and storage, methane is emitted to the atmosphere. Methane is a potent greenhouse gas (GHG) with GWP (global warming potential) of 84 times greater than CO_2 in a 20-year time frame. In spite of its large effect in the atmosphere, methane has a low mean half-life of 9.1 years (that is a low value, in comparison to 100 years reported for CO_2). In pre-industrial area, global methane concentration was 722 ppb (parts per billion) which rose to 1,866 ppb by 2019 (https://en.wikipedia.org/wiki/Atmospheric_methane).

2.2.2 PURIFICATION AND PROCESSING

As motioned above, the crude natural gas is a complex and variable mixture of hydrocarbons (mostly methane) and nonhydrocarbons. Due to difficulties in processing, it should be purified before being used. The refining of

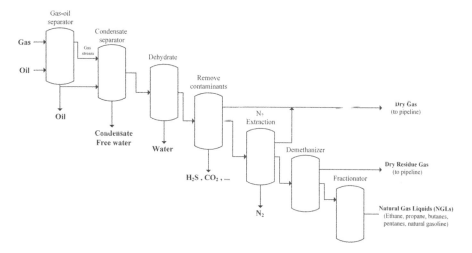

FIGURE 2.1 General steps in natural gas processing (Redrawn from EIA, 2006).

natural gas is a multistep process. The typical steps include removal of liquids (condensates and free water), acid gas (H_2S and CO_2) removal and nitrogen removal (Speight, 2007). Mercury elimination might be necessary. Subsequently, methane is separated from other hydrocarbons (natural gas liquids (NGLs)) in a demethanizer (Figure 2.1).

Sweetening of sour gas is a critical purification step as sulfur compounds (mainly H_2S) are corrosive and poisonous. Their presence in fuels results in emission of SO_x upon combustion which are important and regulated air pollutants. They are also the poisons of many downstream catalysts in methane conversion processes. A common method for desulfurization is chemical absorption process with aqueous alkanolamine solutions; such as monoethanolamine (MEA), diethanolamine (DEA), triethanolamine (TEA), diglycolamine (DGA), di-isopropanolamine (DIPA) and methyldiethanolamine (MDEA). MDEA can absorb H_2S selectively in the presence of CO_2. The use of selective polymer membranes is limited to CO_2 removal and still needs further developments (Faramawy et al., 2016).

2.3 CARBON DIOXIDE

2.3.1 Sources

Carbon dioxide occurs in trace amount in the atmosphere (about 0.04% at present) and also in dissolved form in sea and other natural waters. Its natural sources are volcano and hot spring gases. Dissolution of carbonates in acidic solutions also releases CO_2. It is also produced by biological processes such as fermentation of carbohydrates and aerobic respiration, together with photosynthesis, contributing nature's carbon cycle. Carbon dioxide is a variable part of natural gas.

Carbon dioxide is the end product of combustion of any carbon containing compound. Thus, the widespread use of fossil fuels increased the pre-industrial level (0.028%) to its present value. It has great environmental implications as CO_2 is a greenhouse gas.

It was reported that in 2017, the worldwide CO_2 emissions were approximately 37 Gt. There are different sources for CO_2 emissions; for instance, large sources such as industrial facilities (e.g., cement, steel and power plants) and chemical plants, small to medium sources such as industrial and commercial buildings and smaller sources such as transportation (Muradov, 2014).

In order to reduce GHG emissions, CCS (carbon capture and storage) and CCU (carbon capture and utilization) strategies can be used.

For CCS, there are some economic and technical barriers that must be overcome before its large-scale application. The first economic obstacle is its unprofitability that needs a large capital investment. From technical viewpoints, the leakage rate of CO_2 is uncertain. On the other hand, in some countries (such as Norway, Brazil, UK, India and Singapore) due to limitations in geological storage capacity and subsequent increasing cost of transportation and injection costs, CCS is not a viable option (Khoo et al., 2011).

Recently, CCU is being focused on as it utilizes the waste CO_2 emissions as feedstock to produce valuable chemicals thereby contributing to mitigation in climate change. CCU is advantageous to CCS, as in the former CO_2 is changed to a product that can be sold. Moreover, CO_2 is a renewable, non-toxic and cheap resource (Yu et al., 2008).

The similar aim of CCS and CCU is capturing CO_2 emissions in order to prevent its release to atmosphere, but the final destination of CO_2 is the difference between CCS and CCU; in CCS, the captured CO_2 is transported to a suitable location for its long-term storage, but in CCU, it is charged to a commercial product (Zapp et al., 2012).

2.3.2 CO_2 CAPTURE OPTIONS

The major industrial CO_2 sources are oil refineries, power plants, biogas sweetening, ammonia and ethylene oxide (EO) manufacture, cement production, and steel and iron industries. For instance, electricity production in fossil fuel power plants, caused about 40% of the worldwide CO_2 emissions. Different processes generate CO_2 and a one-size-fit-all technology cannot be applied for its capture. There are different capturing technologies with various maturity levels (Markewitz et al., 2012).

CO_2 capture technologies are categorized to pre-combustion, post-combustion and oxy-fuel combustion (Cuellar-Franca and Azapagic, 2015), which are discussed in detail in below sections.

2.3.2.1 Pre-Combustion Capture

In this strategy, the CO_2 produced via intermediate reactions (reforming/WGS (water-gas shift)) of the fuel is captured before combustion. Power plants and coal gasification are examples of processes that can produce CO_2 as a co-product (Singh et al., 2011).

In pre-combustion capture process, chemical solvent regeneration causes energy penalties which are lower for physical solvents as they need only a reduction in pressure. In case of concentrated CO_2 streams and processes with high operating pressures, physical solvents are more suitable.

2.3.2.2 Post-Combustion Capture

After converting a carbon source into carbon dioxide (e.g., fossil fuel combustion or wastewater sludge digestion), it must be separated from a flue gas, which is the subject of post combustion capture and has application in different industries such as ethylene oxide production, power plants, biogas sweetening, cement, and steel and iron industry (Aresta, 2003).

Different separation technologies can be applied for CO_2 separation, such as absorption, membrane and cryogenic separations, solid sorbent adsorption, and vacuum and pressure swing adsorption. The most common process is absorption by MEA (monoethanolamine), but due to the need to high heat consumption for MEA regeneration, it is not an economic process in all industries (Li et al., 2013; Sung and Suh, 2014).

2.3.2.3 Oxy-Fuel Combustion

Oxy-fuel combustion is applicable to processes involving combustion; such as power plants, iron and steel industries and cement production plants. In these processes, pure oxygen and fuel are burned and a flue gas containing high concentrations of CO_2 without N_2 and NO_x is obtained. In this process, there is no need to separate CO_2 from flue gas but pure oxygen is needed which is expensive and an energy-intensive air separation process is needed for it.

Chemical looping combustion (CLC) and chemical looping reforming (CLR) are two alternatives to oxy-fuel combustion process in which a metal oxide is used to selectively transfer oxygen from air reactor to fuel combustor. CLR has some advantages the most important of which are better sulfur tolerance, higher efficiency in fuel conversion, lower need to steam and the ability to handle dilute CO_2 streams. The main challenge of CLC process is ash handling and solid fuel applications (Najera et al., 2011).

2.3.3 CO₂ SEPARATION TECHNOLOGIES

In CO_2 separation, the main aim is selective removal of CO_2 from process streams or ambient air by which a concentrated CO_2 stream is obtained that can be used in other processes or compressed and transported to other places. Stationary points (process streams) and atmosphere (ambient air) are the major CO_2 sources. Absorption, adsorption and membrane separation are the main separation technologies (Yuan et al., 2016).

2.3.3.1 CO₂ Separation from Stationary Points

Coal power plants (mainly CO_2-N_2 mixtures), gasification plants (mainly CO_2-H_2 mixtures) and natural gas wells (mainly CO_2-CH_4 mixtures) are examples of stationary point sources of CO_2. As each of these cases has definitely different CO_2 partial pressures and temperatures, special capture materials are required for them. For instance, for CO_2-H_2 separation from a coal/biomass gasification plant, if a solvent being resistant to high temperatures is used, there is no need to a cooling unit for decreasing the temperature of syngas (Yuan et al., 2016).

Any material that is selected to be used for CO_2 capture must have selectivity, stability and affinity. Solid sorbents, liquid solvents and membranes can be used for CO_2 separation, which are discussed in details as below (D'Alessandro et al., 2010).

2.3.3.1.1   Solid Sorbents

Solid sorbents are novel materials with specific structure due to which CO_2 molecule can bind selectively from gas mixtures and remove or concentrate it under temperature, pressure or vacuum swing adsorption cycles. Activated carbon, zeolites, alkaline-, amine-enriched solids, metal oxides, lithium zirconates, MOFs (metal organic framework) and COFs (covalent organic framework) are examples of solid sorbents used for CO_2 capture. Molecular sieving, chemisorption and physisorption consist of the following main operations:

- CO_2 is adsorbed on the solid surface from a gaseous stream.
- By means of temperature, pressure or vacuum swing adsorption, the CO_2-rich sorbent is regenerated and CO_2 is liberated.
- For a solid sorbent, CO_2 uptake capacity, stability, selectivity and regenerability are the most important properties.

Interested reader is refereed to Li et al. (2012) and Samanta et al. (2012) which are comprehensive reviews on solid sorbents for CO_2 capture.

For large-scale applications, solid materials should have specific properties, such as favorable adsorption capacity and selectivity, low material cost, desirable kinetics for adsorption, environmentally green synthesis route, mild regeneration conditions, desirable long-term stability during different adsorption/desorption cycles and acceptable tolerance to impurities (Yuan et al., 2016).

2.3.3.1.1.1   Zeolites

Zeolites have well-defined porous structures. They are crystalline aluminosilicates in which the metal atoms are surrounded by four oxygen atoms in a tetrahedron geometry. For CO_2 capture by zeolites, the gaseous stream must be cooled down below 100°C and its impurities must be removed in advance (Yuan et al., 2016).

Zeolite-based CO_2 capture has fast kinetics and the physisorption is reversible which is governed by a thermodynamic equilibrium process. Irreversible chemisorption also occurs on zeolite surface. The kinetic and equilibrium properties are determined by zeolite topology and composition (Si/Al ratio and extra framework cations). Due to stronger electrostatic interactions with carbon dioxide, low Si/Al ratios and extra framework cations are favored. It was reported that zeolites with special structural features such as double eight-ring units and double-crank-shaft chains had high selectivity. For CO_2 capture, different zeolites such as X and Y Faujasites, 4A, 5A and 13X were evaluated (Remy et al., 2013; Spigarelli and Kawatra, 2013).

Large-scale applications of zeolites were postponed, due to following reasons:

- In the presence of moisture and impurities, the adsorption capacity and stability were poor.
- A temperature swing adsorption (that is energy- and time-intensive) is needed for complete regeneration by which adsorption capacity is maintained.

2.3.3.1.1.2 Amine-Grafted Solid Materials Recently, solids impregnated by active functional groups (such as aqueous amine solutions) are being focused as adsorbents. These amine-grafted solid materials have the advantage of aqueous solutions and solid materials for maximizing sorption properties. MCM-41 impregnated by polyethyleneimine is an example of these adsorbents. These functionalized solids need lower operating temperatures and less energy for regeneration. Moreover, these adsorbents have higher CO_2 adsorption capacity and selectivity (Xu et al., 2002).

Amine-grafted porous carbons, polymers and organic resins are examples of different organic supported amines. Among these materials, amine-polymer-silica composite hollow fiber adsorbents are being paid attention. Mesoporous silica/cellulose acetate hollow fibers injected with APS (3-aminopropyltrimethoxysilane), hollow fibers injected with PEI (poly(ethyleneimine)) and PAI (poly(amide-imide))/silica/PEI-glycerol sorbents were synthesized and evaluated for CO_2 capture (Fan et al., 2014).

Due to acceptable mass transfer, negligible influence of CO_2 partial pressure, adsorption kinetics and improved CO_2 capacity at low temperature, amine-grafted solids are attractive for post-combustion CO_2 capture (Yuan et al., 2016).

Before large-scale application of amine-functionalized adsorbents for CO_2 capture, there are some barriers that must be overcome in advance:

- For amine-grafted solid sorbents, long-term stability, scalability and tolerance to impurities must be evaluated.
- As the PSA (pressure swing adsorption) process is energy intensive and TSA (temperature swing adsorption) process has time-consuming regeneration steps, cost and desorption efficiency must be checked.
- Environmental impacts for large-scale production of amine-grafted adsorbents must be evaluated.
- The optimal configuration of amine-grafted adsorbents by which a specific CO_2 capture obtained, must be determined.

2.3.3.1.1.3 Metal Organic Frameworks MOFs, also known as porous coordination polymers, are porous materials composed of metal ions/clusters bridged by organic linkers forming 2D or 3D structures. They have attractive structural properties (such as high surface area, adjustable chemical functionality, high number of active sites, etc.) and are good candidates for selective CO_2 capture. It was reported (Furukawa et al., 2013) that for natural gas purification and coal gasification cases that have elevated pressures, MOF is the best adsorbent, while in the case of low CO_2 partial pressures, zeolites and activated carbon (AC) are suitable adsorbents.

During the years, researchers tried to ameliorate CO_2 capture properties of MOFs by different strategies such as pore-size control, introduction of polar functional groups, open metal sites and incorporation of alkali-metal cations. For exact pore size tuning and creation of desired functionalities, post-synthesis modification and direct assembly are two routes (Sumida et al., 2012).

In order to improve adsorption properties, coordinated solvent molecules are replaced with highly polar ligands. For functionalizing MOFs, aromatic amine derivatives, heterocycle derivatives and alkylamine bearing frameworks were used. In order to discover CO_2–CO_2 interaction, dynamics of the adsorbed carbon dioxide inside MOF must be studied in different ranges of temperature and pressure (Lin et al., 2013).

For large-scale utilization of MOFs for CO_2 capture, real operating conditions must be considered. For most MOFs, moisture is one limiting factor. CO_2 selectivity and adsorption capacity might be limited by impurities present in gaseous stream. In the case of large-scale application of MOF for CO_2 capture, specifically for post-combustion cases, the below issues must be taken into account (Nguyen et al., 2014):

- There are rare reported cases for the application of MOFs with extraordinary adsorption characteristics at mild operating conditions.
- As current MOF synthesis methods use expensive and toxic solvents, they cannot be scaled-up in an environmentally friendly route.
- There is an urgent need to study desorption properties of MOFs. Moreover, multiple adsorption/desorption cycles must be studied to investigate its effect on long-term stability of MOF.

2.3.3.1.1.4 Other Solid Sorbents   Carbon- and graphene-based adsorbents, covalent organic and hyper cross-linked polymers, porous aromatic frameworks, and conjugated microporous polymers are examples of other solid sorbents that can be used for CO_2 capture. Due to their high thermal stability, large adsorption capacity and good adsorption kinetics, activated carbons are good candidates for CO_2 capture in high-pressure pre-combustion scenarios (Slater and Cooper, 2015).

For post-combustion scenarios of CO_2 capture, activated carbon is not a suitable case as, at mild conditions, it has low CO_2 capacity/selectivity and the presence of NO_x and SO_x has negative impact on AC.

For this category of solid sorbents, two topics need further investigations:

- Their recyclability must be evaluated.
- The environmental impact of their large-scale production must be demonstrated.

2.3.3.1.2 Conventional Solvents

Physical absorption and chemical absorption are two main options as conventional methods of CO_2 capture based on solvents. The former relies on pressure or temperature differences and the latter depends on acid-base neutralization reaction (Yuan et al., 2016).

Generally, capture methods based on solvents consist of below steps:

- CO_2 containing gas stream is contacted with the solvent at a relatively low temperature and is absorbed in the absorber column.
- In the desorption column, CO_2-rich solvent is being regenerated by stripping with steam at high temperatures and the absorbed CO_2 is released.
- The CO_2 lean solvent (that was regenerated) is returned to the absorber column.
- Heat exchangers are connected to the absorber and stripper.

As the production costs of absorbents are very high (especially for ionic liquids (ILs)), scale-up of these processes is prohibited. The following points need to be considered in the future (Yuan et al., 2016):

- Chemisorption mechanisms of solvent-based CO_2 capture must be studied in detail in order to improve capture performance and reduce regeneration cost.
- By developing new chemical absorption technologies, long-term stability, impurity tolerance and high CO_2 purity rate could be enhanced significantly.
- Commercial production of absorbents with reasonable price must be achieved.
- Reaction mechanism and kinetics of functionalized ILs that react with CO_2 must be evaluated.

2.3.3.1.2.1 Physical Absorption For physical solvents, absorption capacity is proportional to CO_2 partial pressure and for pre-combustion CO_2 capture, this method is suitable. Some examples of commercial physical absorption technologies are Purisol, Rectisol, Fluor and Selexol. For the application of physical absorption for post-combustion CO_2 capture, an energy-intensive pretreatment step is needed in advance for the flue gas. ILs are good alternative for organic physical solvents, due to their low volatility, non-flammability, high thermal stability and low energy demand for regeneration. Basic IL has low working capacity and high production cost. Hence, functionalized ILs (in which functional groups are incorporated into basic ILs) were developed to improve the CO_2 affinity (Gurkan et al., 2010).

2.3.3.1.2.2 Chemical Absorption In the conventional carbon dioxide capture technologies based on chemical absorption, alkali absorbents, ammonia and aqueous amine/MEA are commonly used as solvents. For natural gas sweetening, MDEA (N-methyldiethanolamine) is used as a tertiary amine and for CO_2 removal from hydrogen and natural gas, MEA is mainly used (Spigarelli and Kawatra, 2013).

The Fluor Econamine FG Plus[SM] (EFG+), KM-CDR and Kerr-McGee/AGG Lummus Crest (KMALC) are three commercial amine-based absorption technologies for carbon dioxide capture. The main advantages of these processes are being mature technologies that are effective at low pressures. However, they are very sensitive to impurities that is a disadvantage (Yuan et al., 2016).

Generally, for primary and secondary amines, the loading capacity of CO_2 is $0.5-1$ mol_{CO2}/mol_{amine}. If the applied amine has less water with higher concentration, the capital cost is reduced (due to decrease in equipment size). Low energy equipment in the stripping stage, low chemical reactivity with flue gas impurities and fast reaction rate can be obtained by the application of mixed amine-based solvents (Yuan et al., 2016).

Piperazine (PZ) is an example of these solvents that can have good performance for post-combustion CO_2 capture. For chemical absorption of carbon dioxide from natural gas and flue gas streams, aqueous ammonia can be used that has high CO_2 loading capacity but due to its high volatility, equipment plugging may occur (Freeman et al., 2010).

2.3.3.1.3 Membranes

Membranes synthesized from (semi)permeable materials are another candidate for selective CO_2 separation. Membrane separation process is based on physical or chemical mechanisms; such as solution-diffusion transport, Knudsen diffusion and molecular sieving. Main advantages of membrane systems are as follows (Powell and Qiao, 2006):

- Simple operation with low energy consumption.
- Compactness without using any hazardous chemicals that have environmental disposal problems.
- Easy installation of membrane modules.

For polymeric membranes, CO_2 permeability and selectivity are important. Commercially, membranes are used for CO_2 removal from natural gas streams. Due to the low pressure of gas streams in post-combustion capture, membranes can have applications in this case as well.

For gas streams with low CO_2 partial pressure and H_2 capture, inorganic, composite, organic and mixed matrix membranes (MMM) can be used.

For large-scale application of membranes for CO_2 separation, inorganic membranes and MMMs are so far used but need more research evaluations. For CO_2 separation from high-pressure gas streams, due to their high CO_2/H_2 selectivities, polymeric membranes are good candidates. For post-combustion CO_2 capture scenarios, these membranes are not suitable because of low CO_2/N_2 selectivities and permeances (Yuan et al., 2016).

2.3.3.1.3.1 Polymeric Membranes

In pre-combustion and post-combustion CO_2 capture scenarios, polymeric membranes are good candidates due to their low synthesis cost. Based upon solution-diffusion mechanism, in polymeric membranes, gas permeation could be enhanced by partial pressure driving force. Traditional polymeric membranes have low selectivity for CO_2/N_2 mixtures (Merkel et al., 2012).

For post-combustion CO_2 capture, a favorable polymeric membrane is the one that have thermal/physical stability, with balanced permeability and selectivity at low pressure drop and being resistance to contaminants.

In natural gas sweetening, commercially, polymeric membranes are used for CO_2 capture. For the majority of these commercial membranes, the CO_2/N_2 selectivity is in the range of 20–30 with low pressure-normalized CO_2 flux. According to data provided in Yuan et al. (2016), a balance is reported for CO_2 permeability and CO_2/N_2 selectivity obtained by hydrophilic polymer-based Polaris membrane (MTR technology). For scale-up of polymeric membranes, there are some challenges, as follows:

- There is a need to appropriately balance CO_2 selectivity and permeability.
- Due to sensitivity to impurities, synthesis procedure must be well controlled and environmentally scalable.

2.3.3.1.3.2 Facilitated Transport Membranes Facilitated transport membranes are of two types, namely liquid-supported membranes and enhanced ion-exchange membranes. Due to acceptable mechanical properties, as well as high stability and selectivity for CO_2/H_2 and CO_2/N_2 separation, these membranes are good candidates for CO_2 capture from high- or low-pressure gas streams (Yuan et al., 2016).

In the facilitated transport membranes, a selective and reversible reaction occurs between CO_2 and the carrier molecules in the membrane, while non-reactive molecules transport by solution-diffusion mechanism. For instance, primary or secondary amines react with CO_2 to form carbonate ion, which diffuses through the membrane and on the permeate side CO_2 is released with a relatively low partial pressure. In these composite membranes, CO_2 capture performance is determined by the selected carrier. Favorable carrier is one that has high absorption capacity, reaction rate and stability. A mixture of pure PVA (poly(vinyl alcohol)) with PVAm (polyvinylamine) or PEI (polyethyleneimine) and hydrophilic mixtures of PVA and PAA (polyallylamine) and PAMAM (polyamidoamine) dendrimers are examples of carriers (Li et al., 2015a). As an example, at 120°C, PVAm composite membrane has a CO_2/H_2 selectivity of more than 200 and a CO_2 permeability of more than 1,000 GPU (Gas Permeance Unit). Before the large-scale application of facilitated transport membranes, some points must be taken into account, as follows:

- The selected membrane must have acceptable and durable CO_2 capture performance under harsh operating conditions.
- Manufacturing process must be economic and environmentally friendly.
- There should be a reasonable relationship between cross-linking degree and membrane thickness (as material properties) and separation performance.

2.3.3.1.3.3 Inorganic Membranes Inorganic membranes are used for CO_2 capture from flue gas streams. Typically, these membranes consist of porous or non-porous materials; such as metal oxides, ceramics, molecular sieves, zeolites, MOFs (ZIFs (zeolitic imidazole frameworks)), graphene oxide and silica. Molecular sieving, solution-diffusion or adsorption-diffusion can promote inorganic membrane-based CO_2 capture. Molecular sieving does not have enough CO_2 separation capability, as CO_2 and N_2 have similar molecular sizes with low permeances. For commercialization of membrane technologies, higher CO_2 separation capability is needed that can be achieved by improving membrane intergrowth, reduced concentration of non-selective pores and controlling membrane thickness (Joshi et al., 2014).

For pre-combustion and post-combustion CO_2 capture scenarios, by inorganic membranes, zeolites are evaluated candidates. For MOF membranes, due to high porosity and high surface area of MOFs, high CO_2 permeance could be obtained (Qiu et al., 2014).

Inorganic membranes have outstanding separation performance at lab-scale but before their scale-up, their long-term stability and reproducibility must be evaluated and their production route must be economic.

2.3.3.1.3.4 Mixed Matrix Membranes   MMM is a combination of bulk continuous polymer phase and inorganic discrete phase (with uniform distribution). In this type of membrane, CO_2 permeability and selectivity are synergetically enhanced and the mechanical and thermal stability is improved. Hence, the selection of inorganic and polymer phase must be optimized. A favorable polymer is one that has good adhesion between inorganic particles and polymer. Generally, polycarbonates, polyimides, polysulfones, polyarylates, poly(arylethers) and poly(arylketones) are used as the polymer phase. For the inorganic fillers, porous and nonporous materials can be used. In order to avoid molecular sieving effects, the material used for inorganic phase must have large pores. Zeolites, silica, carbon nanotubes (CNTs), ZIFs/MOFs and metal oxides are different inorganic candidates as filler in the synthesis of MMMs, among which zeolites have better performance. The main advantage of MOF is that it can easily control the interface morphology between the two phases, by which MMMs stability is affected. The synthesis technology of MMMs is an important aspect. For large-scale production of MMMs, some points must be taken into account (Li et al., 2015b):

• Synthesis process of MMMs must be economically reasonable and can be scaled up.
• The performance of MMMs in the presence of chemical contaminants must be evaluated.
• The long-term stability of MMMs must be investigated.

2.3.3.2 CO$_2$ Separation from Ambient Air

As the presence of carbon dioxide in ambient air and its level is a major concern, its direct capture from ambient air is a very critical research area. Due to the thermodynamic limitations (because of very dilute CO_2 concentration) and energy cost (for transferring large volumes of air through separation process), CO_2 capture from air is very challenging. Cryogenic distillation and membrane-based separation are not economic processes for this separation.

2.3.3.2.1 Traditional Sorbents

For CO_2 separation from ambient air, traditional solid sorbents can be used. Among different available solid sorbents, zeolites and activated carbon are not top alternatives because at atmospheric pressure, the presence of moisture decreases the selectivity toward CO_2. Moreover, because of low heats of adsorption, these sorbents have low CO_2 adsorption (Choi et al., 2009). Calcium- or sodium-based oxides, as chemisorbents, need cheap heat sources with lower overall costs.

Generally, traditional solid sorbents are advantageous due to their high thermal stability (for physisorbents) and high adsorption capacity under high temperatures (for chemisorbents). Their major drawback is the need to high energy for desorption of chemisorbents (Yuan et al., 2016).

2.3.3.2.1.1 Alkaline Solutions   In separation technologies based on alkaline solutions, chemical absorption with $LiOH/NaOH/KOH/Ca(OH)_2$ occurs and metal carbonates with CO_2 are produced which, in turn, recycled back to the hydroxide-based solution. In this cyclic process, CO_2 is captured that can be sequestered or used as a C_1 feedstock. This process needs different equipment; such as open towers, packed columns and pools. In spite of the low cost of alkaline solution process, due to expensive energy-consuming regeneration step, the whole process is expensive (Goeppert et al., 2012).

2.3.3.2.1.2 Amine-Grafted MOFs   At low CO_2 partial pressures, MOFs (with(out) functionalized ligands and/or unsaturated metal centers) do not have acceptable performance for CO_2 capture. A combination of MOFs with (poly)amines has good performance for CO_2 separation from ambient air. In amine-grafted MOFs, a strong chemical bond is formed that improves the affinity toward CO_2 and consequently a high CO_2 selectivity is obtained. For instance, Mg-MOF-74 was modified by ethylenediamine and the CO_2 adsorption capacity was enhanced. A combination of Mg-MOF-74 and N,N-dimethylethylenediamine presented high selectivity and fast kinetics for CO_2 adsorption from air. Generally, amine-grafted MOFs have high CO_2 capacity but their stability is unclear. They are very sensitive to moisture and their synthesis routes are expensive.

2.3.3.2.1.3 Amine-Grafted Oxides   For CO_2 separation from ambient air, solid supported amines/polyamines are promising candidates. In this case, amine loading is important for improving CO_2 adsorption capacity. Different materials such as alumina, silica and carbon were evaluated as support and L-lysine, polyethyleneimine and allylamine were evaluated as deposited components for adsorbent used for CO_2 separation from ambient air. Under the condition of low relative humidity, these amine-functionalized oxides have excellent CO_2 adsorption capacity and stability. Due to their slow adsorption kinetics, most of these studied materials frequently need several hours for achieving saturation (Yuan et al., 2016).

2.3.3.2.1.4 Other Technologies   Membrane-based technologies are another alternative for CO_2 separation from ambient air. Recently, an ion-exchange resin-based membrane was designed, in which positive ions of the resin attract CO_2. This is an acceptable material for operation in dry and warm climates. Capital cost and long-term stability of this approach must be evaluated.

2.3.4 CO₂ Storage Options

After capturing carbon dioxide, it is compressed and shipped to be stored in ocean, ground (namely geological storage) or as a mineral carbonate (Li et al., 2013).

Ocean storage has been studied for more than 25 years, but it has never been evaluated on large scale. Ocean has huge capacity and the injected CO_2 can be stored at great depths.

In geological storage, carbon dioxide is injected into gas and depleted oil reservoirs, coalbed formations and deep saline aquifers, in the depth of 800–1,000 m. According to the temperature and pressure of the reservoirs, CO_2 can be stored as a compressed gas, liquid or in supercritical conditions. For CO_2 storage in large amount of 700–900 Gt, deep saline aquifer formations are possible choice (Song and Zhang, 2013).

In mineral carbonate concept, CO_2 is reacted with Mg and Ca oxides to form carbonates. This is "storage and utilization" option. Ca and Mg are in the form of silicate minerals in the nature such as olivine, wollastonite and serpentine. Finland, Portugal, USA and Australia have large deposits of these minerals. In the mineral carbonation process, stable carbonates are formed in which CO_2 can be stored for long periods (decades to centuries) without the risk of its leakage. Due to its high energy penalties and costs, this technology is not developed at large-scale applications. The cost of mining, transportation and mineral preparation must be considered as well, which reduced overall CO_2 removal efficiency (Khoo et al., 2011; Nduagu et al., 2012).

2.3.5 CO$_2$ UTILIZATION OPTIONS

CO_2 utilization is another alternative to CO_2 storage for the captured carbon dioxide. Generally, CCU is classified to two groups: direct CO_2 use and CO_2 conversion (Huang and Tan, 2014).

Refrigerant fluids, EOR (enhanced oil recovery) and ECBM (enhanced coalbed methane recovery) are examples of direct CO_2 use (Chauvy et al., 2019). Waste streams containing CO_2 with high purity (such as ammonia production) can be used for direct utilization of CO_2. In the food industry, CO_2 is used as the carbonating agent, packaging gas, preservative and solvent for the extraction of flavors. In the pharmaceutical industry, CO_2 is used as an intermediate in drug synthesis or as respiratory stimulant. In EOR and ECBM, CO_2 is used for the extraction of crude oil from an oil field or for the extraction of natural gas from unmineable coal deposits, respectively. For more than 40 years, EOR is used in many oil-producing countries such as Norway, USA and Canada (Yu et al., 2008).

Its direct use is beyond the scope of this book and will not be discussed in detail here. CO_2 utilization for the production of fuels and chemicals is the main subject of this book which will be discussed in detail in the subsequent chapters.

2.3.6 CONCLUSIONS ON TECHNOLOGIES FOR CO$_2$ CAPTURE

Scientists and engineers did their best to develop a CO_2 capture technology which is thermodynamically scalable and efficient. Different technologies were developed and reported in this regard; namely cryogenic distillation, absorption, adsorption and membrane separation. All of them can be used for CO_2 capture from stationary points. For CO_2 separation from ambient air, absorption by alkaline solution, adsorption by zeolites and amine-grafted solids and membrane separation by composite membranes are possible separation methods (Yuan et al., 2016).

For CO_2 capture by solid sorbents and membranes (in comparison to conventional MEA absorption), valuable progresses have been achieved. However, for postcombustion CO_2 capture scenarios, the application of MOFs, MMMs and inorganic membranes needs more research. For large-scale application, polymeric membranes, amine-grafted sorbents and facilitated transport membranes have high potential. The main disadvantage of chemical absorption process is its high energy-intensive regeneration step. As solid sorbents and membrane-based technologies can have acceptable performance, they can be good alternatives to chemical absorption process; after solving the below issues:

- CO_2 separation in the presence of impurities and long-term durability and stability of the membrane in that situation.
- Scalable, cost-effective and environmentally benign procedure for membrane synthesis.

In comparison to CO_2 separation from stationary point source, CO_2 separation from ambient air has higher operating costs. Before large-scale utilization of cost-effective CO_2 separation from ambient air, below points must be taken into account:

- There should be a reasonable relationship between CO_2 selectivity and working capacity, and the required energy for regeneration.
- During cyclic CO_2 adsorption-desorption process, long-term stability must be kept.
- There should be a reasonable relationship between the characteristic of the selected material, operating conditions and CO_2 separation performance.
- The CO_2 separation performance must be evaluated at lab and pilot scales.
- The presented process must be economically and environmentally viable.

2.4 COAL

2.4.1 CHEMICAL COMPOSITION

Coal is a heterogeneous material which is difficult to characterize because it lacks a uniform and defined nature or structure (Olah et al., 2018). Coal is formed by gradual and continuous carburization of wood and other part of plant materials. This process is a combination of biological and thermal processes during which deoxygenation-aromatization occur. Thus, the coals can be classified on different bases such as the original plant (called coal type), the degree of metamorphosis or maturity (called coal rank), impurity level including ash and heteroatoms (called coal grade) and industrial properties such as cocking or agglomeration (Crelling et al., 2006).

Coals can be ranked as lignite, semibituminous, bituminous and anthracite with the increasing order of maturity and thus aromaticity (that is, the ratio of aromatic carbon to total carbon). The H/C molar ratio in bituminous coal and anthracite are about 0.8 and as low as 0.2, respectively. Low rank coals have more oxygenated functional groups attached to the clusters. Bituminous coal consists of aromatic clusters linked by short (-CH$_2$-)$_n$ (n = 1–4), ether, sulfide and biphenyl bridges which readily undergo cleavage under thermochemical conditions (Olah et al., 2018).

Upgrading of coal to petroleum-like liquids (with lower S, N and ash) and gaseous products fuels improves the burning and transport properties and thus the vast coal sources could be used as a source of clean fuels and useful chemicals. The conversion of coke to liquid products (coal liquefaction) can be implemented for most coals (except for anthracite) either directly by hydrogenation under high operating temperatures and pressures along with solvents to create liquids which are highly aromatic in nature (Ahmed et al., 2019) or indirectly via syngas generation by gasification followed by Fischer–Tropsch and methanol, dimethyl ether (DME) synthesis.

2.4.2 GASIFICATION

Gasification is a thermochemical process which transforms a carbon-rich feed to a combustible gas mixture. The primary product of the process, syngas, may be used as a fuel for power generation or as feedstock for the synthesis of chemicals or fertilizers.

A wide variety of feeds including coal, coal char, biomass and petcoke can be processed. According to operation condition, different thermochemical processes can occur with coal and biomass. Combustion is a high-temperature strongly exothermic process which occur in the presence of excess oxygen (or air) forming CO$_2$ and H$_2$O. Gasification is also a high-temperature process but occurs under oxygen-deficient conditions commonly with fossil fuels. Pyrolysis and liquification occur at lower temperatures in the absence of oxidant with or without catalyst.

Gasification, like combustion, is a high temperature process; however, it is under limiting oxygen supply unlike combustion which typically uses excess air. Therefore, unlike gasification, the elements in combustion products are in their highest oxidation states and the product has no heating value (Table 2.1). Soot formation is common to both, primarily as a result of mass transfer limitations. The pollutants (H$_2$S, NH$_3$) in gasification can be separated more easily and thus has environmental benefits compared to combustion.

In evaluation of the solid feeds for gasification, moisture, volatile matter, fixed carbon, ash content, elemental analysis (C, H, O, N, S) are important parameters.

TABLE 2.1

Comparison of Products in Combustion and Gasification

Element	Combustion	Gasification
Carbon	CO$_2$, CO, soot	CO, soot
Hydrogen	H$_2$O	H$_2$
Sulfur	SO$_2$, SO$_3$	H$_2$S, COS
Nitrogen	NO, NO$_2$, N$_2$	NH$_3$, HCN, N$_2$

According to the oxidant (air or oxygen) used, the gasifier is air-blown or oxygen-blown type. In gasification of coal, about one-fifth to one-third of theoretical oxygen (that required for complete combustion) is added (Mishra et al., 2018).

Newer protocols suggest gasification in the presence of supercritical or near critical water (600–750°C) to make use of the advantage of improved transport properties. This method is especially suited for high moisture feeds (such as biomass) The main drawbacks are high installation cost and high energy consumption (Molino et al., 2018).

2.4.2.1 Chemistry and Thermodynamics

In gasification, coal and coal char react with a gasifier (air, oxygen, steam, CO_2, H_2 or their mixture) at elevated temperatures (>1,000°C). Thus, the gasifier may be either oxidizing or reducing gas. The product contains CO, H_2, CH_4, CO_2 and other gases the composition of which depends on feed composition, operating condition (T and P) and gasifier type. Low-rank coals can be gasified at lower temperatures, but the cold gas yield is not high in fixed bed gasifiers (Kang et al., 2017). Similar chemistry is applicable to coke derived from petroleum or other sources.

In thermodynamic analysis of gasification, coal or coal char are assumed as pure carbon (graphite) and considering the most important reactions involved (Ahmed et al., 2019) which is a reasonable approximation. The main reactions occurring during coal gasification are as follows:

$$coal \rightarrow char + gases\left(CO, CO_2, H_2, CH_4, CH_X\right) (endothermic) \tag{2.1}$$

$$2C(char) + 1.5O_2 \rightarrow CO + CO_2 \text{ (exothermic)} \tag{2.2}$$

$$C(char) + H_2O \Leftrightarrow CO + H_2 \text{ (endothermic)} \tag{2.3}$$

$$CO + H_2O \Leftrightarrow CO_2 + H_2 \text{ (mildly exothermic)} \tag{2.4}$$

$$CO + 3H_2 \Leftrightarrow CH_4 + H_2O \text{ (exothermic)} \tag{2.5}$$

$$C(char) + H_2O \Leftrightarrow CO + H_2 \text{ (endothermic)} \tag{2.6}$$

$$CH_X \rightarrow (1 - X/4)C + (X/4)CH_4 \text{ (pyrolysis)} \tag{2.7}$$

$$CH_X + mH_2 \rightarrow \left(1 - (X+2m)/4\right)C + \left((X+2m)/4\right)CH_4 \text{ (hydropyrolysis)} \tag{2.8}$$

The equilibrium composition calculation can be done by using the equilibrium constants of individual thermodynamically independent reactions (stoichiometric approach) or by Gibbs free energy minimization of the system knowing the products (non-stoichiometric model). Unfortunately, the accurate prediction of gas composition is not possible by equilibrium calculations because all reactions may not be in equilibrium. The other reason is the possibility of occurrence of other reactions.

The reactions involving heteroatoms S and N can be exemplified by:

$$H_2S + CO_2 \Leftrightarrow COS + H_2O \quad \Delta H^0_{298} = 6.97 \, kJ/mol \tag{2.9}$$

$$HCN + H_2O \Leftrightarrow NH_3 + CO \quad \Delta H^0_{298} = -49.7 \, kJ/mol \tag{2.10}$$

$$0.5N_2 + CO_2 \Leftrightarrow NO + CO \quad \Delta H^0_{298} = 373 \, kJ/mol \tag{2.11}$$

The composition of these impurities in the product stream depends on S and N content in the feed, gasifier type and operating conditions, but S is mainly as H$_2$S (with lower amounts of carbonyl sulfide, COS) and N as NH$_3$ (but also as HCN, NO$_x$ and N$_2$). Chlorine is found mostly as HCl if present in the feed (Recari et al., 2016).

Minerals in the fuel do not gasify and form a solid ash that, according to operation temperature, leaves the gasifier as a glass-like slag or other forms.

The influence of operating parameters on product composition can be predicted from the set of reactions taking place. Higher temperatures favor endothermic reactions over exothermic reactions. Consequently, the CO production will increase with temperature, whereas CH$_4$ formation will decrease. Higher pressures favor gaseous reactions with decreasing mole such as methane and carbon dioxide formation.

The main issue is supplying the heat of reaction. It can be supplied externally (i.e., allothermal), but the most common method is allowing exothermic reactions to occur in the same vessel (i.e., autothermal). Reactions 2.4 and 2.5 can be used for tailoring the final product composition and thus are often conducted in downstream blocks.

The process flow diagram of different gasification technologies is nearly the same. In the first step coal is prepared by crushing and drying and if necessary is subjected to operations preventing cake formation. Then it is gasified by a mixture of O$_2$ (or air) and steam. The product gas is then cooled and refined from particulates (e.g., coal char) and contaminants (e.g., H$_2$S and CO$_2$).

Catalytic gasification occurs at lower temperatures thereby the syngas composition changes. It is advantageous by reducing the operating temperature and pressure.

2.4.2.2 Gasification Catalysts

Gasification of coal and biomass are examples of non-catalytic gas-solid reactions. The rate of surface reactions can be increased by using appropriate catalysts. Transition metals and especially alkali metals have been used as catalysts in coal gasification (Bell et al., 2011). Kapteijin et al. (1986) tested and screened group IA alkali metals for coal gasification by CO$_2$. For an acid washed steam-activated peat char, the catalytic activity increased with atomic weight of alkali metal in the order:

uncatalyzed < Li < Na < K < Rb < Cs

For group IIA metals, however, the same trend with respect to atomic weight was not observed and the following order of activity was obtained (Kapteijin et al. 1986):

$$uncatalyzed < Be < Mg < Sr \approx Ba < Ca \approx K$$

It has been found that alkali, alkali earth and transition group VIII metals are effective additives, especially K and Ca species. K_2CO_3, KCl and Li_2CO_3 are the most effective additives. For coal and coal char, the activity of Li decreases with CO and CO_2. KCl is not desirable because of the corrosive nature of Cl⁻ anion. The following factors are effective in high effectiveness of K_2CO_3: K_2CO_3 strongly increases the rate, it is mobile and exhibits high dispersion on coke, it is independent of coal rank, and is not affected by addition mode. Its drawback is volatility at higher temperatures which reduce activity and recovery by formation of potassium alumi-nosilicates. Potassium interacts with other components and forms eutectics with low melting points which increase agglomeration of the ash. The proposed solutions were demineralization of coal and adding interaction inhibitors such as calcium (Zhang et al., 2017).

The characterization of coal char structure with and without K_2CO_3 catalyst by X-ray diffraction and laser Raman techniques showed that the catalyst changed the structure of the organic unit in coal char. The mechanism by which the catalyst affects the structure of coal char in pyrolysis was investigated by X-ray photoelec-tron spectroscopy. A direct evidence of electron transfer in catalyst–coal interactions was obtained, and a K-char intermediate formed. This makes it easier for H_2O to attack the K-char intermediate, resulting in increasing gasification reactivity of coal char (Zhang et al., 2016). Potassium carbonate can inhibit the graphitization and condensation of coal char in pyrolysis.

However, incorporation of the catalyst into the process has economic and envi-ronmental impacts. Catalyst recovery increases the operation costs. The unre-covered catalyst causes water and soil pollution. Exxon studied the recovery of catalyst as potassium by water wash and lime treatment, the so-called digestion process (Euker and Reitz, 1981). Using such treatments, catalyst recovery effi-ciencies of about 90% have been achieved for K_2CO_3-catalyzed steam gasification process (Yuan et al., 2019).

2.4.2.3 Kinetics

Coal gasification is a multistage process comprising, coal pyrolysis (rapid), combus-tion (rapid and limited) char gasification (slow) and slag formation.

Among the main steps of gasification, the char gasification step is much slower and thus the rate limiting step (Mishra et al., 2018). Thus, the design and construc-tion of a coal gasifier strongly depends on char gasification. The characteristics and thus the reactivity of char depend on coal precursor and formation conditions such as heating rate and gas-particle dynamics.

The kinetics of coal gasification has been extensively studied and reviewed by Mishra et al. (2018) and Irfan et al. (2011). Two kinetics are commonly applied for high pressure gasification of coke, namely power-law or global rate and

Langmuir–Hinshelwood (LH) rate expression. The former can be written as Eq. (2.12) (Irfan et al., 2011) and is applicable over a limited range of operating conditions:

$$r_g = A \exp\left(\frac{-E_a}{RT}\right) P_g^n \tag{2.12}$$

where r_g is the rate of gasification, E_a is the activation energy and P_g is the partial pressure of the gasifying agent.

In reality, the reaction involves several adsorption-desorption steps. The LH kinetics is mechanism-based and applicable over a wider range of conditions. Thus, for CO_2 as gasifying agent with the following mechanism:

$$C + CO_2 \Leftrightarrow C(O) + CO, \quad k_1, k_2 \tag{2.13}$$

$$C(O) \to CO \quad k_3 \tag{2.14}$$

In which k_1 and k_2 are rate constants for the forward and reverse reactions in Eq. (2.13), k_3 rate constant of Eq. (2.14) and C(O) is surface complex reaction intermediates, the intrinsic rate expression is obtained as Eq. (2.15) (Liu et al., 2017):

$$r_{CO2} = \frac{k_1 [C_t] P_{CO2}}{1 + \dfrac{k_2}{k_1} P_{CO} + \dfrac{k_1}{k_3} P_{CO2}} \tag{2.15}$$

where $[C_t]$ is the overall concentration of active sites. The inhibitory effect of CO as implied by Eq. (2.15) is due to its competition with CO_2 for active sites.

The structural effects of gas-solid reactions, in addition to the intrinsic kinetics, have been accounted for by different models. The volumetric model (VM) is the simplest one, assuming that the reaction occurs within the whole volume of the solid. The conversion rate, X, for a first-order reaction is then

$$r = \frac{dX}{dt} = k_{VM}(1 - X) \tag{2.16}$$

The grain or shrinkage core model (Szekely and Evans, 1970) assumes that reaction occurs only over the surface of a shrinking carbon core with a rate given by Eq. (2.17),

$$\frac{dX}{dt} = k_{GM}(1 - X)^{2/3} \tag{2.17}$$

These models cannot describe the observed maximum reaction rate (Mahinpey and Gomez 2016). Consequently, random pore model (RPM) (Bhatia and Perlmutter, 1980; Bhatia and Vartak, 1996) is considered as the best model.

It assumes an initial increase in accessibility of pores followed by coalescence or overlap of the pore surfaces which reduce the available area for reaction. The resulting rate is

$$\frac{dX}{dt} = k_{RPM}(1-X)\sqrt{1-\psi \ln(1-X)} \qquad (2.18)$$

where the parameter ψ characterizes the influence of pore structure on kinetics of gasification and is defined by Eq. (2.19)

$$\psi = \frac{4\pi L_0(1-\varepsilon_0)}{S_0^2} \qquad (2.19)$$

In which L_0 is the pore length, ε_0 is the solid porosity and S_0 is the pore surface area. In above equations, X is defined as

$$X = \frac{m_0 - m(t)}{m_0 - m_a} \qquad (2.20)$$

where m_0, $m(t)$ and m_a are the initial mass and mass of particle at time t and mass of ash, respectively.

It is generally believed that oxidation reactions involving oxygen are very rapid occurring in short distance from the point of entrance of oxygen and mixing with gases or contacting with the solid. The evolved heat pyrolyzes the coke producing char which reacts with H_2O, CO_2 and other combustion or pyrolysis gases. Thus, the assumption made previously that the solid reactant is carbon (char) is possibly the case, although its type affects gas-solid reaction rates. The overall reaction is perhaps kinetic limited below 1,000°C. Above this temperature, the reaction becomes progressively pore-diffusion limited while at very high temperature it becomes surface-film (external) mass transfer limited as summarized in Figure 2.2.

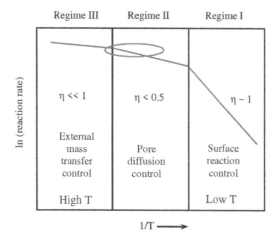

FIGURE 2.2 Reaction regimes versus temperature in combustion and gasification. Technical operation range encircled (Redrawn from Walker et al.1959; Tanner and Bhattacharya 2016).

Therefore, another important parameter is the effectiveness factor, η, the ratio of observed rate per intrinsic rate accounting for mass transfer limitations (Levenspiel, 1999). Accordingly, the reaction regime can be divided into three regimes according to mass transfer limitation as summarized in Figure 2.2. The mass transfer limitations increase with reaction temperature resulting in decrease in effectiveness factor.

Mass transfer limitations affect the global kinetics of gasification. Thus, in low temperature range (region I in Figure 2.2), in which mass transfer effects are negligible, the observed kinetics is the same as microkinetics with the same kinetic parameters being applicable. In region II, the apparent activation energy is about half of the intrinsic value and the order of reaction is about $(n+1)/2$ for an nth order intrinsic kinetics. In region III, the activation energy is very small and the overall reaction is apparently first-order, irrespective of the intrinsic kinetics (Tanner and Bhattacharya 2016). Nevertheless, operation in region I is not of a practical interest because of the low overall reaction rates and also undesirable product selectivity. In practice, higher operation temperatures are used to meet the operational requirements (Figure 2.2).

2.4.2.4 Gasifiers

Because of high operating temperatures, heat transfer, mass transfer and reaction rate play critical roles in gasification. These conditions should be satisfied in a gasifier where the solid reacts with gasifying agent. According to solid-gas contacting patterns, three types of gasifiers are available commercially: namely, fixed bed (or moving bed), fluidized bed and entrained flow gasifiers.

The basic configurations for gasifiers are moving-bed, fluidized bed and entrained flow gasifiers (Figure 2.3).

In fixed-bed (or moving-bed) gasifier, the gasifying agent (steam, oxygen and/ or air) slowly passes through a bed of solids moving slowly under the action of gravity, and so the alternative name moving-bed gasifier. This type is the oldest type which is still widely used. It is further classified according to solid-gas contact as up-draft (with gas entering from bottom, counter current with the solid), down draft (with gas entering from the top, co-current with solid) and cross draft type. Updraft type is one of the oldest and simplest of all designs and is more common due to lower tar formation (Mishra et al., 2018). Coal particles within 5–80 mm are preferred.

A temperature profile is formed within the bed. Thus, the coal is successively preheated, dried, pyrolyzed, gasified and combusted in the combustion zone with a temperature of 1,300°C for dry ash and 1,500–1,800°C for slagging ash.

The two main practical advantages of fixed-bed gasifiers are that they are cost effective for small scale operation and that the produced gas is cleaner with lower particulates and tar contents compared with fluidized bed gasifiers (Ragnarsson et al., 2015). The main drawbacks are poor heat transfer resulting in non-homogeneous temperature profiles and tighter feed specification requirement. Thus, lower moisture contents (<20 wt.%) and more uniform feed particle sizes distribution are required to prevent clogging and channeling (Safarian et al., 2019).

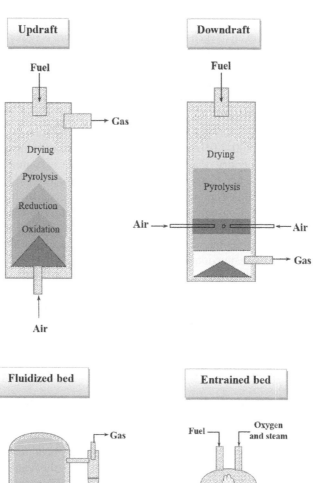

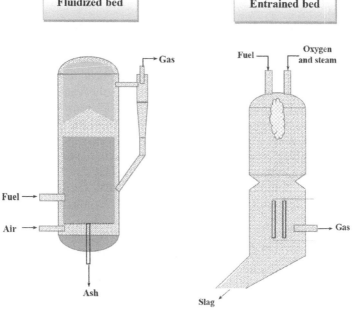

FIGURE 2.3 Gasifier configurations.

In a fluidized-bed gasifier, the solid particles act as a fluid. Particles smaller than 5 nm are suspended in an oxygen-rich gas. The temperature is rather uniform and relatively low within the bed. Therefore, the feed must be highly reactive such as low-grade coals (Speight, 2011). The throughput is higher than fixed bed gasifiers. Solid and liquid fuels can be used in these gasifiers. Due to short gas residence times, in order to ensure about high carbon conversion, the solid feed should be reconstituted into pellets (for biomass) or pulverized into small granules.

Fluidization avoids formation of Clinker and defluidization because of the operating temperature (800–1,050°C) which is below the fusion temperature of the ash. Thus, the ash is either dry or agglomerated. The former is commonly found in low-grade coal while the latter can be found for any coal grade.

The main potential advantage of fluidized bed gasifiers is high solid-gas heat and mass transfer rates. Consequently, a uniform temperature distribution is obtained and the reactor is more flexible to the change in feed quality. The main disadvantage is more dust and particulate content of the product gas (Warnecke, 2000) causing problems in downstream equipment (Safarian et al., 2019).

In an entrained flow gasifier, the pulverized coal particles (<0.1 mm) are suspended in a stream of steam and oxygen at high space velocities in co-current flow. According to the feeding mode, dry (with N_2) or wet, nearly all forms of coals are acceptable. These gasifiers exhibit high carbon conversions as they operate above slagging temperature of coal (1,400–1,600°C). They have large capacity because of a short residence time in order of seconds. The large commercial types are Texaco, Koppers-Totzeck and Shell. The available capacities are larger than those of the previous types.

In addition to the three main types of gasifiers discussed above, hybrid combinations of these gasifiers also exist. The Kellogg Brown and Root (KBR) gasifier, for instance, shares the characters of both fluidized and entrained flow gasifiers.

There are still novel gasifiers designs such as molten metal bath gasifiers.

The state of ash (dry, aggregate and slag) depends on the temperature. In dry ash operations (<1,000°C), the ash is collected as dry without sintering. According to its inorganic constituents, aggregation occurs at 1,000–1,200°C and the particles sinters which are collected under controlled condition for steady operation. Slagging occurs above 1,200°C. Melting properties should be appropriate. Fluxing agents might be necessary (Furimsky, 1999). In moving bed gasifiers where $T = 1,500–1,800°C$, the ash is slag or dry. In fluidized bed gasifiers ($T = 800–1,050°C$), ash is discharged as dry or aggregate. In entrained flow gasifiers (1,400–1,600°C), ash is discharged as slag.

2.4.2.4.1 Gas Cleaning

In addition to CO, H_2 and CH_4, the gas product contains undesirable constituents such as ash, entrained soot, tar, H_2S and trace amounts of NH_3, COS, HCl and HCN which should be avoided or removed. Common methods include addition of lime to fluidized bed solid feed to adsorb H_2S, ex-situ H_2S absorption or adsorption and removal of tar.

Lurgi Fixed Bed Dry Bottom (FBDB™) technology can utilize low grade (high water, high ash) coal by reacting with oxygen and steam. According to downstream or product requirements, the raw syngas can be processed by shift reaction or purified by the Rectisol™ process. The dry syngas yield is around 2,000 Nm³/ton dry coal according to Technology Handbook (2016).

The Rectisol™ syngas purification process removes acid gases (CO_2, H_2S, HCN, COS and also NH_3 and mercaptans) and other impurities typical of gasification of low-grade feeds. Cold methanol is used as the physical solvent for absorption of impurities which is regenerated by flashing and stripping. The co-products include tar acids (phenols), ammonia and sulfur.

2.4.2.5   Commercial Technologies

Gasification is not a new concept. Converting coal to a combustible gas has been practiced commercially since the early 19th century (Ahmed et al., 2019). Commercially, coal gasification may be used for fuel gas production for power generation or syngas as a feed for chemical syntheses as the first step in indirect coal liquification or coal to chemicals.

The fixed-bed technology of British Gas and Lurgi (BGL) developed a slagging bottom to reduce the excess steam used in fixed-bed gasifiers to reduce the temperature below the melting point of ash. This allows the ash to melt drain through a slag trap. The capacity of this gasifier per cross section area is over twice than that of a dry bottom gasifier.

The U-Gas technology employs a fluidized-bed gasifier using agglomeration of ash to eliminate coal agglomeration. Crushed limestone can be injected with coal for capturing sulfur. The char and ash that exit the gasifier with the product gas are recycled to a high temperature agglomeration zone where the fresh coal introduced is pyrolized, gasified and the ash particles are softened.

For large-scale gasification of petroleum refinery residues, coal and petroleum coke, the most widely used gasification type is entrained flow gasifier (Basu, 2013).

An example of dry-fed entrained flow gasifier is the Shell gasifier. It is an oxygen-blown gasifier and is capable of processing a wide range of coals (from brown coal to anthracite). Its drawbacks, being typical of entrained-flow gasifiers, are high oxygen requirements and high duty of waste heat recovery. However, dry-feed delivery slightly reduces oxygen demand compared to slurry feed gasifier at the expense of a more complex coal-feeding system. This technology is marketed by Siemens, Thermal Power Research Institute (TPRI), MHI, Uhde and Shell.

The GE gasifier (formerly Texaco), the East China University of Science and Technology (ECUST) and the CB&I E-Gas gasifier (formerly Conoco-Phillips) are commercial examples of slurry-fed entrained flow gasifiers.

Southern Company and KBR developed a hybrid-type transport bed gasifier also known as the TRIG™ gasifier. This gasifier is an advanced pressurized circulating one operating at pressures up to 40 bar and moderate temperatures (815–1,065°C) and high carbon conversion with low tars and oils formations are achieved. In this gasifier, air or pure oxygen can be used as the oxidant, according to the application (such as chemicals production or power generation) (Ahmed et al., 2019).

In Kellogg coal gasification process (formerly molten salt gasification process), a bath of molten Na and Ca carbonates is used. Oxygen and steam enter from the bottom while coal is feed in the beneath of the surface. Partial oxidation of coal produces CO which is partly converted into hydrogen via water-gas shift reaction

(Cornils et al, 2013). The Rockgas process is a pressure modification (1,000°C, 2 MPa) of Kellogg process being used for the production of low-Btu gas for in-site consumption.

In 2011, 60 MM tpy of coal were gasified worldwide and the coal consumption for chemical purposes is increasing. Syngas obtained from coal is commercially used by some companies (such as in Eastman acetic anhydride process) (Cornils et al, 2013).

According to a study conducted in 2014, there were 686 gasifiers worldwide, 272 of which were commercial operating with a total capacity of 116,000 MW (thermal) (Ahmed et al., 2019).

2.5 HEAVY OIL RESIDUES

Heavy residues (also called residuum, heavy ends, bottoms) are non-volatile oil cuts of low value for fuels and feedstock. They contain high proportions of heteroatoms (S and N) and metals. Thus, they are not really hydrocarbons and their combustion is associated with considerable emission of particulate matters, ash, NO_x and SO_x. Therefore, they should be upgraded by costly processes before being used by catalytic processes such as residue fluid catalytic cracking (RFCC), and thermal processes such as visbreaking, delayed coking and gasification. Catalytic processes are not robust against the increasing metal content (1,000 ppm Ni+V) and rising coke formation tendencies (Coradson carbon) of the current heavy crude oils. Consequently, partial oxidation or thermal process gasification are valuable options. It is highly flexible and practically applicable to all types of high-sulfur heavy residues to produce synthesis gas, hydrogen or power. It can be applied to atmospheric and vacuum residues of distillation in refineries processing heavy or extra heavy crude oils (Bader et al., 2018). Here, Air Liquide Multi-Purpose Gasification, Texaco/GE Gasification or the Shell Gasification Process as entrained-flow gasification technologies can be employed (Higman and van der Burgt, 2008).

Pitch is a concentrate of crude oil asphaltenes in which sufficient oil has been removed from asphalts to give a coal like solid with a high melting point (Cornils et al, 2013). It contains low levels of ash and is brittle, so easily pulverized for gasification e.g., by suspending in water for Texaco gasification process.

Compared to coal and other solid fuels, relatively few studies have been published for liquid fuels such as heavy oils gasification in the literature. The entrained flow gasifier is the most suitable type for heavy oils. The main operating parameters are oil flow rate, oil temperature (affecting appropriate spraying) and oxygen/oil and steam/oil ratio affecting the temperature and efficiency of carbon conversion.

The Shell gasification process is capable of upgrading bottom-of-the barrel to syngas. Gasification is non-catalytic partial oxidation process operation at 1,300°C and 80 bar. The reactor effluent is cooled down to about 350°C by simultaneous 120 bar saturated steam generation and heating the feedstock and oxygen. The solid particulates (ash and soot) are removed by two step water scrubbing (Speight, 2020).

Lurgi MultiPurpose Gasification (MPG™) utilizes heavy residues from refinery (asphalt, bitumen, tar, visbreaker residue, hydrocracker residue, FCC (fluid catalytic cracking) residue, vacuum residue) or other processes (e.g., coal tar) to produce syngas by non-catalytic partial oxidation. Slurries of coal (1 mm) are also acceptable.

Using the proprietary MPG-burner, gasification is performed in the refractory lined reactor at typically 1,200–1,500°C and 30–100 bar (Speight, 2020).

Vaezi et al. (2011) used a thermodynamic equilibrium simulation neglecting char formation. They used only two independent reactions, namely WGS and methanation. The model was validated by comparing the numerical results with the data of extra-heavy oil (Orimulsion™).

2.6 BIOMASS

Biomass is referred to organic materials such as bark, roots, leaves and seeds produced by plants. In boarder sense in may include microbial and animal wastes such as chicken and manure. This term, thus, is intended to be applied for materials with little or no direct food application but are of industrial applicability. The main sources of biomass are agricultural and forestry wastes, municipal wastes and energy crops (rapid growing plants with high biomass yields).

Biomass is a complex and variable mixture of carbohydrates, lipids, proteins and minerals. The main component of plant biomass are carbohydrates (about 75 wt.% dry basis) and lignin (about 25%) which depends on source. The carbohydrates are mostly cellulose and hemicelluloses.

Cellulose is a polymer of glucose linked via β-glycoside bonds. Its high crystallinity impedes its hydrolysis. Hemicellulose is an amorphous polysaccharide with a random structure and lignin is an amorphous polymer composed of metoxylated phenyl propane unit. Lignocellulosic biomass is the most abundant class of biomass which is an inexpensive and renewable raw material. The main processes for biomass conversion are biochemical and thermochemical (Sharma and Sheth, 2016). These processes are accompanied by depolymerization or breaking down biomass molecules under the action of enzymes or heat, respectively.

Biomass is a promising feedstock for chemical industries (e.g., in biorefineries) in the future as it is renewable and, on a net basis, has much lower CO_2 emissions (if any) than fossil fuels. It can be converted into C_1 feedstocks by thermochemical (gasification) and biological processes (fermentation). The main C_1 products are CO, methane and CO_2, respectively.

2.6.1 THERMOCHEMICAL CONVERSIONS

Similar to coal conversion, the main thermochemical conversion of biomass can be classified as combustion, gasification, pyrolysis and liquefaction according to oxidizing potential, reaction temperature and heating rate. However, unlike coal and coal char, biomass has high hydrogen and oxygen content making operating conditions of the respective reactions somewhat different. In contrast, coal has a higher amount of fixed carbon and ash. The ash of coal is mostly silica alumina while that of biomass alkali and alkali earth metal oxides and carbonates (Mallick et al., 2017).

The gasification of biomass to syngas is now considered as a mature and acceptable technology compared to other biomass conversion technologies.

Biomass gasification produces combustible gas by partial oxidation of biomass using air, oxygen, carbon dioxide and/or steam as gasifying agents at high temperatures (800–1,000°C). The product contains carbon monoxide, carbon dioxide, hydrogen, methane and nitrogen. Ethane and propane (as light hydrocarbons), tars (as heavier hydrocarbons that condense at temperatures between 250 and 300°C) and hydrogen chloride (HCl) and hydrogen sulfide (H_2S) (as undesirable gases) can also be present (Molino et al., 2016).

To meet the required specifications, several preparation techniques including drying and size reduction can be used. The high moisture contents of biomass entered into the plant (typically 30%–60%) should be reduced to about 10%–15% by drying (e.g., in rotary dryers) (Krishna et al., 2019). Size reduction (in knife chippers or hammer mills) and classification is necessary for appropriate uniform size.

Alkali metal salts of weak acids such as carbonates (K_2CO_3, Na_2CO_3) and sulfides (K_2S, Na_2S) are capable of catalyzing the carbon-steam gasification reaction. However, their application is costly and their performance decreases over time (Krishna et al., 2019).

2.6.1.1 Gasification Models

The performance of biomass gasification can be estimated by kinetic rate and thermodynamic models.

Kinetic rate models are crucial for designing and optimizing gasifiers. According to the required accuracy and details, they could be computationally intensive (Sharma, 2008). Nevertheless, development of kinetic models for biomass gasification has been the subject of many researches (Puig-Arnavat et al., 2010). These models describe the rate of char gasification based on theoretical and experimental studies and are especially useful when the solid-gas contact times are short. The kinetic models always contain parameters that limit their applicability to different plants.

Thermodynamic equilibrium models, in contrast, are independent of gasifier design. Therefore, they may be more appropriate for investigating the effect of important fuel process parameters. However, at low outlet temperatures range of 750–1,000°C, thermodynamic equilibrium may not be achieved (Bridgwater, 1995).

A comparison of non-stoichiometric equilibrium model using minimizing the Gibbs free energy with fifteen chemical species with experimental data for biomass gasification in downdraft gasifiers highlighted the well-known weakness of equilibrium models (i.e., overestimation of hydrogen and underestimation of methane in the product gas) (Gambarotta et al., 2018).

2.6.1.2 Gasification Technologies for Biomass

The most frequently used type of gasifiers for biomass gasification are moving-bed and fluidized-bed reactors each type with variations (Warnecke, 2000). Among moving bed gasifiers, updraft types show higher hot gas efficiency (90%–95%), higher turn-down ratios (5–10) and accepts higher moisture levels (up to 60% wet basis) in biomass gasification compared to downdraft and cross draft types (Knoef, 2005). However, its tar formation is much higher (30–150 g/Nm³).

In entrained flow gasifiers, the residence time is in the order of seconds or tens of seconds. Thus, to achieve high biomass conversions, high operation temperatures are required which necessitates high O_2 feeding. Furthermore, very small biomass particle sizes ($\leq$100 μm) are required to enhance heat and mass transfer to particles (Higman and Van der Burgt, 2008). These conditions are not suitable for most biomass types.

For coal gasification, a well-known technology is the entrained flow gasifier, but further developments are necessary for its operation with biomass feedstock. This technique uses high temperature and short residence times to achieve high throughputs which rely on finely ground solid particles or slurries. High temperature reduces tar formation which tends to be high for biomass.

Because of highly fibrous nature of biomass, its milling is energy intensive. Torrefaction can loosen the fibrous nature of biomass making it more easily grindable while still possible to be formed into pellets without using binders. These requirements limit the torrefaction temperatures to 200–300°C (Basu, 2013). The other challenge is formation of large amount of molten ash which can corrode the refractories or metal linings due to its high alkaline metal contents.

The gasification of biomass involves several steps, namely drying, pyrolysis, and gasification or partial oxidation.

Prins et al. (2007) studied the effect of fuel composition (C/O and H/O atomic ratios) on the thermodynamic efficiency of gasification systems. A chemical equilibrium model was used to estimate the highest possible gasification efficiency. It was found that exergy loss for wood (O/C$\approx$0.6) is higher than that for coal (O/C$\approx$0.2). For a gasification temperature of 927°C, a fuel with O/C of 0.4 correspond to a lower heating value of 23 MJ/kg was recommended. At higher temperature (1,227°C), a fuel with O/C ratio below 0.3 (LHV$\approx$26 MJ/kg) was considered as the preferable feed. They concluded that enhancing the heating value of solid biofuels with high oxygen contents by thermal pre-treatment or mixing with coal would be an attractive strategy.

2.6.2 BIOCHEMICAL CONVERSION

Biogas is a renewable fuel with major constituent being CH_4 and CO_2. It is a product of anaerobic digestion of biomass (Italiano et al., 2015). A vast variety of organic materials including agricultural waste, manure, municipal waste, plant material, sewage sludge, green waste or food waste can be used as raw material. The composition depends on biomass type and digester condition, but a typical composition is in the range of 50%–75% CH_4, 50%–25% CO_2, 0%–10% N_2 and 0%–3% H_2S (Appari et al., 2014).

The anaerobic digestion involves four biological steps in series each with the aid of special microorganisms (Weiland, 2010):

1. *Hydrolysis*. Hydrolysis of large organic molecules to simpler and soluble ones (e.g., carbohydrates to sugars and proteins to amino acids) by anaerobic bacteria.
2. *Acidogenesis*. Conversion of the simple organic molecules to carbon dioxide, hydrogen, ammonia and organic acids (e.g., propionic acid) by acidogenic bacteria.

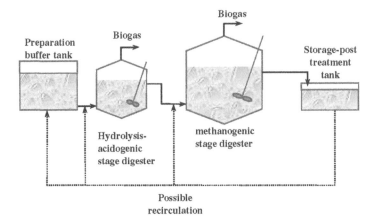

FIGURE 2.4 Two-stage anaerobic digestion (Redrawn from Roš and Zupančič, 2012).

3. *Acetogenesis.* Conversion of organic acids to acetic acid, along with additional hydrogen, ammonia and carbon dioxide by acetogenic bacteria.
4. *Methanogenesis.* Decomposition of the acetic acid to methane and carbon dioxide by methanogenic bacteria.

The overall reaction of anaerobic digestion for cellulosic feed can be simplified as:

$$\left(C_6H_{10}O_5\right)_n + nH_2O \rightarrow 3nCO_2 + 3nCH_4 \tag{2.21}$$

The digester may be classified as plug flow digesters, fixed dome digesters and floating drum digesters. They vary in size from 1 m^3 (household units) to as large as 2,000 m^3 (Luque et al., 2008). Germany is currently the largest biogas producer in Europe.

Because of different growth characteristics of microorganisms in hydrolysis, acidogenesis and methanogenesis, two-stage digesters have been developed to optimize the condition for each phase (Figure 2.4), thereby reducing the hydraulic retention time (HRT, the ratio of reactor volume (V) to influent rate (Q), V/Q) (Zinoviev et al., 2010).

Factors such as pH, temperature, organic loading rate, hydraulic retention time and carbon-to-nitrogen (C/N) ratio play a significant role during the bio-degradation of the solid material (Ramatsa et al., 2014). The digestion process may take several weeks which means that large volumes are necessary.

The main upgrading process is the removal of CO_2 and H_2S by absorption, adsorption, cryogenic or membranes. A review of upgrading methods can be found in Muñoz et al. (2015).

2.7 CONCLUSIONS

C_1 feedstocks can be derived directly from abundant natural sources, waste streams and flue gases or from a great variety of raw materials including fossil fuels, renewables and organic wastes by thermochemical and biological processes. Heavy organic material (coal, biomass, organic wastes and refinery residues) can be converted into syngas by gasification and then to liquid fuels and methanol by Fischer–Tropsch and methanol synthesis, respectively. This application is expected to be the main use of gasification technologies in the future.

Another possibility is conversion of raw C_1 molecules to more reactive ones. Thus, carbon dioxide and methane, as abundant C_1 molecules with low reactivity, can also be converted into syngas, by hydrogenation or partial oxidation, respectively, and then to liquid products in the same way.

In either case, the share of unconventional feedstocks in chemical industries can be increased by which sustainable and greener chemical synthesis processes and thus a circular economy based on reuse of the wastes can be developed.

NOMENCLATURE

AC	activated carbon
APS	3 aminopropyltrimethoxysilane
BGL	British Gas and Lurgi
BTL	biomass to liquids
CCS	carbon capture and storage
CCU	carbon capture and utilization
CLC	chemical looping combustion
CLR	chemical looping reforming
CNT	carbon nanotube
COF	covalent organic framework
CTL	coal to liquids
DEA	diethanolamine
DGA	diglycolamine
DIPA	di-isopropanolamine
DME	dimethyl ether
ECBM	enhanced coal-bed methane recovery
ECUST	East China University of Science and Technology
FBDB	fluid bed dry bottom
EO	ethylene oxide
EOR	enhanced oil recovery
FCC	fluid catalytic cracking
GHG	greenhouse gas
GTL	gas to liquid
GWP	global warming potential
HRT	hydraulic retention time
IL	ionic liquid
KBR	Kellogg Brown and Root
MDEA	N-methyldiethanolamine
MEA	monoethanolamine
MMM	mixed matrix membranes
MOF	metal organic framework
MPG	multipurpose gasification

NGL	natural gas liquid
PAA	polyallylamine
PAI	poly(amide-imide)
PAMAM	polyamidoamine
PEI	poly(ethyleneimine)
ppb	parts per billion
PSA	pressure swing adsorption
PVA	poly(vinyl alcohol)
PVAm	polyvinylamine
PZ	Piperazine
RFCC	residue fluid catalytic cracking
RPM	random pore model
TEA	triethanolamine
TPRI	Thermal Power Research Institute
TSA	temperature swing adsorption
VM	volumetric model
WGS	water-gas shift

REFERENCES

Ahmed, S., Schmalzer, D. K., Van Essendelft, D. T., Shadle, L. J., Siefert, N. S., Link, D., Richards, G. A., Seol, Y., Shekhawat, D., Soeder, D. J., Ramezan, M., Stiegel, G. J., Loftus, P. J., Kemp, I. C., Smith, L., Smith, J. D., Dees, D. W., Energy Resources, Conversion, and Utilization, In Green, D. W., Southard, M. Z., (Eds.) Perry's Chemical Engineers' Handbook, 9th ed., McGraw-Hill, New York (2019).

Appari, S., Janardhanan, V. M., Bauri, R., Jayanti, S., Deutschmann, O., A detailed kinetic model for biogas steam reforming on Ni and catalyst deactivation due to sulfur poisoning, Appl. Catal. A. G., 471 (2014) 118–125.

Aresta, M., Carbon Dioxide Recovery and Utilization, Kluwer Academics Publishers, The Netherlands (2003).

Bader, A., Hartwich, M., Richter, A., Meyer, B., Numerical and experimental study of heavy oil gasification in an entrained flow reactor and the impact of the burner concept, Fuel Proc. Tech., 169 (2018) 58–70.

Basu, P., Biomass Gasification, Pyrolysis, and Torrefaction Practical Design and Theory, 2nd ed., Academic Press, Amsterdam (2013).

Bell, D. A., Towler, B. F., Fan, M., Coal Gasification and its Applications, Elsevier, Amsterdam (2011).

Bhatia, S. K., Perlmutter, D. D., A random pore model for fluid-solid reactions: I. Isothermal and kinetic control, AIChE J., 26 (1980) 379–386.

Bhatia, S. K., Vartak, B. J., Reaction of microporous solids: the discrete random pore model, Carbon 34 (1996) 1383–1391.

Bridgwater, A. V., The technical and economic feasibility of biomass gasification for power generation, Fuel, 74 (1995) 631–653.

Chauvy, R., Meunier, N., Thomas, D., De Weireld, G., Selecting emerging CO_2 utilization products for short- to mid-term deployment, Appl. Energy 236 (2019) 662–680.

Choi, S., Drese, J. H., Jones, C. W., Adsorbent materials for carbon dioxide capture from large anthropogenic point sources., ChemSusChem., 2 (2009) 796–854.

Chong, Z. R., Yang, S. H. B., Babu, P., Linga, P., Li, X. S., Review of natural gas hydrates as an energy resource: Prospects and challenges, 162 (2016) 1633–1652.

Cornils, B., Herrmann, W. A., Wong, C.-H., Zanthoff, H.-W. (Eds.) Catalysis from A to Z, Wiley-VCH, Weinheim (2013).

Crelling, J. C., Hagemann, H. W., Sauter, D. H., Ramani, R. V., Vogt, W., Leininger, D., Krzack, S., Meyer, B., Orywal, F., Reimert, R., Bonn, B., Bertmann, U., Klose, W., Dach, G., Coal, Ullmann's Encyclopedia of Industrial Chemistry, Wiley-VCH Verlag GmbH & Co. KGA, Weinheim (2006).

Cuellar-Franca, R. M., Azapagic, A., Carbon capture, storage and utilization technologies: a critical analysis and comparison of their life cycle environmental impacts, J. CO₂ Util., 9 (2015) 82–102.

D'Alessandro, D. M., Smit, B., Long, J. R., Carbon dioxide capture: prospects for new materials, Angew. Chem. Int. Ed., 49 (2010) 6058–6082.

EIA, Energy Information Administration, Natural gas processing: the crucial link between natural gas production and its transportation to market (2006) 1–11.

Euker, C. A., Reitz, R. A., Exxon catalytic coal gasification process development program. Final Project Report for the U. S. Department of Energy under Contract No. ET-78-C-01-2777 (1981).

Fan, Y., Labreche, Y., Lively, R. P., Jones, C. W., Koros. W. J., Dynamic CO₂ adsorption performance of internally cooled silica-supported poly(ethylenimine) hollow fiber sorbents, AIChE J., 60 (2014) 3878–3887.

Faramawy, S., Zaki, T., Sakr, A. A. E., Natural gas origin, composition, and processing: a review, J. Nat. Gas Sci. Eng., 34 (2016) 34–54.

Freeman, S. A., Dugas, R., van Wagener, D. H., Nguyen, T., Rochelle, G. T., Carbon dioxide capture with concentrated, aqueous piperazine, Int. J. Greenhouse Gas Cont., 4 (2010) 119–124.

Furimsky, E., Gasification in petroleum refinery of 21ˢᵗ century, Oil Gas Sci. Tech. Rev. IFP, 54 (1999) 597–618.

Furukawa, H., Cordva, K. E., O'Keeffe, M., Yaghi, O. M., The chemistry and applications of metal-organic frameworks. Science, 341 (2013) 974–988.

Gambarotta, A., Morini, M., Zubani, A., A non-stoichiometric equilibrium model for the simulation of the biomass gasification process, Appl. Energy, 227-1 (2018) 119–127.

Goeppert, A., Czaum, M., Prakash, G. K. S., Olah, G. A., Air as the renewable carbon source of the future: an overview of CO₂ capture from the atmosphere., Energy Env. Sci., 5 (2012) 7833–7853.

Gurkan, B. E., de la Fuente, J. C., Mindrup, E. M., Ficke, L. E., Goodrich, B. F., Price, E. A., Schneider, W. F., Brennecke, J. F., Equimolar CO₂ absorption by anion-functionalized ionic liquids, J. Am. Chem. Soc., 132 (2010) 2116–2117.

Higman, C., van der Burgt, M., Gasification, Gulf Professional Publishing, Oxford, UK (2nd ed.) (2008).

Huang, C. H., Tan, C. S., A review: CO₂ utilization. Aero. Air Qual. Res., 14 (2014) 480–499.

Irfan, M. F., Usman, M. R., Kusakabe, K., Coal gasification in CO₂ atmosphere and its kinetics since 1948: a brief review, Energy 36 (2011) 12–40.

Italiano, C., Vita, A., Fabiano, C., Lagana, M., Pino, L., Bio-hydrogen production by oxidative steam reforming of biogas over nanocrystalline Ni/CeO₂ catalysts, Int. J. Hyd. Energy, 40 (2015) 11823–11830.

Joshi, R. K., Carbone, P., Wang, F. C., Kravets, V. G., Su, Y., Grigorieva, I. V., Wu, H. A., Geim, A. K., Nair, R. R., Precise and ultrafast molecular sieving through graphene oxide membranes, Science, 343 (2014) 752–754.

Kang, T.-J., Park, H., Namkung, H., Xu, L.-H., Park, J.-H., Heo, I., Chang, T.-S., A study on the direct catalytic steam gasification of coal for the bench-scale system, Korean J. Chem. Eng., 34 (2017) 2597–2609.

Kapteijn, F., Porre, H., Moulijn, J. A., CO₂ gasification of activated carbon catalyzed by earth alkaline elements. AIChE J., 32 (1986) 691–695.

Khoo, H. H., Bu, J., Wong, R. L., Kuan, S. Y., Sharratt, P. N., Carbon capture and utilization: preliminary life cycle CO_2, energy, and cost results of potential mineral carbonation, Energy Proc., 4 (2011) 2494–2501.

Knoef, H. A. M. (Ed.), Handbook Biomass Gasification. BTG Publisher, Enschede, The Netherlands (2005) pp. 239–241.

Krishna, B. B., Biswas, B., Bhaskar, T., Gasification of Lignocellulosic Biomass, In Biofuels: Alternative Feedstocks and Conversion Processes for the Production of Liquid and Gaseous biofuels, Elsevier Amsterdam (2019).

Latimer, A. A., H. Aljama, A. Kakekhani, J. S. Yoo, A. Kulkarni, C. Tsai, M. Garcia-Melchor, F. Abild-Pedersen, J. K. Nørskov, Mechanistic insights into heterogeneous methane activation, Phys. Chem. Chem. Phys., 19 (2017) 3575–3581.

Levenspiel, O, Chemical Reaction Engineering, 3rd ed., Wiley, USA (1999).

Li, J. R., Sculley, J., Zhou, H. C., Metal-organic frameworks for separations, Chem. Rev., 112 (2012) 869–932.

Li, L., Zhao, N., Wei, W., Sun, Y., A review of research progress on CO_2 capture, storage, and utilization in Chinese Academy of Sciences, Fuel, 108 (2013) 112–130.

Li, R., Wang, Z., Liu, Y., Zhao, S., Wang, J., Wang, S., A synergistic strategy via the combination of multiple functional groups into membranes towards superior CO_2 separation performances, J. Memb. Sci., 476 (2015a) 243–255.

Li, X., Ma, L., Zhang, H., Wang, S., Jiang, Z., Guo, R., Wu, H., Cao, X. Z., Yang, J., Wang, B., Synergistic effect of combining carbon nanotubes and graphene oxide in mixed matrix membranes for efficient CO_2 separation., J. Memb. Sci., 479 (2015b) 1–10.

Lin, L. C., Kim, J., Kong, X., Scott, E., McDonald, T. M., Long, J. R., Reimer, J. A., Smit, B., Understanding CO_2 dynamics in metal-organic frameworks with open metal sites, Angew. Chem. Int. Ed., 52 (2013) 4410–4413.

Liu, L., Cao, Y., Liu, Q., Yang, J., Experimental and kinetic studies of coal-CO_2 gasification in isothermal and pressurized conditions, RSC Adv., 7 (2017) 2193–2201.

Luque, R., Herrero-Davila, L., Campelo, J. M., Clark, J. H., Hidalgo, J. M., Biofuels: a technological perspective, Energy Env. Sci., 1 (2008) 542–564.

Mahinpey, N., Gomez, A., Review of gasification fundamentals and new findings: reactors, feedstock, and kinetic studies, Chem. Eng. Sci., 148 (2016) 14–31.

Mallick, D., Mahanta, P., Moholkar, V. S., Co-gasification of coal and biomass blends: Chemistry and engineering, Fuel, 204 (2017) 106–128.

Markewitz, P., Kuckshinrichs, W., Leitner, W., Linssen, J., Zapp, P., Bongartz, R., Schreiber, A., Muller, T. E., Worldwide innovations in the development of carbon capture technologies and the utilization of CO_2, Energy Env. Sci. 5 (2012) 7281–7305.

Merkel, T. C., Zhou, M., Baker, R. W., Carbon dioxide capture with membranes at an IGCC power plant, J. Memb. Sci., 389 (2012) 441–450.

Mishra, A., Gautam, S., Sharma, T., Effect of operating parameters on coal gasification, Int. J. Coal Sci. Technol., 5 (2018) 113–125.

Molino, A., Chianese, S., Musmarra, D., Biomass gasification technology: the state-of-the-art overview, J. Energy Chem., 25 (2016) 10–25.

Molino, A., Larocca, V., Chianese, S., Musmarra, D., Biofuels production by biomass gasification: a review, Energies, 11 (2018) 811–841.

Muñoz, R., Meier, L., Diaz, I., Jeison, D., A review on the state-of-the-art of physical/chemical and biological technologies for biogas upgrading, Rev. Env. Sci. Bio Tech., 14 (2015) 727–759.

Muradov, N., Liberating energy from carbon: introduction to decarbonization, Springer, New York (2014).

Najera, M., Solunke, R., Gardner, T., Veser, G., Carbon capture and utilization via chemical looping dry reforming, Chem. Eng. Res. Des., 89 (2011) 1533–1543.

Nduagu, E., Bergerson, J., Zevenhoven, R., Life cycle assessment of CO_2 sequestration in magnesium silicate rock – a comparative study, Energy Conv. Manag., 55 (2012) 116–126.

Nguyen, N. T., Furukawa, H., Gandara, F., Nguyen, H. T., Cordova, K. E., Yaghi, O. M., Selective capture of carbon dioxide under humid conditions by hydrophobic chabazite-type zeolitic imidazolate frameworks., Angew. Chem. Int. Ed., 53 (2014) 10645–10648.

Olah, G. A., Molnár, Á., Prakash, G. K. S., Hydrocarbon Chemistry, 3rd ed, Wiley, New York (2018) 123–325.

Powell, C. E., Qiao, G. G., Polymeric CO_2/N_2 gas separation membranes for the capture of carbon dioxide from power plant flue gases, J. Memb. Sci., 279 (2006) 1–49.

Prins, M. J., Ptasinski, K. J., Janssen, F. J. J. G., From coal to biomass gasification: Comparison of thermodynamic efficiency, Energy 32 (2007) 1248–1259.

Puig-Arnavat, M., Bruno, J. C., Coronas, A., Review and analysis of biomass gasification models, Renew. Sust. Energy Rev. 14 (2010) 2841–2851.

Qiu, S., Xue, M., Zhu, G., Metal-organic framework membranes: from synthesis to separation application., Chem. Soc. Rev., 43 (2014) 6116–6140.

Ragnarsson, B. F., Oddsson, G. V., Unnthorsson, R., Hrafnkelsson, B., Levelized cost of energy analysis of a wind power generation system at Burfell in Iceland, Energies, 8 (2015), 9464–9485.

Ramatsa, I. M., Akinlabi, E. T., Madyira, D. M., Huberts, R., Design of the bio-digester for biogas production: a review, Proceedings of the World Congress on Engineering and Computer Science (2014) Vol II.

Recari, J., Berrueco, C., Abelló, S., Montané, D., Farriol, X., Gasification of two solid recovered fuels (SRFs) in a lab-scale fluidized bed reactor: influence of experimental conditions on process performance and release of HCl, H_2S, HCN and NH_3, Fuel Proc. Tech., 142 (2016) 107–114.

Remy, T., Peter, S. A., Tendeloo, L. V., der Perre, S. V., Lorgouilloux, Y., Kirschhock, C., Baron, G. V., Denayer, J. F. M., Adsorption and separation of CO_2 on KFI zeolites: effect of cation type and Si/Al ratio on equilibrium and kinetic properties., Langmuir, 29 (2013) 4998–5012.

Roš, M., Zupančič, G. D., Organic farm waste and municipila sludge, In Kole, C., Joshi, C. P., Shonnard, D. R. (Eds.) Handbook of Bioenergy Crop Plants, CRC Press (2012).

Safarian, S., Unnþorsson, R., Richter, C., A review of biomass gasification modelling, Renew. Sust. Energy Rev., 110 (2019) 378–391.

Samanta, A., Zhao, A., Shimizu, G. K.H., Sarkar, P., Gupta, R., Post-combustion CO_2 capture using solid sorbents: a review., Ind. Eng. Chem. Res., 51 (2012) 1438–1463.

Sharma, A. K., Equilibrium and kinetic modeling of char reduction reactions in a downdraft biomass gasifier: a comparison, Solar Energy, 82 (2008) 918–928.

Sharma, S., Sheth, P. N., Air-steam biomass gasification: experiments, modeling and simulation, Energy Con. Manag. 110 (2016) 307–318.

Singh, B., Strømman, A. H., Hertwich, E. G., Comparative life cycle environmental assessment of CCS technologies, Int. J. Greenhouse Gas Cont. 5 (4) (2011) 911–921.

Slater, A. G., Cooper, A. I., Function-led design of new porous materials, Science, 348 (2015) 988.

Song, J., Zhang, D., Comprehensive review of Caprock-Sealing mechanisms for geologic carbon sequestration, Env. Sci. Tech. 47 (2013) 9–22.

Speight, J. G., Natural Gas: A Basic Handbook, Gulf Publishing Company, Houston, TX (2007).

Speight, J. G., The Refinery of the Future, Elsevier, Amsterdam (2011).

Speight, J. G., Shale Gas Production Processes: Gulf Professional Publishing, USA (2013).

Speight, J. G., Handbook of Gasification Technology Science, Processes, and Applications, Wiley (2020).

Spigarelli, B. P., Kawatra, S. K., Opportunities and challenges in carbon dioxide capture, J. CO₂ Util., 1 (2013) 69–87.

Sumida, K., Rogow, D. L., Mason, J. A., McDonald, T. M., Bloch, E. D., Herm, Z. R., Bae, T. H., Long, J. R., Carbon dioxide capture in metal-organic frameworks. Chem. Rev., 112 (2012) 724–781.

Sung, S., Suh, M. P., Highly efficient carbon dioxide capture with a porous organic polymer impregnated with polyethylenimine, J. Mater. Chem. A., 2 (2014) 13245–13249.

Szekely, J., Evans, J. W., A structural model for gas-solid reactions with a moving boundary, Chem. Eng. Sci., 25 (1970) 1091–1107.

Tanner, J., Bhattacharya, S., Kinetics of CO₂ and steam gasification of Victorian brown coal chars, Chem. Eng. J., 285 (2016) 331–340.

Technology Handbook (Ed.), Air Liquide Global E&C Solutions US Inc. (2016).

Vaezi, M., Passandideh-Fard, M., Moghiman, M., Charmchi, M., Gasification of heavy fuel oils: a thermochemical equilibrium approach, Fuel 90 (2011) 878–885.

Walker Jr., P. L., Rusinko Jr., F., Austin, L. G., Gas Reactions of Carbon, In: Eley, D. D., Selwood, P. W., Weisz, P. B., (Eds.), Advances in Catalysts, vol. XI, Academic Press, New York (1959) pp. 133–221.

Warnecke, R., Biomass and bioenergy, gasification of biomass: comparison of fixed bed and fluidized bed gasifier, Biomass Bioenergy, 18 (2000) 489–497.

Weiland P. Biogas production: current state and perspectives. Appl. Microbiol. Biotech., 85 (2010) 849–860.

Xu, X. C., Song, C. S., Andresen, J. M., Miller, B. G., Scaroni, A. W., Novel polyethylenimine-modified mesoporous molecular sieve of MCM-41 type as high-capacity adsorbent for CO₂ capture, Energy Fuels, 16 (2002) 1463–1469.

Yu, K. M., Curcic, I., Gabriel, J., Tsang, S. C., Recent advances in CO₂ capture and utilization, ChemSusChem 1 (2008) 893–899.

Yuan, Z., Eden, M. R., Gani, R., Toward the development and deployment of large-scale carbon dioxide capture and conversion processes, Ind. Eng. Chem. Res., 55 (2016) 3383–3419.

Yuan, X., Lee, K. B., Kim, H. T., Investigation of Indonesian low rank coals gasification in a fixed bed reactor with K2CO3 catalyst loading, 92 (2019) 904–912.

Zapp, P., Schreiber, A., Marx, J., Haines, M., Hake, J. F., Gale, J., Overall environmental impacts of CCS technologies – a life cycle approach, Int. J. Greenhouse Gas Cont. 8 (2012) 12–21.

Zhang, J., Zhang, R., Bi, J., Effect of catalyst on coal char structure and its role in catalytic coal gasification, Catal. Comm., 79(2016) 1–5.

Zhang, J., Li, J., Mao, Y., Bi, J., Zhu, M., Zhang, Z., Zhang, L., Zhang, D., Effect of CaCO₃ addition on ash sintering behavior during K₂CO₃ catalyzed steam gasification of a Chinese lignite, Appl. Thermal Eng., 111 (2017) 503–509.

Zinoviev, S., Müller-Langer, F., Das, P., Bertero, N., Fornasiero, P., Kaltschmitt, M., Centi, G., Miertus, S., Next-generation biofuels: survey of emerging technologies and sustainability issues, ChemSusChem, 3 (2010) 1106–1133.

3 C_1 Interconversions

3.1 INTRODUCTION

Methane and carbon dioxide are the most abundant C_1 compounds (Olah et al., 2018). Therefore, their utilization as C_1 building blocks for the production of chemicals and fuels is gaining interest. However, their use as chemical feedstock is limited; due to their low reactivity resulted from high stability of the molecules. Methane and carbon dioxide are the fully reduced and fully oxidized forms of carbon, respectively (Lee et al., 2015). The portfolio of the products by heterogeneous catalysis is, therefore, limited and many potential processes are still under development.

An alternative route is to convert these compounds into intermediate more reactive C_1 building blocks (platform C_1 compounds) such as CO and methanol which are subsequently transformed into the desired products. Thus, most of the commercialized conversions are via an indirect route through cascade reactions in which methane and carbon dioxide are first converted into more reactive CO, CH_3OH and then converted into the desired product. This, however, will increase the production cost compared to the direct routes.

C_1 compounds can be converted into each other in reactions not involving C–C bond couplings. The conversion of abundant poorly reactive compounds such as methane and CO_2 to more reactive compounds such as CO and methanol are considered as a promising route for indirect conversion of the former compounds.

This chapter will present examples of processes that have been used to convert these two C_1 compounds into more versatile and reactive C_1 compounds. There are large numbers of other possible C_1 interconversions typically useful for energy transport/storage. Examples are methanol reforming to CO_2, which tend to be used for hydrogen or energy transfer and is not of interest in C_1 chemistry. Carbon dioxide hydrogenation to methane (Power to Gas, PTG) is developed for storage of renewable energies such as wind and solar (Sahebdelfar and Takht Ravanchi, 2021).

3.2 METHANE CONVERSIONS

3.2.1 REFORMING OF METHANE

In reforming of methane, methane is converted into a mixture of CO and H_2 (synthesis gas) by partial oxidation. The product may also contain variable amounts of CO_2. According to the oxidizing agent, steam, CO_2 and oxygen, the process is termed as steam reforming (Eq. 3.1), dry reforming (Eq. 3.2) and partial oxidation (POX) (Eq. 3.3) of methane, respectively:

$$CH_4 + H_2O \Leftrightarrow CO + 3H_2 \qquad \Delta H_{298}^0 = 206 \, \text{kJ/mol}$$
$$\Delta G^0 = 70317 - 75.16T \, \text{kJ/mol} \tag{3.1}$$

DOI: 10.1201/9781003279280-3

43

$$CH_4 + CO_2 \Leftrightarrow 2CO + 2H_2 \qquad \Delta H_{298}^0 = 248\,kJ/mol$$
$$\Delta G^0 = 61770 - 67.32T\,kJ/mol \tag{3.2}$$

$$CH_4 + 0.5O_2 \rightarrow CO + 2H_2 \qquad \Delta H_{298}^0 = -35.6\,kJ/mol$$
$$\Delta G^0 = 70317 - 75.16T\,kJ/mol \tag{3.3}$$

Where T is in kelvins. The above reactions are basic reforming processes. They differ in product syngas composition and heat of reaction. According to the catalyst and operation condition, several side reactions accompany methane reforming. These include water-gas shift (WGS) (Eq. 3.4):

$$CO + H_2O \Leftrightarrow H_2 + CO_2 \qquad \Delta H_{298}^0 = -41.2\,kJ/mol$$
$$\Delta G^0 = 8545 - 7.84T\,kJ/mol \tag{3.4}$$

And coke formation reactions such as methane decomposition (Eq. 3.5), Boudouard (or CO disproportionation) reaction (Eq. 3.6) and carbon monoxide reduction by water (Eq. 3.7):

$$CH_4 \Leftrightarrow C + 2H_2 \qquad \Delta H_{298}^0 = 75\,kJ/mol$$
$$\Delta G^0 = 21960 - 26.45T\,kJ/mol \tag{3.5}$$

$$2CO \Leftrightarrow C + CO_2 \qquad \Delta H_{298}^0 = -172\,kJ/mol$$
$$\Delta G^0 = -39810 + 40.87T\,kJ/mol \tag{3.6}$$

$$CO + H_2 \Leftrightarrow C + H_2O \qquad H_{298}^0 = -131\,kJ/mol$$
$$\Delta G^0 = -31263 - 33.399T\,kJ/mol \tag{3.7}$$

Most of the above-mentioned reactions are equilibrium limited. From Gibbs free change of reaction–temperature relations given above, one deduces that the endothermic Eq. (3.5) is favored above 555°C, while the exothermic reactions (3.6) and (3.7) are thermodynamically favorable below 700 and 670°C, respectively (Jafarbegloo et al., 2015).

Coke formation tendency in POX is lower than steam reforming of methane (SRM) and dry reforming of methane (DRM) since oxygen is a stronger oxidant than water and CO_2 and thus can oxidize and remove the coke deposited in situ.

3.2.1.1 Steam Reforming of Methane

Steam reforming of methane (SRM, Eq. 3.1) is the oldest and most widely used methane reforming reaction for production of synthesis gas (or syngas) and hydrogen. It is used in Haber-Bosch process for ammonia synthesis for producing hydrogen from natural gas. The reverse of this reaction (methanation) occurs at much lower temperatures and is used for eliminating carbon oxides in ammonia plants and also for producing synthetic natural gas (SNG) from coal or biomass.

The SRM reaction is highly endothermic and thus is thermodynamically favored at higher temperatures. Under stoichiometric feed composition, coke formation is favored; therefore, a higher H_2O/CH_4 ratio is used. Although high pressures reduce equilibrium conversions, pressures above 20 atm are used to avoid excessive compressions because the product needs high pressures for the subsequent conversion. Thus, the typical industrial operating conditions are 750–950°C and 30–35 bar and $H_2O/CH_4 = 2.5$–3.0 mol/mol.

The reaction is catalyzed by metals of groups 8–11. Noble metals exhibit the best performance but their high price limits their application. Among base metals, Ni is the preferred one because of a combination of high activity and stability. Fe and Co are also active, but they are prone to oxidation under reaction conditions. The order of activity is Rh > Ru ≫ Ni > Pt > Pd > Re, however, it may differ somewhat in different references due to variation in dispersion among catalysts (Bartholomew and Farrauto, 2006).

The commercial catalyst is Ni (15–25 wt.%) supported on α-Al_2O_3, $CaAl_2O_4$, $MgAl_2O_4$ spinel or MgO. Prior to startup, the nickel oxide should be reduced by the feed or preferably hydrogen at about 600°C.

The effectiveness factor for industrial catalysts is estimated in the range of 0.02–0.05, meaning that the reaction rate per catalyst weight is a few percent of that for small particles. The low effectiveness factor implies a high diffusion limitation, that is, the reaction rate is much larger than that for entering and leaving catalyst pellets (van der Oosterkamp, 2003). Thus, within pores, the reacting species might be in chemical equilibrium while in bulk phase not. Therefore, to evaluate the coke forming tendency, one must consider both bulk and equilibrium composition of the gases. To increase particle surface area, the commercial catalysts are typically finned.

Coke formation and sintering are the main causes of catalyst deactivation under industrial operation conditions. The main source of carbon is methane decomposition (Eq. 3.5) or Boudouard reaction (Eq. 3.6). Whisker carbon is formed through decomposition of methane adsorbed on Ni surface (Usman and Wan Daud, 2015). Nevertheless, carbon deposition can be suppressed on Ni catalysts by careful selection of alkali promoters like MgO, La_2O_3 and Ce_2O_3 which increase the adsorption of CO_2 and steam on the catalyst surface and thereby enhancing the rate of carbon removal (Rostrup-Nielsen et al., 2002).

Sulfur compounds are catalyst poisons at concentrations as low as 50 ppm. Furthermore, the low melting point of NiS enhances surface mobility and thus sintering (Albertazzi et al., 2007).

To supply the required heat of reaction, SRM is operated in fire-heated fixed-bed multitubular reactors in commercial practice (Figure 3.1). The catalyst is loaded in vertical tubes surrounded by furnaces. It can be said that steam reforming is "heat flux limited" which means that the reaction is limited by heat transfer and not by kinetics (van der Oosterkamp, 2003). The basic configurations of reformer furnaces are top fired, bottom fired, side fired and trace-wall (Figure 3.1).

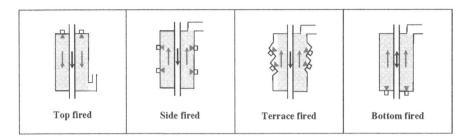

FIGURE 3.1 Basic types of radiation zones in steam reformers (Redrawn from Häring, 2008).

The process gas enters the reformer at 450–600°C, while the product outlet temperature is 700–950°C. High wall temperatures are one of the problems of steam reforming, in which the vertical tubes in a furnace of special construction can attain skin temperatures exceeding 950°C (Le Page 1987).

Despite improvements over years, SR is still very energy demanding (Kondratnko and Baerns, 2003). SRM is very capital and energy intensive because of a need for a large volume of superheated steam and endothermic nature of the reaction. Because of the simultaneous occurrence of WGS, the actual H_2/CO ratio of SRM is larger than the stoichiometric value (3). Thus, SRM gives a hydrogen-rich syngas which is often suitable for hydrogen production. For chemical synthesis, a lower ratio is more desirable.

SRM is a well-established technology, but further improvements are needed for reformer tube material, controlling carbon limits and developing more effective catalysts with higher sulfur and carbon tolerance.

3.2.1.2 Methane Dry Reforming

For the synthesis of oxygenate compounds, a lower H_2/CO ratio is desirable. The carbon dioxide (or dry) reforming of methane (Eq. 3.2) utilizes two greenhouse gases to produce CO-rich syngas. A potential advantage of using CO_2 in reforming feed is that the process could be used for low-quality CO_2-rich natural gas without the need for separation.

Dry reforming of methane (DRM) is more endothermic than SRM and thus higher reaction temperature are necessary which increase the danger of catalyst deactivation.

Supported noble metals (e.g., Ru, Rh, Pd, Ir and Pt) and other transition metals (e.g., Fe, Co, Ni and Cu) have been investigated for DRM. Noble metals are very active and tolerant to carbon formation (Rezaei et al., 2006). Their high activity has been proven theoretically by first principles calculations as well (Lee et al., 2015). However, commercial use of noble metals in catalyst formulation is limited due to their high cost. Ni and Co show comparable activity and are also cost effective, but deactivate rapidly due to coke formation and metal sintering (Jang et al., 2019). Most of the studies on DRM catalysts have been focused on metal-promoted or bimetallic supported catalysts.

The support should provide a high surface area for dispersion of active metal and show high thermal stability under prominent high reaction temperature. Oxide support such as SiO_2, CeO_2, ZrO_2, Al_2O_3, MgO, CaO, TiO_2 and La_2O_3 are most commonly investigated. Certain physicochemical properties of the support such as acidity/basicity (CaO, La_2O_3 and MgO), oxygen storage capacity (CeO_2, CeO_2-ZrO_2 and TiO_2) and reducibility (CeO_2 and ZrO_2) could influence intrinsic activity and coke formation behavior of the catalyst (Abdulrasheed et al., 2019).

Catalysts with mixed oxide supports have also been investigated. A series of Pt/ZrO_2-SiO_2 catalysts with varying ZrO_2 to SiO_2 ratios were prepared by deposition precipitation and tested for DRM (Reddy et al., 2010). ZrO_2 was precipitated in colloidal silica solution with ammonia. The following order in activity and stability was observed: 4: 1 (ZrO_2: SiO_2)> pure ZrO_2>3: 1(ZrO_2: SiO_2)>2: 1 (ZrO_2: SiO_2)> pure SiO_2. The difference was explained by Pt-ZrO_2 interfacial sites. A small amount of SiO_2 would increase the interfacial area between Pt and ZrO_2 resulting in increased activation of CO_2 compared to pure ZrO_2 or SiO_2. The Pt-ZrO_2 interfacial sites were also considered to be active in oxidation of carbon and reducing coke formation.

Within the temperature range of 560–700°C, carbon formation is thermodynamically favorable by both decomposition of methane (Eq. 3.5) and Boudouard reaction (Eq. 3.6). Therefore, dry reforming of methane is usually carried out at temperatures above 750°C to reduce carbon formation (Jafarbegloo et al., 2015). These conditions also reduce the reactor size by increasing the reaction rate.

Carbon formation is a major problem in DRM and should be minimized for commercialization of the technologies. Two technologies based on this reaction have been developed, both of which address coke formation and operate at high temperatures (>900°C) (Arora and Prasad, 2016). The Calcor process has been used for in-site production of high purity CO to avoid the problems associated with CO transport to manufacturing plants. Carbon formation is controlled by using excess CO_2 in a desulfurized feed, which goes through reformer tubes filled with Ni catalysts of different activities and shapes at high temperature and low pressure. In SPARG process (sulfur-passivated reforming), steam is partly replaced by CO_2. Natural gas is first pre-reformed and then passed over sulfur-passivated nickel catalyst. The undercoordinated sites that can promote carbon nucleation are preferentially blocked by the H_2S added to the feed gas. The chemisorption of H_2S to the catalytic sites is thermodynamically favored over carbon growth. Meanwhile, the catalyst still remains sufficiently active to give high conversions of methane. The low H_2/CO ratio can be obtained by adjusting the CO_2 concentrations in the feed gas (Lee et al., 2015). The SPARG process still needs to separate the H_2S for downstream use of the syngas product, as H_2S can poison most catalysts.

3.2.1.2.1 Mechanism and Kinetics

The commonly agreed mechanism for DRM involves activation of CH_4 on metal and activation of CO_2 on the acidic or basic support. The activation of CO_2 leads to a surface formate or carbonate on acidic or basic supports, respectively. Over inert supports, the formation of carbon by methane dehydrogenation will limit the CO_2 activation by the support leading to deactivation. The metal-support interaction is relatively weak on inert supports. Therefore, catalysts with inert supports, such as

SiO_2, exhibit inferior activity in dry reforming of methane (Lee et al., 2015). These catalysts follow a mono-functional pathway by which both CH_4 and CO_2 are activated on active metal (Pakhare and Spivey, 2014).

The mechanism of SRM is well established and the simplified steps are as follows:

$$CH_4 + (n+1)S \Leftrightarrow CH_{4-n} - S + nS - H \tag{3.8}$$

$$CH_x - S + S - O \Leftrightarrow CO + x/2H_2 + 2S \tag{3.9}$$

$$CH_x - S \Leftrightarrow S - C \rightarrow \text{whisker carbon} \tag{3.10}$$

$$H_2O + S \Leftrightarrow S - O + H_2 \tag{3.11}$$

For dry reforming, the supplementary steps are:

$$CO_2 + S \Leftrightarrow S - O + CO \tag{3.12}$$

$$CO_2 + S - C \Leftrightarrow 2CO + S \tag{3.13}$$

The catalytic steam and dry reforming of methane are known to be structure sensitive (Vogt et al., 2020). This means that not all atoms of a metallic nanoparticle have the same catalytic activity. By changing nanoparticle size, the fraction of surface-active sites will change. Smaller nanoparticles have lower coordinated (step and edge) sites compared to larger ones which have flatter and terrace sites. In steam and dry reforming of methane activity, selectivity and stability could be affected by the metal particle size of the supported catalyst. Activation of methane, recombination of C and O and coupling of C–C to form carbon nanofibers all might be structure sensitive in both reactions.

Xu and Froment (1989) proposed the following kinetic expressions for SRM and accompanying reactions on a $Ni/MgAl_2O_4$ catalyst based on a Langmuir-Hinshelwood mechanism

$$-r_1 = k_1 \left(\frac{P_{CH4}P_{H2O}}{P_{H2}^{2.5}} - \frac{P_{CO}P_{H2}^{0.5}}{K_1} \right) / \text{DEN}^2 \tag{3.14}$$

$$-r_2 = k_2 \left(\frac{P_{CO}P_{H2O}}{P_{H2}} - \frac{P_{CO2}}{K_2} \right) / \text{DEN}^2 \tag{3.15}$$

$$-r_3 = k_3 \left(\frac{P_{CH4}P_{H2O}^2}{P_{H2}^{3.5}} - \frac{P_{CO2}P_{H2}^{0.5}}{K_2} \right) / \text{DEN}^2 \tag{3.16}$$

In which $\text{DEN} = 1 + K_{CO}P_{CO} + K_{H2}P_{H2} + K_{CH4}P_{CH4} + K_{H2O}P_{H2O}/P_{H2}$. Also, P is the partial pressure, k kinetic rate constants and K is the equilibrium constant and 1, 2 and 3 refer to SRM, WGS and steam reforming to CO_2 reactions, respectively.

For dry reforming, power-law rate expressions are commonly used as

$$-r = kP_{CH4}^m P_{CO2}^n \qquad (3.17)$$

Mechanism-based kinetic models have also been proposed (e.g., Verykios, 2003; Quiroga and Luna, 2007).

Despite its apparent economic and environmental benefits, dry reforming of methane is not regarded as an industrially mature technology. This is largely due to the rapid deactivation of conventional reforming catalysts by coke deposition and active metal sintering (Jang et al., 2019).

3.2.1.3 Partial Oxidation of Methane

Partial oxidation (POX) of methane to synthesis gas has attracted much attention, and a large number of studies have been published. The main advantage of this concept is the possibility for the conversion of all types of hydrocarbons.

The reaction occurs in refractory-lined reactors with burning fuel under limited oxygen supply. In catalytic partial oxidation (CPOX), the product gases pass a catalyst bed to achieve equilibrium composition before leaving the reactor (Figure 3.2). CPOX has the advantage of lower reaction temperature (800–900°C versus 1,200–1,500°C for noncatalytic), but it is still under development.

Both noble metals and non-noble metals were reported to be active catalysts (Tsang et al, 1995). Noble metals show high activity and selectivity in CPOX. Among non-noble metals, Ni, Co, Fe and Cu exhibited good catalyst performance. However, compared to noble metals, they have the problem of high coke formation and low stability. Ni is very active but deactivates rapidly. The performance could be improved by using a suitable support and promoters.

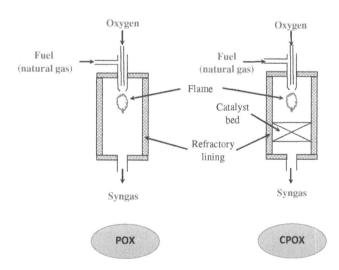

FIGURE 3.2 Reactor designs for POX and CPOX (Innovative catalytic reforming).

Reduced Ni/Al$_2$O$_3$ is the commonly used catalyst in the partial oxidation of methane to synthesis gas. However, the inactive NiAl$_2$O$_4$ phase is often formed during calcination in air. Complete oxidation of methane to CO$_2$ and H$_2$O occurred at a low temperature over the oxidized NiO/Al$_2$O$_3$ or NiAl$_2$O$_4$, but syngas was produced over Ni/Al$_2$O$_3$ catalyst reduced at high (>750°C) temperatures (Nakagawa et al., 1999).

Under conditions relevant to industrial operations; that is, high operating temperatures and pressures, homogeneous gas-phase reactions are promoted. This can lead to increased side reaction, thermal runaway and explosion.

The coke propensity is ordered as Ni > Pd > Rh > Ru > Pt ≈Ir.

High-pressure versions are known as HP POX processes (Cornils et al., 2013). The licensors are Shell and Texaco, which use high flame temperatures (1,400–1,500°C) to avoid soot formation in the flame (van der Oosterkamp, 2003).

The disadvantages of partial oxidation are requiring a flow of pure oxygen with related plant cost and risk of explosion if not the hydrocarbon-oxygen mixture properly controlled. Coke should be separated from the product by scrubbing.

3.2.1.4 Combined Reformings

The choice of a reforming reaction depends on ultimate use of the syngas. When hydrogen is the desired product, steam reforming is preferred. From a C$_1$ chemistry standpoint, an H$_2$/CO molar ratio of 2 is desirable as it is close to the stoichiometry of the synthesis of higher alcohols and liquid hydrocarbons. The desirable H$_2$/CO is adjustable by combining reforming reactions where operation under autothermal mode is also possible. In recent years, co-reforming of methane has been proposed in a single reactor (Liu et al., 2013b). A series of combinations along with their used are outlined below.

Autothermal reforming (ATR) combines exothermic partial oxidation of methane and endothermic steam reforming of methane in a single vessel. ATR was developed by Haldor Topsøe in late 1950s for methanol and ammonia synthesis. Later in 1990s, it was further developed for gas to liquid (GTL) by many companies.

The reformer is a refractory-lined pressure vessel that operates adiabatically (Figure 3.3). It comprises a thermal zone where combustion of methane or natural gas with oxygen or enriched air gas in a sub-stoichiometric flame occurs and a catalytic zone where equilibration of the syngas over Ni-based catalyst occurs. The key elements of ATR technology are a specially designed burner and the catalytic syngas production. The "autothermicity" is achieved by proper control of the combustion by the oxygen concentration. The reactions involved, in addition to Eqs. (3.1), (3.3) and (3.4), are the highly exothermic methane combustion reactions, for example:

$$CH_4 + 1.5O_2 \rightarrow CO + 2H_2O \quad \Delta H^0_{298} = -520\,\text{kJ/mol} \qquad (3.18)$$

The substoichiometric feed (H$_2$O/C = 0.5–3.5, O$_2$/C = 0.4–0.6) is burned at about 2,500°C where homogeneous reaction occurs. The product gas passes a catalyst bed (mostly Ni/MgAl$_2$O$_4$ spinel) where reforming reaction occurs and leaves the reactor in equilibrium at 850–1,100°C (Horn and Schlögl, 2015). The product is free of higher hydrocarbons and soot. The high reaction temperature reduces catalyst deactivation by poisoning and coke formation.

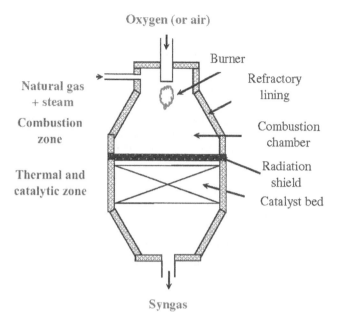

FIGURE 3.3 Schematic representation of an autothermal reactor.

The composition of syngas product can be adjusted by proper selection of the process conditions and also by recycling of CO_2.

3.2.1.4.1 Bi-Reforming

The close relation between steam and dry reforming has been noted in the literature. *Bi-reforming* of methane (BRM) is the catalytic combination of SRM and DRM by which a specific ratio of methane, steam, and CO_2 produces a syngas with essentially hydrogen to carbon monoxide ratio of 2/1 (Eq. 3.19), tentatively called "metgas" to differentiate it from the other syngas mixtures of varying H_2/CO ratio.

$$3CH_4 + 2H_2O + CO_2 \Leftrightarrow 4CO + 8H_2O \qquad (3.19)$$

This specific H_2/CO ratio of about 2 is close to stoichiometric feed for the manufacture of methanol and Fischer–Tropsch hydrocarbon products, with complete utilization of hydrogen. In most literature, Ni has been introduced as the catalyst. The typical catalyst is Ni/MgO (Olah et al, 2013). The Ni catalyst promoted with K_2O over MgO/LDH (layer double hydrotalcite), for example, with alkaline features and ability of CO_2 adsorption has shown good performance in BRM (Cunha et al., 2020).

3.2.1.4.2 Tri-Reforming Process

The addition of oxygen to reforming feed can reduce the problems related to coke formation of DRM and SRM, and also can make the system more energy efficient by in situ energy generation from methane oxidation. The tri-reforming of methane (TRM) is thus a synergetic combination of SRM, DRM, and POX of methane in a single fixed-bed reactor. The catalyst plays an important role and supported Ni is considered as an effective catalyst. For a coke-resistant catalyst, re-oxidation can cause catalyst deactivation. Thus, not all SRM catalysts are suitable for TRM, e.g., NiO/Al$_2$O$_3$ (Majewski and Wood, 2014).

The CO$_2$ may be captured from flue gases. The process is appropriate for low-quality CO$_2$-rich natural gas.

The H$_2$/CO ratio is typically between 1.5 and 2 and the temperature is around 700–850°C. The activity in terms of CO$_2$ conversion follows the order Ni/MgO > Ni/CeO$_2$ ~ Ni/ZrO$_2$ ~ Ni/Al$_2$O$_3$ > Ni/CeZrO at 850°C (Song and Pan, 2004).

3.2.2 PARTIAL OXIDATION TO C$_1$ OXYGENATES

The direct conversion of methane to C$_1$ oxygenates (methanol and formaldehyde) is of great interest because the intermediate methane reforming in indirect routes is highly energy intensive. Therefore, a single-step partial oxidation route would be of great economic interest. These reactions, however, are great challenges in heterogeneous catalysis (Beznis et al., 2010). The favorable operating temperature range is 400–600°C, above which C$_2$ hydrocarbons and syngas are successively produced (Arutyunov, 2018).

3.2.2.1 Methane to Formaldehyde

Formaldehyde is commercially produced by oxidative dehydrogenation of methanol, which in turn is produced from syngas obtained from reforming of methane. The single-step preparation via partial oxidation reaction with molecular oxygen can be shown as:

$$CH_4 + O_2 \Leftrightarrow HCHO + H_2O \quad \Delta H_{298}^0 = -276\,kJ/mol \tag{3.20}$$

The reaction is thermodynamically favorable but much less favorable than the competitive or consecutive oxidation to carbon oxides. Kinetic studies showed that formaldehyde is the primary product and CO and CO$_2$ are consecutive products (Berndt et al., 2000). It is very difficult to control the subsequent over-oxidation kinetically because formaldehyde is much more reactive than methane and thus more susceptible to oxidation. Thus, reducing diffusion limitation (e.g., by introducing porosity or reducing catalyst size) in the catalyst and reducing contact time in the reactor could reduce carbon oxides formation.

Various catalysts including silica-supported metal oxides (VO$_x$/SiO$_2$, MoO$_x$/SiO$_2$ and FeOx/SiO$_2$) and composite metal oxides (FePO$_4$, Fe(MnO$_4$)$_3$) have been developed for partial oxidation of methane to formaldehyde. VO$_x$ and MoO$_x$ exhibited comparatively higher activities (Shimura and Fujitani, 2019). For these active components, the inert SiO$_2$ has been recognized to be better than acidic or basic supports (e.g., Al$_2$O$_3$, TiO$_2$ and SnO$_2$). Furthermore, high dispersion of the active component is essential for high formaldehyde yield as the aggregated species promote oxidation of formaldehyde to carbon oxides.

Silica-supported vanadium and molybdenum oxide catalysts have traditionally been used in the reactions (McCormick et al., 2002). Vanadium oxide is more active whereas molybdenum oxide is more selective to formaldehyde (Du et al., 2006). Applying meso-porous silica support favors dispersion of active sites and site isolation thereby increasing formaldehyde selectivity in vanadium oxide catalysts (Nguyen et al., 2006).

As a consequence, two main approaches are commonly used: one is using high surface area mesoporous silica with well-ordered structure (SBA-15 and MCM-41) instead of amorphous silica as the support. Over conventional silica supports such as Cab-O-sil, bulk particles rather than monomers are formed for more than 5 wt.% loading (Berndt et al., 2000). The other is improving the preparation method (e.g., using sol-gel, ion exchange and grafting) because the conventional impregnation method is not always effective in site isolation. Nevertheless, the carbon-based form-aldehyde yield is still very low (3%–4%).

Additional measures are addition of promoter(s). Shimura and Fujitani (2019) added 14 metal oxides (MO$_x$) to their base VO$_x$/SiO$_2$ catalyst. Ga$_2$O$_3$, In$_2$O$_3$, Y$_2$O$_3$ and Al$_2$O$_3$ resulted in significant improvements, while MgO, MoO$_x$, TiO$_2$, CeO$_2$ and CaO were less effective. The preparation method and promoter loading were also effective. The best result was obtained with VO$_x$/Ga$_2$O$_3$/SiO$_2$ (V loading 2–3 and Ga loading 0.1–0.5 wt%) prepared by sequential impregnation. Under 590°C, GHSV = 60,000 h^{-1}, CH$_4$/O$_2$ = 2, the formaldehyde yield was twice that of unpro-moted catalyst. XRD and H$_2$-TPR results illustrated that it was possibly due to improved reducibility.

Over VMCM-41 catalysts prepared from different tetra-valent vanadium precur-sors (vanadyl acetate and vanadyl sulfate) by hydrothermal method through a mild gelation, the sample prepared form V(acac) (acac = acetylacetone anion) showed pre-dominantly monomer with slight oligomers and showed the highest formaldehyde yield (5.3 kg/kg.h) at 600°C (Dang et al., 2018).

The high dispersion of active species in FeO$_x$/SiO$_2$ catalysts is also essential. By increasing Fe loading in the 0.01–3.2 wt% range, EFR analysis showed differ-ent dispersions and Fe^{3+} coordination symmetries, that is, isolated Fe^{3+} ions, small Fe$_2$O$_3$ clusters and Fe$_2$O$_3$ particles in FeO$_x$/SiO$_2$ catalysts with relative abundance depending on loading (Parmaliana et al., 2002). The highest yield (1.5 × 10^{-3} g/m^2.s) occurred for 0.05–0.1 Fe atom/nm^2 at 650°C. The three Fe^{3+} types mentioned above exhibit different coordination, reducibility and activity. The best performance is seen for isolated Fe^{3+}.

Isolated V^{5+} species are considered as active sites (Loricera et al., 2017).

3.2.2.2 Methane to Methanol

The direct methane to methanol (DMTM) conversion by partial oxidation has received much attention:

$$CH_4 + 0.5O_2 \Leftrightarrow CH_3OH \quad \Delta H^0_{298} = -126 \, kJ/mol \tag{3.21}$$

The reaction can be performed non-catalytically with low conversion using high CH$_4$/O$_2$ ratios at 350–500°C (Tomkins et al., 2017). In sum, 7%–8% yield with 60% selectivity of methanol, for instance, has been achieved with CH$_4$/O$_2$ = 10 mol/mol at 430–470°C (Zhang et al., 2008). The reaction can be also preceded by a catalyst but significant improvement is not often observed.

It is generally believed that the process is controlled by gas-phase radical reactions with catalyst acting mostly as radical quencher thereby inhibiting overoxidation by blocking or reducing gas-phase radical chain reactions (Zhang et al., 2002; Sanchez-Sanchez and Lercher, 2019).

The oxidation of methane to methanol occurs in nature by methanotrophe bacteria by their methane monooxygenase (MMO) enzymes. There are two types of MMO, one water soluble containing di-iron centers as active sites; the other is particulate and bounded to membrane (pMMO) with copper centers.

Compared to methane to formaldehyde, much progress has been achieved for methanol selectivity in recent years (Horn and Schlogl, 2015). The best approach is to mimic copying the concept from homogeneous or biocatalysts. These efforts have been partly successful. Fe-zeolite catalysts, for example, have mono- and dinuclear Fe = O centers analogous to MMO enzymes. Cu(II)-O-Cu(II) configuration over Cu-ZSM5 catalyzes methane oxidation under mild reaction conditions (Al-Shihri et al., 2020).

The conventional supported metal oxides catalysts for methane oxidation are active for DMTM temperatures above 500°C but the methanol yields are typically low (Chorkendorff and Niemantsverdriet, 2017).

An interesting class of catalysts is zeolites containing transition metals such as Cu, Fe, Ni and Co. Cu-HZSM catalysts have been more extensively studied and their yield has been increased more than other catalysts in the past decade. While Fe-zeolite catalysts need N_2O as the oxidant for activation, Cu-zeolite and other transition metal zeolites can be activated by molecular oxygen. Many papers have been published for Cu-zeolites especially due to their ability to activate a variety of oxidants such as O_2, N_2O, H_2O_2 and H_2O (Mahyuddin et al., 2019). Therefore, Cu-zeolite catalysts are the candidate for commercialization. N_2O and other oxidants cannot compete with O_2 which is free and environmentally friendly.

The identification of active sites in Cu-zeolite is still controversial. A bis(μ-oxo) dicopper nucleus has been proposed as active centers from UV-vis studies (Smeets et al., 2005). It has been questioned later and a bent mono(μ-oxo) dicopper species was proposed based on resonance-enhanced Raman spectroscopy (Woertink et al., 2009). On the other hand, a recent theoretical study based on EXAFS proposed trinuclear copper-oxo cluster. Two isolated [Cu-O-Cu]$^{2+}$ has been proposed (Ravi et al., 2017).

Methanol selectivity and productivity has been correlated with the amount of Brønsted acid sites. In situ FTIR spectroscopy showed that Brønsted acid sites improve the stability of the methanol formed and inhibit its overoxidation to carbon oxides. Over copper-exchanged mordenite with high Brønsted acid site, a productivity of 0.18 mol (MeOH)/mol (Cu) with 95% selectivity was obtained (Sushkevich and van Bokhoven, 2018).

The influence of oxidant species (O_2, H_2O or N_2O) in DMTM reaction has been studied over $FePO_4$ catalysts synthesized by an ammonia gel method. The resulting catalyst structures were then revealed by different characterization techniques and the influence of oxidants on the structure and phase of $FePO_4$ during the reaction was examined. A high methanol yield was observed with N_2O as the oxidizing agent. The methane conversion with O_2 and N_2O increased steeply with temperature but increased much slower when H_2O was applied as an oxidant (Dasireddy et al., 2019).

Co/ZSM-5 catalysts have been reported as effective catalysts. Cobalt can be incorporated by different methods: wet ion-exchange (Pierella et al., 2008), incipient wetness impregnation, solid-state ion-exchange and sublimation (Wang et al, 2000). Cobalt-loaded ZSM-5 samples may contain at least two types of cobalt species: Co^{2+} in ion-exchange position and cobalt oxidic species. These cobalt species exhibit different catalytic properties (Beznis et al., 2010).

The preparation method determines the cobalt species present in materials. Krisnandi et al. (2015) studied the effect of Co species and pore size of the zeolite in oxidation of methane to methanol at 150°C. The results illustrated that catalysts prepared by impregnation exhibit higher performance compared to the ion-exchange method. Hierarchical zeolites showed better performance compared to microporous zeolites.

The oxidation of methane to methanol over metal exchanged zeolites could be performed either continuously (catalytic) or step-wise (stoichiometric chemical looping) although the latter has been reported more due to higher yield and selectivity (Mahyuddin et al., 2019). Chemical looping avoids direct contact of methane and oxidant, thereby decreasing the chance of over-oxidation.

The chemical looping approach, as the name implies, involves cyclic expose of a material with redox properties to an oxidant and methane under different temperatures and then extracting the product. This method was used for Cu-exchanged zeolites with MFI structure and since then received attention (Sushkevich and van Bokhoven, 2019). Activation of Cu compound is typically done over 400°C, while the reaction of methane and methanol desorption (by steam) needs lower temperatures (200°C) to inhibit overoxidation which results in a swing in the reaction cycle. A single turn-over is achieved for each active site. Methanol yields of 0.47 mol/mol Cu with remarkable 98% selectivity have been obtained with CuMOR and CuCHA by chemical looping for methanol synthesis from methane.

To date, no synthetic catalyst is capable of converting methane into methanol with high yields using oxygen as the sole oxidant in a single step (Dinh et al., 2018).

3.2.3 METHANE HALIDES

One method for activation of methane is to use low activation energy methane halogenation to methyl halide intermediates under mild operating conditions (Zhang et al., 2011). CH_3X are isostructural with methanol and versatile platform molecules which can be easily converted into a large spectrum of products by dehalogenation as HX. Attempts have been made for halogen-facilitated oxidative coupling of methane with oxidation-reduction cycles of the halogen with special emphasis on bromine (Mesters, 2016). The methyl halide is subsequently converted into higher hydrocarbons by condensation.

Currently, CH_3Cl is manufactured commercially by the reaction of methanol with HCl in the liquid phase over $ZnCl_2$ catalysts or in gas-phase (350°C) over alumina catalysts.

Functionalization of methane can be performed by halogenation or oxyhalogenation (Alvarez-Galvan et al., 2011). As the first route of methane activation, the reaction of seemingly unreactive methane with molecular halogen was surprising. The kinetic study of thermal and photochemical halogenation resulted in proposing chain–radical mechanism reaction which was extended to all halogens and paraffins. The halogen radical (X·) is first formed by homolytic dissociation of X_2 and recovery from alkyl radical reaction in which X_2 plays an important role (Scheme 3.1).

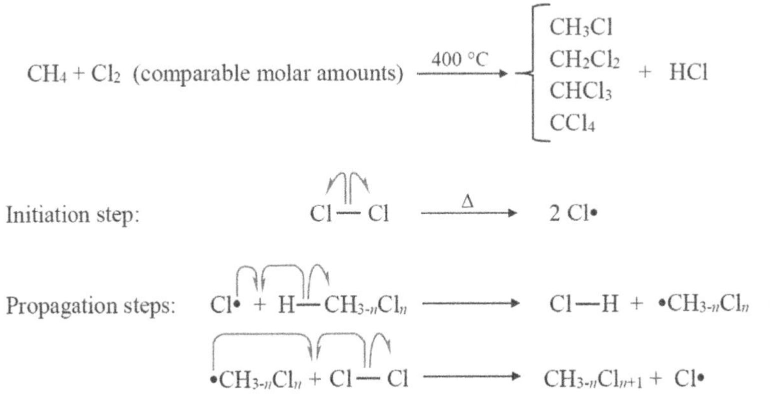

$$CH_4 + Cl_2 \text{ (comparable molar amounts)} \xrightarrow{400\ °C} \left\{ \begin{array}{l} CH_3Cl \\ CH_2Cl_2 \\ CHCl_3 \\ CCl_4 \end{array} \right\} + HCl$$

Initiation step: $Cl-Cl \xrightarrow{\Delta} 2\ Cl\bullet$

Propagation steps: $Cl\bullet + H-CH_{3-n}Cl_n \longrightarrow Cl-H + \bullet CH_{3-n}Cl_n$

$\bullet CH_{3-n}Cl_n + Cl-Cl \longrightarrow CH_{3-n}Cl_{n+1} + Cl\bullet$

SCHEME 3.1 Mechanism of free-radical chlorination of methane (Redrawn from Bruckner, 2002).

The order of radical stability is F· < Cl· < Br· < I·. Fluorination is highly exothermic and explosive. The next two are moderately exothermic whereas reaction with iodine is endothermic and equilibrium limited. The difference in radical reactivity results in different selectivities because the resulting holomethane participates in successive reactions in the same way, resulting in polyhalogenation.

As the reaction proceeds via free radical formation, substitution of each chlorine renders the product more active successively and thus increases the reaction rate. Consequently, the order of chlorination rate is $CH_4 < CH_3Cl < CH_2Cl_2 < CHCl_3$ (Podkolzin et al., 2007).

Unlike fluorination which results in total H substitution, chlorine gives monosubstituted product at high Cl/C (>10) ratios while bromine acts similarly at less excess bromine (Br/C>5) and bromination is limited to CH_2Br_2 (Lin 2017).

Typically, bromine has been used in halogen-mediated methane activation. Chlorine is less favorable because bromination of methane is much less exothermic, more selective and sufficiently rapid to go to completion (Osterwalder and Stark, 2007). The discovery of mono-bromination with high selectivity by Olah et al. (1985) increased attention to this route.

Halogens react readily with methane. However, the usual free radical halogenation is not selective and thus high CH_4/X_2 ratios are necessary to minimize higher substitutions (Olah et al., 2018). Mono-halogenated (CH_3X with X = Cl, Br) methane and dihalogenated methane (CH_2X_2) are the major products under controlled conditions.

A breakthrough in methane halogenation was made by Olah in 1980s, who proposed disruptive halogenation route. It is based on the activation of C-H via electrophilic substitution by polarizing X_2 with a superacid (Olah et al., 1985). This route shows an intrinsic tendency to CH_3X because the halogen with strong electron donation destabilizes the carbonium ion and thus prevents further substitution (Scheme 3.2). Furthermore, the catalytic route allows operation at lower temperatures.

This concept was proved with SbF_7, but for practical space velocities, Olah used noble metals supported on solid acids. Zelolites showed rapid activity loss and extensive dealumination.

$$CH_4 + Cl_2 \xrightarrow[\text{1 atm, 100-250 °C}]{Pt/Al_2O_3} CH_3Cl \quad (\text{selectivity} = 92 \sim 100\,\%)$$

Pathway at the surface of the catalyst:

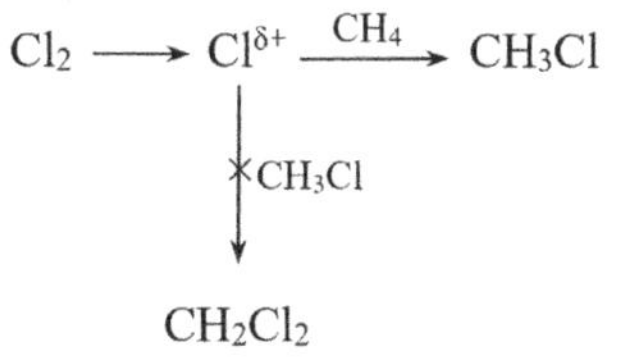

$$Cl_2 \longrightarrow Cl^{\delta+} \xrightarrow{\ CH_4\ } CH_3Cl$$

$$\downarrow CH_3Cl$$

$$CH_2Cl_2$$

SCHEME 3.2 Monochlorination of methane with Cl_2 (Redrawn from Lu and Zhou, 2017).

3.2.3.1 Oxidative Halogenation

The use of halide acid (HX) as a halogen source is advantageous as in the next reaction step (CH_3X conversion to the final product), it can be recovered. Oxidative halogenation uses hydrogen halide in the presence of oxygen as the halogen source:

$$2CH_4 + 2HX + O_2 \rightarrow 2CH_3X + 2H_2O$$

Oxidative chlorination chemistry is used commercially for the synthesis of vinyl chloride monomer (VCM) from ethylene via ethylene dichloride (EDC):

$$CH_2 = CH_2 + 0.5O_2 + 2HCl \xrightarrow{220-240°C,2-4\,bar} CH_2Cl - CH_2Cl + H_2O \quad \Delta H_{298}^0 = -239\,\text{kJ/mol} \tag{3.22}$$

No free chlorine is formed under the reaction conditions.

The catalysts are based on a reducible metal, commonly supported $CuCl_2$. The supports often contain activators and stabilizers such as chlorides of the rare earths and alkali metals (Weissermel and Arpe, 1997). The catalytic cycle is believed to be via reduction of $CuCl_2$ to $CuCl$ over the surface by ethylene and reoxidation to $CuCl_2$ by HCl and O_2. Many efforts have been done to apply this catalyst to oychlorination of methane which has only limited success (Podkolzin et al., 2007) because the reducible metal catalyzes the formation of chlorine gas by Deacon reaction (Eq. 3.23) as well:

$$2HCl + 0.5O_2 \rightarrow Cl_2 + H_2O \tag{3.23}$$

Consequently, gas-phase chlorination also occurs. Furthermore, the catalysts are unstable and volatile under the reaction temperature required for methane activation. The performance of copper catalysts could be improved with different promoters. It has been reported that KCl reduces the melting point of copper chloride and also promotes the redox step (Lin et al., 2017). To inhibit melt segregation of copper and potassium chloride at high temperatures, LaCl₃ has been added as a stabilizing promoter (Garcia and Resasco, 1990).

An important challenge for commercialization of the process is the oxidation of methane to carbon oxides being promoted possibly due to enhanced formation of oxygen vacancies in the presence of the halogen (Zichittella et al., 2017).

Researchers in Dow Chemicals showed that oxychlorination can be done over non-reducible and stable La compounds, especially LaOCl (lanthanum oxychloride) and its derivatives for methane, ethylene and paraffins (Podkolzin et al., 2007). LaCl₃ (full chlorinated form) itself is a promoter of Cu-based catalysts. They found that in contrast to the accepted mechanism that active sites are based on reduction, the oxidation state of La³⁺ does not change during the reaction as shown by XPS. They proposed a mechanism based on oxidation-reduction of the surface chlorine. According to Lercher and coworkers (Figure 3.4), transient ClO⁻ anion, resulting from the oxidation of Cl⁻ in LaOCl and LaCl₃ with molecular oxygen, is proposed as the active site for the initial step of methane activation. Activity measurements and spectroscopy analysis showed that La-based catalysts (LaOCl, LaCl₃ and La with middle chlorine) are very active and stable (for three-week runs) (Podkolzin et al., 2007).

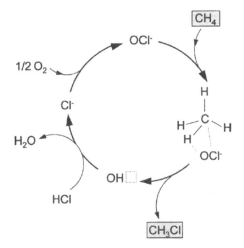

FIGURE 3.4 Lercher's proposal for catalytic cycle in oxychlorination (Redrawn from Peringer et al., 2006).

Zichittella et al. (2017) compared oxychlorination and oxybromination of methane over a variety of catalysts to elucidate the effect of hydrogen halide on catalyst performance. The steady-state activity was in the order RuO_2 > Cu-K-La-X > CeO_2 > VPO > TiO_2 > $FePO_4$. The strongly oxidizing RuO_2 showed also high combustion activity while the least reducible $FePO_4$ showed the highest selectivity to halomethanes (CH_3X, CH_2X_2). Cu-K-La-X and CeO_2 were found to be selective in chloromethane formation but formed CO_2 in oxybromination. In contrast, vanadyl pyrophosphate (VPO) and TiO_2 were selective in oxybromination but exhibited high CO selectivity.

3.2.3.2 Processes

Free radical halogenation was commercialized soon in early 20th century (by Hoechst company in 1923). In chlorination of methane to methyl chloride, methane and chlorine mixed with the recycle gas are fed to the reactor with CH_4/Cl_2 molar ratio of 4–5 at 400–450°C and slightly above atmospheric pressure (Weissermel and Arpe, 1997). Asahi Glass, Dow, Hiils, Montecatini and Scientific Design have developed various modifications for industrial operation differing in heat management of the highly exothermic reaction.

In thermal chlorination high-purity methane and chlorine are required (Olah et al., 2018). Water vapor hydrolyzes the chlorinated product and causes corrosion problems. Oxygen and sulfur compounds terminate the radical chain. Even inert gases such as nitrogen are built up with recycling.

Direct oxychlorination of methane has not yet been used commercially; however, a process which can be viewed as an indirect oxychlorination was developed by Lummus and put on stream in a 30,000 tonne-per-year unit by Shinetsu in Japan in 1975 (Weissermel and Arpe, 1997).

The researchers at UC Santa Barbara and Gas Reaction Technologies (GRT, Inc.) have developed a process for the bromination of CH_4 to CH_3Br by Br_2 and subsequent conversion of CH_3Br to higher hydrocarbons, in particular liquid fuels (Figure 3.5) (McFarland, 2012; Lorkovic et al., 2004).

There are technological problems such as a handling large amount of corrosive bromine and regeneration of the halogen.

The Deacon process (Eq. 3.23) was formerly used to generate chlorine by oxidation of HCl using a supported $CuCl_2$ catalyst at about 430°C. The process (invented in 1868) has been obsoleted for over a hundred years. However, the reaction has gained considerable importance in another field, in producing ethylene dichloride (EDC), a precursor for vinyl chloride monomer (VCM). Recently, the Deacon process has become increasingly interesting as an eco-efficient process for the recovery of chlorine from HCl-containing waste streams, e.g., over shaped RuO_2/SnO-alumina catalysts (Cornils et al., 2013).

3.2.4 SULFURATED METHANES

Methane can be sulfurated to CS_2 (Eq. 3.24) which can be subsequently converted into hydrocarbons.

$$CH_4 + 2S_2 \xrightarrow{\text{oxide catalyst}} CS_2 + 2H_2S \quad \Delta H_{298}^0 = 75\,\text{kJ/mol} \quad (3.24)$$

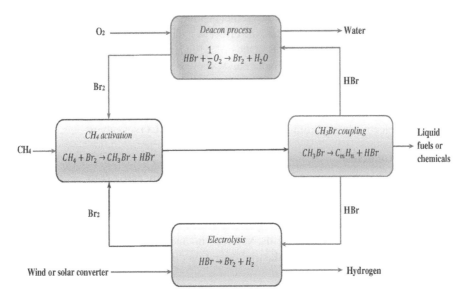

FIGURE 3.5 Conversion of CH_4 to CH_3Br followed by subsequent coupling of CH_3Br to produce higher hydrocarbons (Redrawn from McFarland, 2012).

This step can be performed by reacting preheated CH_4 with vaporized elemental sulfur at 580–635°C and pressure of 2.5–5 atm in an adiabatic reactor (Folkins process) (Taifan and Baltrusaitis, 2016).

Other sulfur allotropes such as S_6 and S_8 can also exist and participate in the sulfuration reaction:

$$CH_4 + 2/3S_6 \xrightarrow{\text{oxide catalyst}} CS_2 + 2H_2S \tag{3.25}$$

$$CH_4 + 1/2S_8 \xrightarrow{\text{oxide catalyst}} CS_2 + 2H_2S \tag{3.26}$$

With lower sulfur in the feed, hydrogen can also be produced

$$CH_4 + S_2 \xrightarrow{\text{oxide catalyst}} CS_2 + 2H_2 \tag{3.27}$$

$$CH_4 + 1/3S_6 \xrightarrow{\text{oxide catalyst}} CS_2 + 2H_2 \tag{3.28}$$

$$CH_4 + 1/4S_8 \xrightarrow{\text{oxide catalyst}} CS_2 + 2H_2 \tag{3.29}$$

Figure 3.6 shows the Gibbs free energy of the reactions versus temperature. The equilibrium conversion of Eqs. (3.24–3.26) is practically complete above 500°C at atmospheric pressure.

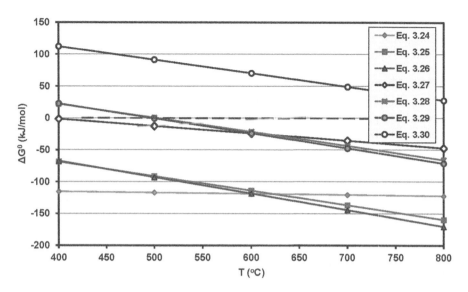

FIGURE 3.6 Gibbs free-energy change of methane sulfuration reactions versus temperature (data from Thacker and Miller, 1944).

CH_4 reforming with H_2S has been proposed as an alternative route.

$$CH_4 + 2H_2S \Leftrightarrow CS_2 + 4H_2, \qquad \Delta H_r^o (298K) = 232\,kJ/mol$$
$$\Delta G_r^o (298K) = 184.94\,kJ/mol \tag{3.30}$$

This reaction is thermodynamically similar to steam reforming and much less favorable than Eq. (3.24) (Figure 3.6). Fair equilibrium conversions (>50%) can be obtained above 800°C at atmospheric pressure for stoichiometric feeds. Sulfuration reactions typically become thermodynamically more favorable at higher temperatures.

With an excess of H_2S, temperatures above 1,000°C are necessary to avoid solid carbon formation (Huang and T-Raissi, 2008). This puts a tremendous constraint on the design of the catalyst to accelerate the kinetics of this process.

This reaction is much less studied than Eq. (3.24) above. The reaction has been studied in laboratory scale over metals with or without support. $CrMo/La_2O_3$-ZrO_2 is an effective catalyst (Taifan et al., 2017a).

$CH_4 + H_2S + CO_2$ are the main components of sour natural gas. Such conversions are useful for the utilization of low-grade natural gases without needing sweetening (Taifan et al., 2017b).

3.2.5 METHANE TO HYDROGEN CYANIDE

Hydrogen cyanide (or hydrocyanic acid, HCN) is a raw material and intermediate in the synthesis of important materials such as polymers and pharmaceuticals and thus can be considered as a C_1 building block. It is a byproduct of the manufacture of acrylonitrile via the ammoxidation of propene. However, because of its volatility (boiling point: 26°C) and high toxicity, HCN transportation is hazardous and in-site production is preferred.

3.2.5.1 Chemistry of Reaction

HCN can be produced by the oxidative or dehydrogenative reaction of NH_3 with hydrocarbons, preferably methane (Weissermel and Arpe, 1997). The ammoxidation of methane is the oxidative route (Eq. 3.31):

$$CH_4 + NH_3 + 1.5O_2 \rightarrow HCN + 3H_2O \quad \Delta H^0_{298} = -473 \, kJ/mol \quad (3.31)$$

The reaction was discovered by Andrussow in 1927 and thus the name of the commercial process.

Side reactions are oxidation of ammonia to N_2, NO and H_2O as well as partial oxidation of methane, giving CO and trace of CO_2 as by-products. The catalyst is usually Pt in the form of gauze or supported with additives such as rhodium to increase the mechanical strength of Pt. The operating conditions are 1,000–1,200°C and somewhat above 1 atm. To avoid HCN decomposition, short residence times (0.0003 s) with rapid quenching are applied.

Bodke et al. (2000) studied the effect of the addition of hydrogen in ammoxidation of methane over Rh and Pt catalysts and found that the addition of small amount of H_2 results in increasing HCN selectivity by about 10% based on NH_3, but it was not significant based on methane. However, the conversion of NH_3 decreased. Therefore, the addition of H_2 is useful if NH_3 is separated from the products and recycled.

The ammo(n)dehydrogenation processes for coupling of CH_4 and NH_3 are performed in the absence of an oxidant (Eq. 3.32):

$$CH_4 + NH_3 \rightarrow HCN + 3H_2 \quad \Delta H^0_{298} = +251 \, kJ/mol \quad (3.32)$$

Unlike Eq. (3.31), Eq. (3.32) is highly endothermic and needs high temperatures (>1,200°C) and external heat supply.

From mass spectroscopy-based experiments, a key role has been suggested for methanimine (CH_2NH) intermediate, the presence of which has been further confirmed by in situ photoionization studies (Schwarz, 2011). According to one mechanistic scenario, the formation of HCN occurs by two competing routes: a surface-bound pathway via Pt aminocarbenes and a gas phase channel involving the subsequent dehydrogenation of methanimine to HCN (Cornils et al., 2013).

The mechanism of the reaction is assumed to be Langmuir-Hinshelwood in which CH_4 acts both as the reactant and poison. The reaction comprises several steps, that is, dissociation of CH_4 and NH_3 on the catalyst surface and the reaction of surface carbon with the produced N fragments or N radical or ammonia itself. Ultra-high vacuum (UHV) studies by Auger electron spectroscopy also revealed that the catalyst surface is covered with a carbon atom monolayer which is responsible for preventing total decomposition of NH_3 to N_2 (Hasenberg and Schmidt, 1985; Hasenberg and Schmidt, 1986).

The oxygen-free production leads to higher selectivity toward HCN and a CO$_2$-free product (easier separation). Furthermore, H$_2$ is a valuable by-product. However, a more complicated reactor system is required.

3.2.5.2 Processes

HCN could be synthesized almost from any system including the three constituting elements H, C and N if sufficient energy is supplied. Currently, only NH$_3$/hydrocarbon systems are of commercial importance. An alternative carbon source such as CO, formaldehyde and methanol has received attention since 1980s. A review of the processes can be found elsewhere (Maxwell, 2004; Gail et al., 2012)

The Andrussow technology for ammoxidation, developed by BASF, is the most widely used method. The yield is relatively low compared to nonoxidative routes. The selectivity is 88% based on methane and 90% based on ammonia (Weissermel and Arpe, 1997).

Desulfurized natural gas with other reactants near-stoichiometric composition is used as the feed. The reaction is carried out over Pt-10%Rh or Pt-Ir catalysts in a gauze reactor. Gauze reactors use catalysts in the form and shape of gauze that ensures short contact times. The reactor effluent is rapidly quenched to less than 400°C in a waste-heat boiler that is directly below the catalyst to avoid the decomposition of HCN (Maxwell, 2004). The product gases are freed from unreacted ammonia by scrubbing with sulfuric acid. The hydrogen cyanide is then absorbed in water or diethanolamine.

The Andrussow ammoxidation process has been modified by DuPont, ICI and other companies. The DuPont HCN process, for instance, uses the process for the manufacture of HCN over Pt-Rh catalysts. Direct microwave heating ensures optimal flow distribution across the catalyst.

The drawback of this method is the need for high-purity CH$_4$ because higher hydrocarbons cause carburization of the Pt catalyst. Sulfur and phosphorous compounds are poisons of the catalyst (Gail et al., 2012).

In Degussa BMA (Blausäure-Methan-Ammoniak, or HCN methane-ammonia) process for ammo(n)dehydrogenation, heat is provided by passing the gases through externally heated gas-fired ceramic tube bundles. Sintered corundum tubes coated with Pt, Ru or Al layer (~15 microns thick) are used with a gauze catalyst or monolith (Maxwell, 2004). Model reactor studies showed that a special temperature profile is required within the reactor (Koberstein, 1973). The selectivity is 90%–91% based on CH$_4$ and 83%–84% based on ammonia at 1,300°C. The downstream is similar to the Andrussow process. The advantages of the process are a CO$_2$-free product and valuable H$_2$ waste gas. Due to high endothermicity, high investment and high maintenance costs for the converter are disadvantages of this process especially in larger scales. It is less important than the Andussow process, although it gives more HCN.

The Shawinigan HCN process, also called the Fluohmic process, was developed by Shawinigan Chemicals, now a division of Gulf Oil Canada. The chemistry is similar to the methane ammonia process but no catalyst is used and the operating temperature is higher. It involves reaction of a hydrocarbon (usually propane or butanes) with NH$_3$ at 1,300–1,600°C in an electrically heated fluidized bed of finely divided coke as catalyst. The feed is rapidly heated to 1,300°C at normal pressures.

To avoid carbon black deposition NH_3/CH_4 of 1.01–1.08 is employed. The selectivity to HCN is approximately 87%. The low ammonia concentration in the off-gas eliminates the need for recovery. Because of high electric energy consumption (~6 kWh/kg HCN), it is attractive only in locations where low-cost electricity is available (Maxwell, 2004).

A study by Stroebe et al. (2001) showed that BMA exhibits higher efficiency than the Andrussow process whereas the Shawinigan process is worst in terms of environmental effects.

3.3 CARBON DIOXIDE CONVERSIONS

3.3.1 REVERSE WGS

A natural indirect route for utilization of CO_2 is its conversion into CO followed by CO conversion into the desired product. This can be accomplished by the reduction of CO_2 by methane (dry reforming of methane, Eq. 3.2) or hydrogen (reverse water-gas shift (RWGS), Eq. 3.33), among others:

$$CO_2 + H_2 \Leftrightarrow CO + H_2O \quad \Delta H_{298}^0 = 41.1\,kJ/mol \tag{3.33}$$

The reaction is endothermic and equilibrium limited; therefore, it is favored at higher temperatures. Figure 3.7 shows the equilibrium conversion of CO_2 as a function of temperature for different initial H_2/CO_2 molar ratios. It shows that high CO_2

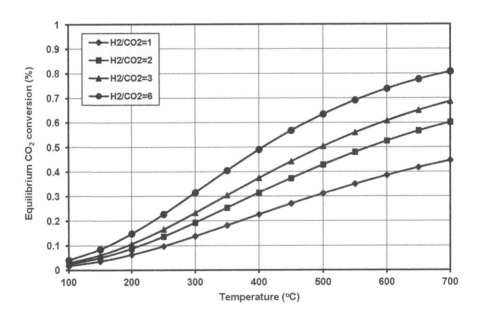

FIGURE 3.7 Equilibrium conversion of CO_2 versus temperature for RWGS with different feed compositions ($P = 0.1$ MPa).

conversions can be achieved only above 600°C. Furthermore, increasing hydrogen in the initial feed increases the equilibrium conversion of CO_2.

Further hydrogenation of CO can produce oxygenates and hydrocarbons as undesirable by-products. At lower temperatures, the main competitive reaction is the thermodynamically more favorable methanation reaction.

RWGS occurs in many reactions in which CO_2 and H_2 coexist and is believed to be a key step in CO_2 hydrogenation reactions. When RWGS and subsequent reaction occur in separate reactors, the water produced in the first step should be removed before being sent to the next reactor as it is harmful to methanol and Fischer–Tropsch synthesis catalysts.

3.3.1.1 Catalysts

The catalysts active in WGS reaction should be also active in RWGS reaction. The traditional low-temperature shift conversion (LTSC) and high-temperature shift conversion (HTSC) catalysts, however, exhibit poor performance in RWGS because of different reaction condition requirements. Most notably, RWGS is to be operated at higher temperatures. This will enhance catalyst deactivation.

The catalysts used for RWGS can be classified as supported metal catalysts, metal oxides and metal carbides.

The Cu-based LTSC catalyst (e.g., Cu-ZnO/Al$_2$O$_3$) has the advantage of no methane formation activity, but it is susceptible to sintering. Therefore, thermal stabilizers such as Fe have been added to catalyst formulation. Alkali metals have been used both in Cu-based WGS and RWGS catalysts to enhance the selectivity (Chen et al., 2003).

Two mechanisms have been proposed for RWGS over Cu-based catalysts, namely redox and the formate decomposition mechanisms (Porosoff et al., 2017). The former can be simplified as the following reaction sequence (Eqs. 3.34 and 3.35):

$$CO_2 + 2Cu^0 \Leftrightarrow Cu_2O + CO \tag{3.34}$$

$$Cu_2O + H_2 \Leftrightarrow 2Cu^0 + H_2O \tag{3.35}$$

The formate formation mechanism suggests that CO could form via a formate intermediate (HCOO*) formed by the hydrogenation of CO_2 followed by its decomposition to CO and a surface OH* via C=O bond cleavage.

Noble metals are active in RWGS (Goguet et al., 2004) but are not suitable for commercial application because of high price. Ni-based catalysts such as Ni/CeO$_2$-ZrO$_2$ have been used for RWGS. A drawback of Ni as active species of RWGS catalysts is its high methanation activity. It has been shown that the RWGS selectivity increases by decreasing metal loading (<3 wt%.) or reducing NiO particle size below 5 nm. On the contrary, by using higher loadings or larger particles (>30 nm), methanation is enhanced (Lu and Kawamoto, 2014).

Many metal oxides have been proposed as substitute to metals. Metal oxides are typically more tolerant to sintering and poisons.

The commercial Fe$_2$O$_3$/Cr$_2$O$_3$ HTSC catalysts also show poor performance under the RWGS reaction condition because of the reduction of iron oxide to Fe and coke formation.

ZnO-based catalysts such as ZnO/Al_2O_3 or $ZnAl_2O_4$ are active in RWGS reaction but are unstable due to volatilization of ZnO at high temperatures (400–700°C).

In_2O_3 and Ga_2O_3 are other active oxides with order of activity $In_2O_3 > Ga_2O_3$ (Sun et al., 2014). CeO_2 promotes both catalysts by enhancing adsorbing CO_2 and generation of oxygen vacancies (Zhao et al., 2012; Wang et al., 2016). Mixed oxides like barium zirconate pervosike are another class of catalysts employed with reasonable performance at high temperatures (Kim et al., 2014).

Transition metal carbides (TMC) are a novel class of catalysts with electron hybridization configuration similar to noble metals. They have been used at higher temperatures which do not favor methanol formation. Mo_2C is more active than other carbides which is attributed to its facilitated oxygen transfer on the surface. Mo_2C exhibits dual properties of H_2 dissociation and cleavage of C=O bonds and acts as reducible oxides. This catalyst breaks C=O bonds and form surface CO* and O* species. CO* can be desorbed as CO while O* in the form of oxycarbide is substituted with H_2 (Su et al., 2017). The incorporation of alkali metals, especially K, increases CO selectivity in RWGS (from 75% to 90% at lower temperatures, Porosoff et al., 2017). The use of an appropriate support further improves the performance.

Mo_2C is also active in methane and methanol formation. It can be used as catalyst support in hydrogenation reaction.

The main drawback of TMC catalysts are their low surface area and high production cost. These catalysts are prepared by progressive reduction and carburization of oxide precursors (e.g., MoO_3) by a mixture of hydrogen and hydrocarbon (Tuomi et al., 2016).

3.3.2 METHANOL SYNTHESIS (CONVENTIONAL)

Methanol (methyl alcohol) was first discovered by destructive distillation of wood and hence the historical name "wood alcohol". Within 1825–1930, it was produced by this method. However, it is currently produced from syngas (CO+H_2 and some CO_2) through catalytic hydrogenation of carbon monoxide. More recently, CO_2 has received much attention as an alternative carbon source for methanol synthesis.

3.3.2.1 Chemistry

Currently, methanol is produced from syngas (with H_2/CO ratio ≈2, called "*metgas*") at industrial scales (Eq. 3.36). Metgas is obtained by steam reforming of methane and it contains approximately 3 vol.% CO_2. Metgas passes over $Cu/ZnO/Al_2O_3$ catalyst at 5–10 MPa and 200–300°C (Dasireddy et al., 2018):

$$CO + 2H_2 \Leftrightarrow CH_3OH \quad \Delta H^0_{298} = -90.8\,kJ/mol \qquad (3.36)$$

The reaction is exothermic and equilibrium limited.

Spectroscopic studies illustrated that CO_2 is the principal source of methanol while CO serves as producing CO_2 by WGS (Eq. 3.4) (Olah et al., 2018). Thus, the overall equation for methanol synthesis from syngas is a combination of two exothermic equilibrium limited reactions (Eqs. 3.4 and 3.37):

$$CO_2 + 3H_2 \Leftrightarrow CH_3OH + H_2O \quad \Delta H^0_{298} = -49.5\,kJ\,/\,mol \qquad (3.37)$$

Therefore, the distinction between CO and CO_2 in methanol synthesis is somewhat arbitrary. The stoichiometry of the feed is then defined by the so-called stoichiometric number (*SN*) as

$$SN = \frac{\left(y_{H_2} - y_{CO_2}\right)}{\left(y_{CO} + y_{CO_2}\right)} \tag{3.38}$$

where y is the mole fraction of components in the feed. The desirable value of stoichiometric number for methanol synthesis is slightly larger than 2 (Takht Ravanchi and Sahebdelfar, 2021).

3.3.2.2 Catalysts

The first effective catalyst for methanol synthesis was ZnO developed by BASF in 1920s. It operates under rather drastic conditions (320–380°C, 25–35 MPa), and hence names high pressure methanol (HPM) synthesis. ZnO is an active catalyst but it deactivates rapidly by recrystallization under methanol synthesis conditions. Cr(III) exhibits poor catalytic activity but stabilizes ZnO and allows its application for commercial use. The ZnO/Cr_2O_3 catalyst is prepared by the precipitation and exhibits good tolerance to the poisons found in syngas such as sulfur compounds.

The development of low-pressure methanol (LPM) synthesis process (5–10 MPa) using a copper-based catalyst by ICI was a breakthrough in methanol synthesis with the advantage of milder operating conditions and higher methanol yields. It shows very high methanol selectivity (99.9%) under commercial operation close to thermodynamic equilibrium. However, the catalyst is susceptible to syngas poisons and thus the syngas feed should be carefully purified. The typical catalyst is three metallic Cu-ZnO/Al_2O_3 or Cu-ZnO/Cr_2O_3. Most of the modern plants and all megaplants are based on low-pressure methanol synthesis. The catalyst is prepared by co-precipitating of the metals by alkali solution under controlled pH near neutral pH. Typical composition of commercial catalysts is CuO 35–40 wt.%, ZnO 45–50 wt% and Al_2O_3 10–20 wt.%.

In these catalysts, copper is the active component. ZnO is the structural and chemical promoter. As small crystallites (2–10 nm), it facilitates formation of small Cu crystallites (4–8 nm) on its surface (Bartholomew and Farrauto, 2006). It stabilizes Cu crystallites from sintering. ZnO protects Cu from poisoning by chlorides and sulfur compounds by scavenging them. Furthermore, ZnO neutralizes Al_2O_3 acid sites, otherwise catalyzing methanol dehydration to dimethyl ether (DME). Al_2O_3 stabilizes Cu and ZnO against sintering.

The conventional method for the synthesis of Cu-ZnO-Al_2O_3 catalysts is co-precipitation that consists of three steps (Ghosh et al., 2019):

- Precursor precipitation in the form of hydroxycarbonates
- Calcination of these hydroxycarbonate precursors in a controlled manner to produce CuO-ZnO species (with high dispersion) and some residual carbonates to keep the porosity and surface area high.
- Reduction of the oxidized phases to have active catalyst as Cu^o or Cu^+ nanoparticles that decorates ZnO or ZnO_x.

In catalyst synthesis, the key stage is synthesis of hydroxycarbonate precursors; as final catalyst activity strongly depends on catalyst properties that were fixed during early stages of catalyst synthesis. A high Cu surface area (that depends on meso-porosity and small particle size), minimum contamination from alkali metals (Na⁺), good interaction between Cu and ZnO and formation of Cu-Zn partial oxidized mixed oxides are the required characters of the final catalyst (Behrens, 2016; Kondrat et al., 2017).

In industrial operation, where catalyst poisons are avoided by guard beds, the main cause of catalyst deactivation is slow sintering of copper. It can be inhibited by operation below 250–260°C. The typical catalyst lifetime is 2–4 years.

3.3.2.3 Processes

The most important part of methanol manufacture process is the synthesis reactor or converter. Due to high exothermicity of the reaction and sensitivity of the catalyst to high temperatures, the primary goal is precise temperature control. The extensively used technologies use gas- or liquid-phase converters (Tijm et al., 2001).

Several processes have been developed for methanol synthesis from syngas. The BASF process employs Zn/Cr₂O₃ catalyst. The ICI (now Synetix) process is an LPM process developed in 1966 using Cu-ZnO/Al₂O₃ below 300°C and 5–10 MPa.

The Haldor–Topsoe process is a medium pressure (10–15 MPa, 230–260°C) process. The Lurgi (now Air Liquide) process uses Cu-ZnO catalyst in multitubullar reactors with boiling water in the shell resulting in good temperature control (250°C, 5–8 MPa) with reduced catalyst consumption.

The liquid-phase LPMEOH process uses which Cu-based catalysts in slurry reactors (250°C, 5–10 atm) was developed by Air Products and Chemicals.

3.3.3 Methanol from CO₂ Hydrogenation

CO_2 is thermodynamically a stable molecule ($\Delta G°_f = -394.38$ kJ/mol) and kinetically an inert molecule; hence, its activation is difficult. Being a linear non-polar molecule, CO_2 has two reactive sites, carbon and oxygen. As carbonyl carbon has electron deficiency, CO_2 has strong affinity towards electron-donating and nucleophile reagents, whilst oxygen atom has opposite behavior. Due to the fact that CO_2 conversion is kinetically limited, an efficient catalyst and an external energy input is needed for its conversion to methanol. Nowadays, four processes were developed for methanol synthesis by CO_2 hydrogenation, namely, heterogeneous catalysis, homogeneous catalysis, photocatalysis and electrocatalysis, among which heterogeneous catalysis is the subject of the present section. For the other three approaches, the interested reader is referred to Li et al. (2014a).

Carbon dioxide hydrogenation to methanol (Eq. 3.37) is an attractive route for conversion of CO_2 to valuable products and energy storage (i.e., power to methanol). Compared with methanol synthesis from syngas, methanol synthesis from CO_2 is less exothermic but produces a higher amount of water that deactivates the catalyst (Ud Din et al., 2019).

In the direct process of CO_2 hydrogenation to methanol, syngas production as an energy-consuming endothermic step is eliminated and H_2 and CO_2 are provided from other sources. As syngas generation step has half of the total capital investment cost, its elimination reduces the total plant cost. For coal-based plants, the capital cost for syngas step is even higher (approximately 70%–80%). Of course, hydrogen generation from water electrolysis is still an energy-intensive step.

Methanol yield is influenced by temperature, pressure, H_2/CO_2 ratio and space velocity. Increasing space velocity, pressure and H_2/CO_2 ratio is in favor of methanol production. Thermodynamically, high pressure is favorable for CO_2 conversion. Furthermore, reactor volume, plant area and capital cost are effectively reduced for a high-pressure process. The reactor type has an important effect on CO_2 conversion into methanol. In these processes, a simple tube-cooled reactor is sufficient for controlling reaction temperature; hence, the whole process has a lower cost (Saeidi et al., 2014; Marlin et al., 2018).

As in methanol production process from CO_2, water is produced as a by-product, the crude methanol has 30–40 wt.% water which is higher than that in a syngas-based process. On the other hand, crude methanol obtained from a CO_2-based process has higher purity (less oxygenates) that reduces the overall separation cost of the whole process (Zhong et al., 2020)

3.3.3.1 Catalysts

Homogeneous and heterogeneous catalysts were studied for CO_2 hydrogenation to methanol. For direct CO_2 hydrogenation to methanol, some homogeneous catalysts have been reported in the literature which are reviewed in detail by Li et al. (2014b) and Goeppert et al. (2014). Homogeneous catalysts are not desirable for large-scale applications due to contaminating the products and complex separation. Based upon their reasonable cost, ease of handling and separation, high stability and possibility of reuse, heterogeneous catalysts are preferable.

The conventional Cu-ZnO catalyst for the synthesis of methanol from syngas can also be applied for CO_2 hydrogenation to methanol but low activity and high deactivation was observed. In particular, higher water production is a disadvantage as it accelerates the crystallization of Cu and ZnO particles and causes rapid sintering and deactivation and also production of unwanted higher alcohols and hydrocarbons (as unfavorable by-products) (Jadhav et al., 2014).

Hence, since 1990s, researchers have focused on developing more effective catalysts. Different catalyst formulations have been developed and applied to methanol synthesis from CO_2 (Guil-López et al., 2019a).

3.3.3.1.1   Cu-ZnO-Al$_2$O$_3$ Catalysts

In Cu-ZnO-Al$_2$O$_3$ catalysts, ZnO favors Cu^{2+} reducibility and dispersion of the reduced Cu, thereby increasing active site number. The active Cu^+ sites are stabilized on the surface of ZnO. For catalyzing CO_2 hydrogenation, ZnO basic sites must be in close contact with Cu-metal sites. Hence, Cu-ZnO inter-phase formation is important for methanol synthesis from CO_2. ZnO facilitates the adsorption of CO_2 for its subsequent hydrogenation to methanol. The main role of Al$_2$O$_3$ is improving stabilization and exposition of Cu-active centers under reaction conditions (Tisseraud et al., 2016; Dang et al., 2019).

In the process of CO_2 hydrogenation to methanol over Cu-ZnO-Al$_2$O$_3$ catalysts, active sites are partially or completely reduced Cu with synergic contact with ZnO or partially reduced ZnO_x (Tisseraud et al., 2015).

3.3.3.1.1.1 Modifications for Conventional Cu-ZnO-Al$_2$O$_3$ Catalysts: Promoters and Improvements in Synthesis Promoters are added to Cu-ZnO-Al$_2$O$_3$ catalyst to modify Cu active sites and physicochemical properties of the catalyst (such as basicity or reducibility). Researchers focused on the application of noble metals such as Pt, Pd, Au, or Rh for the formation of Cu-Me alloy phases. The industrial use of these promoters is not advised as they increase the final price of the catalyst and their recovery from the spent catalyst must be taken into account (Pasupulety et al., 2015; Liu et al., 2019).

For modifying basicity and physicochemical properties of the catalyst, zirconium, fluoride and gallium can be used as the promoter to have interaction with alumina. At all three levels, Zr act as a promoter (Dasireddy and Likozar, 2019; Guil-Lopez et al., 2019b):

- Catalyst basicity is increased by Zr addition
- During precipitation, hydroxycarbonate precursor formation is modified by Zr as it favors hydrotalcite precursor formation
- In the presence of ZrO_2, Cu^+ species are stabilized and their deactivation is avoided

For Cu-Zn-Al-Zr catalysts prepared from hydrotalcite-like precursors with Zr^{4+}/(Al^{3+}+Zr^{4+}) ratio of 0–07, carbon dioxide conversion was a function of Cu dispersion and surface area while methanol selectivity was a function of basic sites distribution (Gao et al., 2013). The best performance was obtained for Zr^{4+}/(Al^{3+}+Zr^{4+}) = 0.3 (22.5% CO_2 conversion, 47.4 MeOH selectivity, 137 g$_{MeOH}$ kg$_{cat}$ h^{-1} CH$_3$OH yield at 240°C and 2 MPa). Hydrotalcite-like precursors produce smaller Cu clusters.

For Ga-promoted Cu-Zn-Al catalysts, the concentration of Cu^0/Cu^+ species is regulated by the application of different Ga loadings and consequently, its catalytic performance is modified (Toyir et al., 2001). On the other hand, in Cu/ZnO catalyst co-modified with small loadings of Ga^{3+} and Al^{3+} (Ga^{3+} +Al^{3+} = 3 atom %), Ga insertion into malachite precipitates increased with Ga content (Guil-Lopez et al., 2020). For Al/Ga = 1, catalyst activity was considerably improved (15% CO_x conversion, 1,700 g$_{MeOH}$ kg$_{cat}$ h^{-1} MeOH yield at 250°C and 3.0 MPa) which was better than individually loaded Ga and Al samples. This could be related to change in ZnO defects with Zn^{2+}/Al^{3+}/Ga^{3+} substitution.

The role of fluoride ion as a promoter is decreasing the acidity of alumina species. Due to the fact that incorporation and stabilization of fluoride ions in structure is allowed by hydrocarbonates in the form of hydrotalcites, fluoride ion is incorporated during precursor precipitation. In the presence of fluoride ion, CO_2 adsorption is increased that is favored for selective hydrogenation of methanol (Gao et al., 2016).

Temperature, pH, aging time and complete Na^+ removal are operating parameters that must be controlled for selective synthesis of hydroxycarbonate precursors.

There are different hydroxycarbonate phases, namely hydrotalcite ($Cu_{1-x-y}Zn_yAl_x(OH)_2(CO_3)_{x/2}$), aurichalcite ($(Cu,Zn)_5(CO_3)_2(OH)_6$), amorphous zincian georgeite phase ($(Cu,Zn)(CO_3)(OH)_2$) and zincian malachite ($(Cu,Zn)_2(CO_3)(OH)_2$). It was reported that in the presence of zincian malachite phase in the hydroxycarbonate precursor, the best catalytic performance was obtained. Recently, researchers concluded that the presence of amorphous zincian georgeite as a cooperative phase improved the catalytic performance (Mota et al., 2018; Guil-Lopez et al., 2019b).

The reduction step of the calcined precursor is also important, as it must be controlled in a way to avoid sintering of Cu particles and control the amount of Cu^+ in the final catalyst.

The use of aluminum hydrous oxide sol prepared by peptization of aluminum hydroxide for preparation of the catalyst resulted in good performance in the synthesis of methanol form CO_2-rich syngas (Bahmani et al., 2016). The catalyst showed stronger inter-dispersion between Cu and ZnO and well-dispersed Cu nanoparticles with improved stability and activity in methanol synthesis from syngas compared to those prepared from true solutions.

3.3.3.1.2 Cu-ZnO Supported Catalysts

For Cu-ZnO catalysts, another modification is substituting Al_2O_3 from the catalyst with tri- or tetra-valent metal oxides such as Ga_2O_3 and ZrO_2 or with other supports. In contrast to alumina, zirconium oxide has a weak hydrophilic character that enhances copper stability and dispersion. Furthermore, by substituting Al_2O_3 with ZrO_2, the basicity of the final catalyst is increased which is favorable for methanol selectivity as CO_2 adsorption and its subsequent hydrogenation is higher on basic sites. In Cu-ZnO catalysts supported on Al_2O_3, ZrO_2 and CeO_2 prepared by sol-gel and co-precipitation methods, the best performance was obtained for ZrO_2-supported catalyst prepared by co-precipitation (23.2% CO_x conversion, 33% MeOH selectivity, 331 g_{MeOH} kg_{cat} h^{-1} MeOH yield at 280°C, 5MPa). CeO_2 had no favorable effect and reduced Cu dispersion (Angelo et al., 2015; Dong et al., 2016).

As the interaction of copper and zirconia particles stabilizes microcrystalline copper particles during reduction, the presence of ZrO_2 is favorable for the formation of oxygen vacancies on surface. ZrO_2 has a superior promotional effect that is due to fine tuning capacity of reduced Zr^{3+} species at Cu/ZnO interface, by which key reaction intermediates (such as *CO, *CO_2, *H_2CO and *HCO) are bind in a moderate way and methanol formation is facilitated (Kattel et al., 2016).

For Cu/ZnO catalysts, Ga_2O_3 can also be used instead of Al_2O_3. Thermal reduction of ZnO can be facilitated in the presence of Ga^{3+} and, consequently, $Cu-ZnO_x$ nanoparticles with high activity could be produced. Due to the formation of $ZnGa_2O_4$ (gallium spinel) that creates electronic heterojunction with ZnO, $Cu-ZnO_x$ nanoparticles formation is facilitated. During the process of hydroxycarbonate precursor synthesis, some modifications occur in the presence of Ga^{3+}, as it facilitates precursor formation with hydrotalcite structure. Catalysts obtained from hydrotalcite precursors containing gallium have better Cu dispersion and, consequently, are suitable catalysts for CO_2 hydrogenation to methanol (Li et al., 2018).

SBA-15 as a mesoporous material with high specific surface area can also be used as catalyst support. The role of SBA-15 is confinement of Cu/ZnO particles in its structure that enhances its ability for the interaction of CO_2 and H_2 and, consequently, better catalytic performance is obtained. Due to this confinement, Cu/ZnO particles are stabilized and an optimum inter-particle spacing with uniform distribution is obtained by which Cu/ZnO-SBA-15 catalysts are stabilized. Cu/ZnO loading and Cu/Zn molar ratio are two parameters that influence the catalytic behavior of Cu/ZnO-SBA-15. There is a direct correlation between these parameters and dispersion and morphology of Cu/ZnO particles in SBA-15 structure. When Cu/ZnO particles are formed as a thin homogeneous amorphous layer in the channels, best catalytic performance is obtained (Mureddu et al., 2019).

Carbon nanotubes (CNTs) and graphene oxide aerogels are different carbonaceous materials that can be used as supports for Cu/ZnO particles. In the case of CNTs, Cu/ZnO particles are deposited inside and in the outer walls of CNTs. With CNT supports, a smaller particle size and lower interaction between active phase and support is obtained and the final catalyst has greater reducibility. On the other hand, methyl formate is the main product of these catalysts. In the case of application of graphene oxide aerogel as catalyst support, a catalyst with a high surface area of 458 m^2/g was obtained that produced methanol with a high production rate of 2,950.4 $\mu mol_{CH3OH}/g_{cat}$.h at 250°C and 1.5 MPa (Deerattrakul et al., 2018).

3.3.3.1.3   Other Cu-Based Catalysts

For methanol production from CO_2, Cu nano-particles can be used with other oxides as stable effective catalysts. ZrO_2 is an example in this regard. In CO_2 hydrogenation to methanol, CO_2 is adsorbed on ZrO_2 of the Cu/ZrO_2 catalyst after which a formate species is formed on Cu surface and dissociative adsorption of H_2 occurs. In the next step, H atoms are transferred to formate species and consequently to methoxy species on ZrO_2, which are, in turn, hydrogenated to methanol product (Larmier et al., 2017).

Ag or In_2O_3 are modifiers for Cu/ZrO_2 catalyst for improving its efficiency. Ag increases the surface area, formation of Ag-Cu alloy and concentration of partially reduced ZrO_x sites. In the presence of Ag$^+$ in CuO-ZrO_2 catalysts, the meso-structure and surface area of the catalyst were decreased but the amount of partially reduced ZrO_x sites was increased (Tada et al., 2017).

Cerium oxide (CeO_2) is another oxide to be combined with Cu and used as an effective catalyst for CO_2 hydrogenation to methanol. Due to the interaction between Cu nano-particles and CeO_2 support, Cu catalysts supported on CeO_2 have better catalytic performance in methanol synthesis from CO_2 at 200–300°C and 3 MPa. Due to the presence of oxygen vacancies on Cu/CeO_2 catalyst, this catalyst has high selectivity to methanol that is because of the formation of carbonate intermediates (Wang et al., 2020).

Magnesium oxide (MgO) is another oxide to be used as the support for Cu particles on which small metallic Cu particles with high dispersion can be formed with improved catalytic performance in CO_2 hydrogenation to methanol. Due to the basicity of MgO, CO_2 adsorption was facilitated and reaction pathways was modified (Liu et al., 2016).

Titanium oxide (TiO_2) is also used for the support of Cu catalysts in methanol production from CO_2. The creation of sites with oxygen vacancy is facilitated by the redox properties of TiO_2 and consequently CO_2 activation was improved. Moreover, due to the high surface area of TiO_2, reactivity and dispersion of Cu sites were improved (Tosoni and Pacchioni, 2019).

3.3.3.1.4 Non-Cu Catalysts

3.3.3.1.4.1 Noble Metals Palladium (Pd) and gold (Au) are other metals that have high reactivity for CO_2 hydrogenation to methanol (Bahruji et al., 2016; García-Trenco et al., 2018).

Palladium is very active for CO_2 hydrogenation, and the support type and promoters determine its selectivity to methanol. Supporting Pd on ZnO, bimetallic Pd-Zn alloy was formed as the active phase for selective methanol formation. Gallium oxide (Ga_2O_3) is also a suitable support for Pd. Based upon the interaction between Pd and Ga_2O_3, after reduction, Pd-Ga alloy and Pd-Ga inter-metallic compounds were formed. Pd_2Ga intermetallic compound has improved activity and methanol selectivity, due to the fact that atomic hydrogen is provided to the surface of Ga_2O_3 that retards methanol decomposition and CO production. The combination of Pd and In_2O_3 is another example of Pd inter-metallic compounds that are active in methanol synthesis from CO_2. In sum, a 70% higher methanol rate was reported for Pd-In catalyst in comparison to conventional $Cu/ZnO/Al_2O_3$ catalyst (Fiordaliso et al., 2015).

For the use of gold (Au) as an active catalyst in CO_2 hydrogenation to methanol, different oxides can be used as catalyst support; including Al_2O_3, ZnO, TiO_2, CeO_2 and Fe_2O_3. Vourros et al. (2017) reported that CeO_2 and ZnO are good supports for gold nano-particles, and at temperatures lower than 250°C, respectively, 82% and 90% methanol selectivity was obtained. For Au/ZnO catalyst, mechanistic studies showed that methanol formation from CO and CO_2 proceeds through various pathways that are independent and in the course of CO_2 hydrogenation, CO is not an intermediate. When TiO_2 and Fe_2O_3 were used as supports for Au nano-particles, a high CO_2 conversion (40% and 27%, respectively) was obtained. CeO_2 and ZrO_2 are other oxides studied as catalyst supports (Hartadi et al., 2015).

3.3.3.1.4.2 Non-Noble Metals and Oxides Cen et al. (2016) reported that transition metals (such as Fe, Cu and Co) supported on Mo_2C are active catalysts for selective CO_2 hydrogenation to methanol. This support is simultaneously a co-catalyst in CO_2 hydrogenation to methanol. When Cu was added to Mo_2C, methanol production improved, and when Fe and Co were added, the C_{2+} hydrocarbon production increased.

As indium oxide (In_2O_3) is capable of creating oxygen vacancy on its surface, it can be a good catalyst for methanol synthesis by which 3,690 $\mu mol_{CH3OH}/g_{cat}.h$ can be produced. The composite of In_2O_3 and ZrO_2 is also a good catalyst with 100% methanol selectivity at temperatures higher than 300°C and has high stability for 1,000 h. Another example of a composite catalyst is In_2O_3-Ga_2O_3 ($Ga_xIn_{2-x}O_3$) that is highly active at high temperatures (320–400°C) for CO_2 hydrogenation to methanol (Akkharaphatthawona et al., 2019).

3.3.3.2 Mechanism

Formate, formaldehyde and formyl have been observed during synthesis of methanol from CO_2 (Figure 3.8). Fourier transform infrared spectroscopy (FTIR) studies

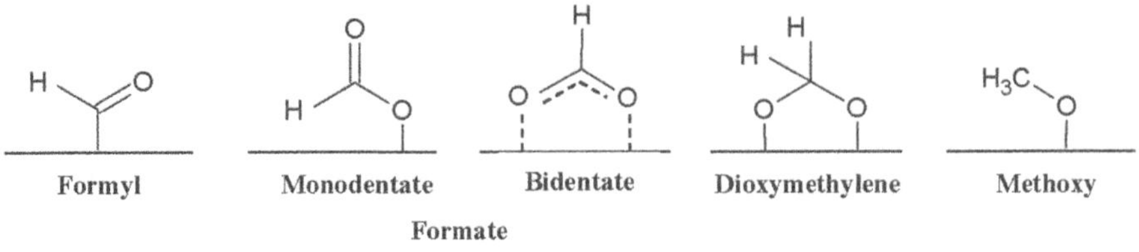

FIGURE 3.8   Surface species found as intermediates in methanol synthesis.

showed that both monodentate and bidenatate surface formate can exist on Cu/SiO₂ catalysts with the latter being more dominant (Yang et al., 2010).

For CO₂ hydrogenation to methanol, the main proposed reaction routes are summarized in Figure 3.9. The mechanisms differ in active sites (metal and surface OH groups) and intermediates.

3.3.3.2.1 The HCOO Mechanism

The conventional formate pathway is the favorable one by majority of researchers. In the beginning mechanisms, a surface formate (HCOO*) is formed from CO₂ reduction with pre-adsorbed surface atomic H by an Eley-Rideal (ER) or Langmuir-Hinshelwood (LH) mechanism. In the next step, dioxomethylene (H₂COO*), formaldehyde (H₂CO*), methoxy (CH₃O*) and CH₃OH are produced. Experimental and DFT calculations confirmed the formate pathway (Kim et al., 2016).

In this mechanism pathway, the CO produced by the RWGS reaction accumulates as the product. On the other hand, by CO hydrogenation, a formyl (HCO*) is produced that is an unstable product and dissociates back to CO and H atoms (Zhong et al., 2020).

Under reaction conditions, the surface is occupied by bidentate formate species that is the most stable intermediate during methanol formation (Liu et al., 2013a).

Researchers reported the formate mechanism for different catalysts; such as Cu surface, Cu clusters and Cu nano-particles supported on metal oxides or metal carbides, alloy surfaces (e.g., Zn-deposited Cu) and Pd-based catalysts (Posada-Perez et al., 2016; Wang et al., 2019).

Kinetic studies have shown that in the formate route, the rate determining step (RDS) is HCOO*/H₂COO* hydrogenation or the corresponding C–O bond dissociation (Kunkes et al., 2015).

3.3.3.2.2 The Revised HCOO Mechanism

In the revised formate (r-HCOO) pathway, CO and CO₂ hydrogenation pathways are active and 2/3 of methanol is produced from CO₂ hydrogenation. Preferentially, HCOO* is hydrogenated to formic acid (HCOOH*) instead of being hydrogenated to dioxymethylene (H₂COO*). In the next step, formic acid is hydrogenated to H₂COOH*, and for the generation of H₂CO* and OH, the C–O bond in H₂COOH* is split. Meanwhile, via a methoxy (CH₃O*) intermediate, H₂CO* is hydrogenated to methanol (Zhong et al., 2020).

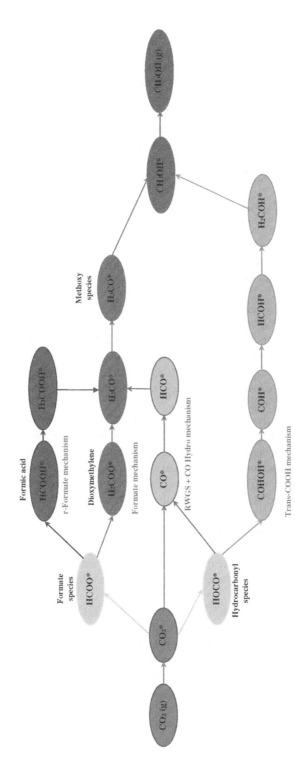

FIGURE 3.9 Proposed reaction mechanisms for hydrogenation of CO_2 to methanol (Redrawn from Zhong et al., 2020).

3.3.3.2.3   The RWGS+CO Hydrogenation Mechanism

This mechanism involves RWGS reaction by which CO_2 is first changed to CO through a carbonyl (COOH*) intermediate, and in the next step, CO is hydrogenated to methanol with different intermediates; such as formaldehyde (H_2CO), formyl (HCO*) and methoxy (H_3CO*). CO* and formyl hydrogenation are the RDSs. This mechanism effectively describes CO byproduct formation and is suggested for Cu-based catalysts (such as Cu/CeO_x, Cu/ZrO_2, $Cu/ZnO/Al_2O_3$ and Cu/TiO_2) at very high pressures (360 bar) and Au-based catalysts (such as $Au/CeOx/TiO_2$) (Kattel et al., 2016).

According to Fisher et al. (1997), formate is the common intermediate for methanol synthesis and RWGS reaction.

3.3.3.2.4   The Trans-COOH Mechanism

According to the *trans*-COOH mechanism, hydrocarboxyl (HOCO*) species is the first hydrogenated species. The hydrocarboxyl pathway is an H_2O-mediated mechanism; H_2O provides the H atoms to which adsorbed CO_2 reacts and HOCO* is formed. In the next steps, HOCO* is hydrogenated to dihydroxycarbene (COHOH*), hydroxymethylidyne (COH*) is formed from COHOH dissociation and CH_3OH (methanol) is formed *via* H_2COH (hydroxymethyl) and HCOH (hydroxylmethylene) intermediates (Yang et al., 2015).

Under the dry H_2 condition, direct hydrogenation of bidentate formate (HCOO*) cannot produce methanol on metallic Cu catalysts, but when Cu catalysts are oxidized with N_2O or O_2, methanol and water are significantly produced; hence, co-adsorbents related to water-derived species or surface oxygen facilitate the formation of methanol (Yang et al., 2013).

Hence, according to the two basic formyl and formate reaction mechanisms, an ideal Cu-based catalyst for CO_2 hydrogenation to methanol is one that easily hydrogenate dioxomethylene and moderately bond CO, being weak enough to prevent CO poisoning and being strong enough for CO hydrogenation rather than its desorption. In this manner, methanol production from both pathways and mechanisms can be facilitated.

It has been reported (Lei et al., 2015) that methanol is mainly produced via the formate route. Although the reaction rate of CO_2 hydrogenation by RWGS is faster than that of CO_2 hydrogenation by the formate pathway, the reaction rate of methanol synthesis from syngas decreases in the presence of CO_2. Hence, new catalysts must be designed in a way to promote the suppression of the active sites for RWGS reaction and the robustness of methanol synthesis sites for inhibiting water production (Tisseraud et al., 2015).

For a Cu-based catalyst combined with ZnO and ZrO_2, Bianchi et al. (1995) reported adsorbed hydrogen spillover from Zr to Cu. Generally, in this catalytic process, two active centers are involved and hydrogen adsorption and dissociation occur on the Cu site and CO_2 adsorption as bicarbonate occurs on the ZrO_2 site. Atomic hydrogen transfers from the Cu surface to ZrO_2 surface by spillover and hydrogenates the adsorbed carbon-containing species in a stepwise manner to methanol desorbed from the surface.

Rasmussen et al. (1994) proposed a micro-kinetic model based on 13 individual steps in which the rate limiting step is dioxomethylene hydrogenation.

3.3.3.3 Commercial Plants

Since the beginning of the 20th century, researchers are focused on methanol production from CO_2 and H_2. In the earliest production plants in USA, CO_2 and H_2 were obtained from by-products of fermentation processes (Chorkendorff and Niemantsverdriet, 2017).

Carbon Recycling International (CRI) operated the first commercial CO_2 to CH_3OH plant in Iceland, in 2012, namely "George A. Olah Renewable Methanol Plant". This demo plant with a capacity of 4,000 t/y is based on CO_2 conversion with the available local geothermal energy sources (steam and heat). Water electrolysis produces the required hydrogen. The produced methanol (namely Volcanol™) is mixed with gasoline. This plant recycles 5,600 tCO_2/y (Zhong et al., 2020).

In Canada, Blue Fuel Energy is planned to use hydroelectricity and concentrated CO_2 emissions from natural gas processing for methanol production (http://bluefuelenergy.com).

It is reported that for a methanol production plant based on CO_2 hydrogenation, the capital investment cost is the same as that of a conventional syngas-based plant. CO_2 and H_2 availability and their price, and necessary electrical energy are limiting factors for the scale-up of CO_2-based methanol production plants (Goeppert et al., 2014).

There are two-step routes for methanol production from CO_2. At first, by RWGS reaction, CO_2 is changed to CO. After water removal, a mixture of CO_2/CO/H_2 is converted into methanol. The Korean Institute of Science and Technology (KIST) developed this process as CAMERE (carbon dioxide hydrogenation to methanol *via* RWGS reaction). One advantage of this two-step route is a reduction in methanol synthesis reactor size. Moreover, this process has higher efficiency and lower operating costs. In order to have a feasible CO_2 conversion to CO (> 60%), the RWGS reactor must be operated at $T > 600°C$ with a zinc aluminate catalyst. In the second reactor, $Cu/ZnO/ZrO_2/Ga_2O_3$ catalyst is used. In Korea, a pilot plant based on the CAMERE process with 100 kg_{CH3OH}/d capacity was built (Joo et al., 2004).

Catalyst deactivation (due to thermal sintering), recycling unconverted gas (due to thermodynamic equilibrium limitations) and low CO_2 conversion in one-pass (due to large release of heat of reaction) are main disadvantages for conventional gas-phase process. Air Products and Chemicals developed a liquid-phase methanol (LPMEOH) synthesis process from syngas, in the late 1970s. The demo unit was operated at Eastman's Chemicals-from-coal in Tennessee (USA) in 1997. A slurry bubble column reactor (SBCR) is used in this process. For absorbing the exothermal heat of the reaction, a high boiling-point mineral oil is used in the reactor. By cooling, water and methanol are separated from solvent and it is recycled back to the reactor. Efficient heat transfer, good reactor temperature cooling, high conversion per pass and mild process temperature are advantages of this liquid-phase process (Din et al., 2019).

RITE (Research Institute of Innovative Technology) and NIRE (National Institute for Resources and Environment) used Cu/ZnO catalyst in hydrocarbon oil (as a hydrophobic solvent) for liquid-phase methanol production. As 95% methanol yield is obtained by this process, there is no need to recycle the un-reacted gas. There is a need to develop catalysts with high activity at milder temperature for liquid-phase processes (Lee and Sardesai, 2005).

In 1993, Lurgi announced the first worldwide demo plant for direct methanol production from CO_2. Their process was based on $Cu/ZnO/Al_2O_3$ catalyst developed by Clariant (formerly Süd-Chemie). In 2010, ALFE (Air Liquide Forschung und Entwicklung GmbH) and Lurgi GmbH reported a long-term methanol production evaluation at pilot scale. Their pilot plant consists of a loop reactor equipped with crude methanol condensation and a separation recycle gas compressor. In order to increase methanol yield and carbon efficiency, a recycle compressor was used to recycle back the unconverted gas to the synthesis loop. The commercial $Cu/ZnO/Al_2O_3$ catalyst was used in this pilot plant at GHSV of 10,500 h^{-1} at 80 bar and 250°C. The CO_2 per pass conversion of 35%–45% was obtained. Within the first 100 h of reaction, the catalyst had slight deactivation, and for the rest of 600 h, it had a stable operation. The space-time-yield (STY) of 0.6 $kg/L_{cat}.h$ was reported for methanol (Zhong et al., 2020).

In Japan, since 1990, NIRE and RITE (now recognized as AIST (Advanced Industrial Science and Technology)) started a joint R&D project for direct methanol production from CO_2. In 1996, a bench-scale plant based on $Cu/ZnO/ZrO_2/Al_2O_3/SiO_2$ catalyst at 250°C and 5 MPa with recycling un-reacted gases and production capacity of 50 kg_{CH3OH}/d with 99.9 wt.% purity was constructed. In 2008, in Osaka, Japan's Mitsui Chemicals Inc. constructed a pilot plant for methanol production from CO_2 and H_2 with an annual capacity of 100 t/y that is used as the feedstock for olefin and aromatic production. In this plant, hydrogen was generated from photochemical water splitting by solar energy (Zhong et al., 2020).

In 2018, a cooperative agreement was signed between Lanzhou New Area Petrochemical Industry Investment Group, DICP (Dalian Institute of Chemical Physics) and Suzhou Gaomi New Energy for direct methanol production from CO_2 and H_2. In 2019, DICP and Petro China produced methanol at pilot-scale with 70% methanol selectivity and 20% CO_2 single-pass conversion. The produced methanol has purity higher than 99.9% (Saeidi et al., 2014).

In EU, the MefCO₂ consortium at Niederaussem had a pilot plant with 500 t/y methanol at 230°C. CO_2 is provided from the flue gas of a power plant and H_2 is obtained from water electrolysis. The produced methanol is separated and by a recycle compressor, unconverted reactants are recycled back.

3.4 CONCLUDING REMARKS

The interconversions of C_1 molecules provide the opportunity of using more abundant and less reactive C_1 compounds to more active and versatile intermediates. Currently, these conversions are the main routes for valorization of natural gas to chemicals and fuels and also chemical fixation of CO_2.

Despite the wide and large-scale use of some of these conversions such as reforming of methane, these processes impact the economy of the whole process and direct routes for CH_4 and CO_2 to chemicals remain attractive for further research and development.

For CO_2 conversions, the supply of hydrogen is critical and sustainable development requires the use of renewable energy sources.

Carbon dioxide chemical recycling for the production of methanol as a carbon neutral renewable fuel is a feasible alternative to overcome the problems of global climate change and fossil fuel depletion.

Technical and thus economic improvement might be expected in the future by using novel concepts such as membrane reactors.

NOMENCLATURE

acac	acetylacetone anion
AIST	Advanced Industrial Science and Technology
ALFE	Air Liquide Forschung und Entwicklung GmbH
ATR	autothermal reforming
BMA	Blausaure-Methan-Ammoniak
BRM	bi-reforming of methane
CAMERE	carbon dioxide hydrogenation to methanol *via* RWGS reaction
CNT	carbon nanotube
CPO	catalytic partial oxidation
CRI	Carbon Recycling International
DICP	Dalian Institute of Chemical Physics
DME	dimethyl ether
DRM	dry reforming of methane
DMTM	direct methane to methanol
EDC	ethylene dichloride
ER	Eley-Rideal
FTIR	Fourier transform infrared spectroscopy
GRT	gas reaction technologies
GTL	gas to liquid
HPM	high pressure methanol
HTSC	high-temperature shift conversion
KIST	the Korean Institute of Science and Technology
LDH	layer double hydrotalcite
LH	Langmuir-Hinshelwood
LPM	low pressure methanol
LPMEOH	liquid-phase methanol
LTSC	low-temperature shift conversion
MMO	methane monooxygenase
NIRE	National Institute for Resources and Environment
POX	partial oxidation
PTG	power to gas
RDS	rate determining step
RITE	Research Institute of Innovative Technology
RWGS	reverse water-gas shift
SBCR	slurry bubble column reactor
SRM	steam reforming of methane

SN	stoichiometric number
SNG	synthetic natural gas
SPARG	sulfur-passivated reforming
STY	space-time-yield
TMC	transition metal carbides
TRM	tri-reforming of methane
UHV	ultra-high vacuum
VCM	vinyl chloride monomer
VPO	vanadyl pyrophosphate
WGS	water gas shift

REFERENCES

Abdulrasheed, A., Jalil, A. A., Gambo, Y., Ibrahim, M., Hambali, H. U., Hamid, M. Y. S., A review on catalyst development for dry reforming of methane to syngas: recent advances, Renew. Sust. Energy Rev., 108 (2019) 175–193.

Akkharaphatthawona, N., Chanlekc, N., Cheng, C. K., Chareonpanicha, M., Limtrakule, J., Witoon, T., Tuning adsorption properties of $Ga_xIn_2\text{-}xO_3$ catalysts for enhancement of methanol synthesis activity from CO_2 hydrogenation at high reaction temperature, Appl. Surf. Sci., 489 (2019) 278–286.

Albertazzi, S., Basile, F., Fornasari, G., Trifiro, F., Vaccari, A., Thermal Biomass Conversion, In Centi, G., van Santen, R. A. (Eds.) Catalysis for Renewables, Wiley, Weinheim (2007).

Al-Shihri, S., Richard, C. J., Al-Megren, H., Chadwick, D., Insights into the direct selective oxidation of methane to methanol over ZSM-5 zeolites in aqueous hydrogen peroxide, Catal. Today, 353 (2020) 269–278.

Alvarez-Galvan, M. C., Mota, N., Ojeda, M., Rojas, S., Navarro, R. M., Fierro, J. L. G., Direct methane conversion routes to chemicals and fuels, Catal. Today, 171 (2011) 15–23.

Angelo, L., Kobl, K., Martínez-Tejada, L. M., Zimmermann, Y., Parkhomenko, K., Roger, A. C., Study of $CuZnMO_x$ oxides (M = Al, Zr, Ce, Ce-Zr) for the catalytic hydrogenation of CO_2 into methanol. Comptes. Rendus. Chim., 18 (2015) 250–260.

Arora, S., Prasad, R., An overview on dry Reforming of methane: Strategies to reduce carbonaceous deactivation of catalysts, RSC Adv., 6 (2016) 108668–108688.

Arutyunov, V., Direct Methane to Methanol: Historical and Kinetics Aspects, Methanol, Elsevier (2018).

Bahmani, M., Vasheghani Farahani, B., Sahebdelfar, S., Preparation of high performance nano-sized $Cu/ZnO/Al_2O_3$ methanol synthesis catalyst via aluminum hydrous oxide sol, Appl. Catal. A. G., 520 (2016) 178–187.

Bahruji, H., Bowker, M., Hutchings, G., Dimitratos, N., Wells, P., Gibson, E., Jones, W., Brookes, C., Morgan, D., Lalev, G., Pd/ZnO catalysts for direct CO_2 hydrogenation to methanol, J. Catal., 343 (2016) 133–146.

Bartholomew, C. H., Farrauto, R. J., Fundamentals of Industrial Catalytic Processes, John Wiley & Sons, New Jersey (2006).

Behrens, M., Promoting the synthesis of methanol: understanding the requirements for an industrial catalyst for the conversion of CO_2, Angew. Chem. Int. Ed., 55 (2016) 14906–14908.

Berndt, H., Martin, A., Bruckner, A., Schreier, E., Muller, D., Kosslick, H., Wolf, G. U., Lucke, B., Structure and catalytic properties of VO_x/MCM materials for the partial oxidation of methane to formaldehyde, J. Catal., 191 (2000) 384–400.

Beznis, N. V., Weckhuysen, B. M., Bitter, J. H., Partial oxidation of methane over Co-ZSM-5: tuning the oxygenate selectivity by altering the preparation route, Catal. Lett., 136 (2010) 52–56.

Bianchi, D., Chafik, T., Khalfallah, M., Teichner, S. J., Intermediate species on zirconia supported methanol aerogel catalysts: V. Adsorption of methanol, Appl. Catal., A., 123 (1995) 89–110.

Bodke, A. S., Olschki, D. A., Schmidt, L. D., Hydrogen addition to the Andrussow process for HCN synthesis, Appl. Catal. A. G., 201 (2000) 13–22.

Bruckner, R., Advanced Organic Chemistry: Reaction Mechanisms, Elsevier (2002).

Cen, Y., Choi, S. M., Thompson, L. T., Low temperature CO_2 hydrogenation to alcohols and hydrocarbons over Mo_2C supported metal catalysts, J. Catal., 343 (2016) 147–156.

Chen, C. S., Cheng, W. H., Lin, S. S., Study of reverse water gas shift reaction by TPD, TPR and CO_2 hydrogenation over potassium-promoted Cu/SiO_2 catalyst, Appl. Catal. A. G., 238 (2003) 55–67.

Chorkendorff, I., Niemantsverdriet, J. W., Concepts of Modern Catalysis and Kinetics, WILEY-VCH, Weinheim (2017) 301–348.

Cornils, B., Herrmann, W. A., Wong, C. H., Zanthoff, H. W. (Eds.) Catalysis from A to Z: A Concise Encyclopedia, WILEY-VCH, Weinheim (2013).

Cunha, A. F., Morales-Torres, S., Pastrana-Martínez, L. M., Martins, A. A., Mata, T. M., Caetano, N. S., Loureiro, J. M., Syngas production by bi-reforming methane on an Ni–K-promoted catalyst using hydrotalcites and filamentous carbon as a support material, RSC Adv., 10 (2020) 21158–21173.

Dang, T. T. H., Seeburg, D., Radnik, J., Kreyenschulte, C., Atia, H., Vu, T. T. H., Wohlrab, S., Influence of V-sources on the catalytic performance of VMCM-41 in the selective oxidation of methane to formaldehyde, Catal. Commun., 103 (2018) 56–59.

Dang, S., Yang, H., Gao, P., Wang, H., Li, X., Wei, W., Sun, Y., A review of research progress on heterogeneous catalysts for methanol synthesis from carbon dioxide hydrogenation. Catal. Today, 330 (2019) 61–75.

Dasireddy, V. D. B. C., Neja, S. S., Blaz, L., Correlation between synthesis pH, structure and $Cu/MgO/Al_2O_3$ heterogeneous catalyst activity and selectivity in CO_2 hydrogenation to methanol, J. CO_2 Util., 28 (2018) 189–199.

Dasireddy, V. D. B. C., Hanzel, D., Bharuth-Ram, K., Likozar, B., The effect of oxidant species on direct, non-syngas conversion of methane to methanol over an $FePO_4$ catalyst material, RSC Adv., 9 (2019a) 30989–31003.

Dasireddy, V. D. B. C., Likozar, B., The role of copper oxidation state in $Cu/ZnO/Al_2O_3$ catalysts in CO_2 hydrogenation and methanol productivity, Renew. Energy, 140 (2019b) 452–460.

Deerattrakul, V., Puengampholsrisook, P., Limphirat, W., Kongkachuichay, P., Characterization of supported Cu-Zn/graphene aerogel catalyst for direct CO_2 hydrogenation to methanol: effect of hydrothermal temperature on graphene aerogel synthesis, Catal. Today, 314 (2018) 154–163.

Din, I. U., Shaharun, M. S., Alotaibi, M. A., Alharthi, A. I., Naeem, A., Recent developments on heterogeneous catalytic CO_2 reduction to methanol, J. CO_2 Util., 34 (2019) 20–33.

Dinh, K. T., Sullivan, M. M., Serna, P., Meyer, R. J., Dincă, M., Román-Leshkov, Y., Viewpoint on the partial oxidation of methane to methanol using Cu- and Fe-exchanged zeolites, ACS Catal., 8 (2018) 8306–8313.

Dong, X., Li, F., Zhao, N., Xiao, F., Wang, J., Tan, Y., CO_2 hydrogenation to methanol over $Cu/ZnO/ZrO_2$ catalysts prepared by precipitation-reduction method, Appl. Catal. B Env., 191 (2016) 8–17.

Du, G., Lim, S., Yang, Y., Wang, C., Pfefferle, L., Haller, G. L., Catalytic performance of vanadium incorporated MCM-41 catalysts for the partial oxidation of methane to formaldehyde, Appl. Catal. A: G., 302 (2006) 48–61.

Fiordaliso, E. M., Sharafutdinov, I., Carvalho, H. W. P., Grunwaldt, J. D., Hansen, T. W., Chorkendor, I., Wagner, J. B., Damsgaar, C. D., Intermetallic GaPd$_2$ nanoparticles on SiO$_2$ for low-pressure CO$_2$ hydrogenation to methanol: catalytic performance and in-situ characterization, ACS Catal., 5 (2015) 5827–5836.

Fisher, I. A., Woo, H. C., Bell, A. T., Effects of zirconia promotion on the activity of Cu/SiO$_2$ for methanol synthesis from CO/H$_2$ and CO$_2$/H$_2$, Catal. Lett., 44 (1997) 11–17.

Gail, E., Gos, S., Kulzer, R., Lorösch, J., Rubo, A., Sauer, M., Kellens, R., Reddy, J., Steier, N., Hasenpusch, W., Cyano Compounds, Inorganic, In Ullmann's Encyclopedia of Industrial Chemistry, Wiley, Germany (2012).

Gao, P., Li, F., Zhan, H., Zhao, N., Xiao, F., Wei, W., Zhong, L., Wang, H., Sun, Y., Influence of Zr on the performance of Cu/Zn/Al/Zr catalysts via hydrotalcite-like precursors for CO$_2$ hydrogenation to methanol., J. Catal., 298 (2013) 51–60.

Gao, P., Yang, H., Zhang, L., Zhang, C., Zhong, L., Wang, H., Sun, Y., Fluorinated Cu/Zn/Al/Zr hydrotalcites derived nanocatalysts for CO$_2$ hydrogenation to methanol, J. CO$_2$ Util., 16 (2016) 32–41.

Garcia, C. L., Resasco, D. E., High-temperature oxychlorination catalysts: role of LaCL$_3$ as an inhibitor of the segregation of active species during heating-cooling cycles., J. Catal., 122 (1990) 151–165.

García-Trenco, A., Regoutz, A., White, E. R., Payne, D. J., Shaer, M. S. P., Williams, C. K., PdIn intermetallic nanoparticles for the hydrogenation of CO$_2$ to methanol, Appl. Catal. B. Env., 220 (2018) 9–18.

Ghosh, S., Uday, V., Giri, A., Srinivas, S., Biogas to methanol: a comparison of conversion processes involving direct carbon dioxide hydrogenation and via reverse water gas shift reaction, J. Clean. Prod., 217 (2019) 615–626.

Goeppert, A., Czaun, M., Jones, J. P., Prakash, G. K. S., Olah, G. A., Recycling of carbon dioxide to methanol and derived products – closing the loop, Chem. Soc. Rev., 43 (2014) 7995–8048.

Goguet, A., Meunier, F., Breen, J. P., Burch, R., Petch, M. I., Ghenciu, A. F., Study of the origin of the deactivation of a Pt/CeO$_2$ catalyst during reverse water gas shift (RWGS) reaction., J. Catal., 226 (2004) 382–392.

Guil-López, R., Mota, N., Llorente, J., Millán, E., Pawelec, B., Fierro, J. L. G., Navarro, R. M., Methanol synthesis from CO$_2$: a review of the latest developments in heterogeneous catalysis, Materials 12 (2019a) 3902.

Guil-Lopez, R., Mota, N., Llorente, J., Millan, E., Pawelec, B. G., Fierro, J. L. G., Navarro, R. M., Data on TGA of precursors and SEM of reduced Cu/ZnO catalysts co-modified with aluminium and gallium for methanol synthesis, Data Br., 24 (2019b) 104010.

Guil-Lopez, R., Mota, N., Llorente, J., Millan, E., Pawelec, B. G., Fierro, J. L. G., Navarro, R. M., Structure and activity of Cu/ZnO catalysts co-modified with aluminium and gallium for methanol synthesis, Catal. Today, 355 (2020) 870–881.

Hartadi, Y., Widmann, D., Behm, R. J., CO$_2$ Hydrogenation to methanol on supported Au catalysts under moderate reaction conditions: Support and particle size effects, Chem. Sus. Chem., 8 (2015) 456–465.

Häring, H.-W., (ed.), Industrial Gases Processing, Wiley-VCH, Weinheim (2008).

Hasenberg, D., Schmidt, L. D., HCN synthesis from CH$_4$ and NH$_3$ on clean Rh, J. Catal., 91 (1985) 116–131.

Hasenberg, D., Schmidt, L. D., HCN synthesis from CH$_4$ and NH$_3$ on platinum, J. Catal., 97 (1986) 156–168.

Horn, R., Schlögl, R., Methane Activation by Heterogeneous Catalysis, Catal. Lett., 145 (2015) 23–39.

Huang, C., T-Raissi, A., Liquid hydrogen production via hydrogen sulfide methane reformation, J. Power Sources. 175 (2008) 464–472.

Jadhav, S. G., Vaidya, P. D., Bhanage, B. M., Joshi, J. B., Catalytic carbon dioxide hydrogenation to methanol: a review of recent studies, Chem. Eng. Res. Des., 92 (2014) 2557–2567.

Jafarbegloo, M., Tarlani, A., Wahid Mesbah, A., Sahebdelfar, S., Thermodynamic analysis of carbon dioxide reforming of methane and its practical relevance, Int. J. Hydrogen Energy, 40 (2015) 2445–2451.

Jang, W. J., Shim, J. O., Kim, H. M., Yoo, S. Y., Roh, H. S., A review on dry reforming of methane in aspect of catalytic properties, Catal. Today, 324 (2019) 15–26,

Joo, O. S., Jung, K. D., Jung, Y., CAMERE process for methanol synthesis from CO$_2$ hydrogenation, Stud. Surf. Sci. Catal., 153 (2004) 67–72.

Kattel, S., Yan, B., Yang, Y., Chen, J. G., Liu, P., Optimizing binding energies of key intermediates for CO$_2$ hydrogenation to methanol over oxide-supported copper, J. Am. Chem. Soc., 138 (2016) 12440–12450.

Kim D. H., Park J. L., Park E. J., Kim Y. D., S. Uhm, Dopant effect of barium zirconate-based perovskite-type catalysts for the intermediate-temperature reverse water gas shift reaction, ACS Catal., 4 (2014) 3117–3122.

Kim, Y., Trung, T. S. B., Yang, S., Kim, S., Lee, H., Mechanism of the surface hydrogen induced conversion of CO$_2$ to methanol at Cu(111) step sites, ACS Catal., 6 (2016) 1037–1044.

Koberstein, E., Model reactor studies of the hydrogen cyanide synthesis from methane and ammonia, Ind. Eng. Chem. Process Des. Develop., 12 (1973) 444–448.

Kondrat, S. A., Smith, P. J., Carter, J. H., Hayward, J. S., Pudge, G. J., Shaw, G., Spencer, M. S., Bartley, J. K., Taylor, S. H., Hutchings, G. J., The effect of sodium species on methanol synthesis and water-gas shift Cu/ZnO catalysts: utilizing high purity zincian georgeite, Faraday Discuss., 197 (2017) 287–307.

Kondratenko, E. V., Baerns, M., Synthesis Gas Generation, In Horváth, I. T. (Ed.), Encyclopedia of Catalysis, Wiley, Germany (2003) 442–456.

Krisnandi, Y. K., Putra, B. A. P., Zaharaa, M. B., Abdullah, I., Howe, R. F., Partial oxidation of methane to methanol over heterogeneous catalyst Co/ZSM-5, Procedia Chem. 14 (2015) 508–515.

Kunkes, E. L., Studt, F., Abild-Pedersen, F., Schlogl, R., Behrens, M., Hydrogenation of CO$_2$ to methanol and CO on Cu/ZnO/Al$_2$O$_3$: is there a common intermediate or not, J. Catal., 328 (2015) 43–48.

Larmier, K., Liao, W. C., Tada, S., Lam, E., Verel, R., Bansode, A., Urakawa, A., Comas-Vives, A., Copéret, C., Reaction intermediates and role of the interface in the CO$_2$-to-CH$_3$OH hydrogenation on ZrO$_2$-supported Cu nanoparticles, Angew. Chem. Int. Ed., 56 (2017) 2318–2323.

Lee, S., Sardesai, A., Liquid phase methanol and dimethyl ether synthesis from syngas, Top. Catal., 32 (2005) 197–207.

Lee, E., Cheng, Z., Lo, C. S., Present and future prospects in heterogeneous catalysts for C1 chemistry, Catalysis, 27 (2015) 187–208.

Lei, H., Nie, R., Wu, G., Hou, Z., Hydrogenation of CO$_2$ to CH$_3$OH over Cu/ZnO catalysts with different ZnO morphology, Fuel, 154 (2015) 161–166.

Le Page, J. F., Applied Heterogeneous Catalysis: Design, Manufacture, Use of Solid Catalysts. Éditions Technip, Paris (1987).

Li, Y. N., He, L. N., Diao, Z. F., Homogeneous hydrogenation of carbon dioxide to methanol, Catal. Sci. Tech., 4 (2014a) 1498–1512.

Li, Y. N., Ma, R., He, L. N., Diao, Z. F., Homogeneous hydrogenation of carbon dioxide to methanol, Catal. Sci. Technol., 4 (2014b) 1498–1512.

Li, M. M. J., Chen, C., Ayvall, T., Suo, H., Zheng, J., Teixeira, I. F., Ye, L., Zou, H., O'Hare, D., Tsang, S. C. E., CO$_2$ hydrogenation to methanol over catalysts derived from single cationic layer CuZnGa LDH precursors, ACS Catal., 8 (2018) 4390–4401.

Lin, R., Amrute, A. P., Pérez-Ramírez, J., Halogen-mediated conversion of hydrocarbons to commodities, Chem. Rev., 117 (2017) 4182–4247.

Liu, L., Jiang H., Liu, H., Li, H., Recent Advances on the Catalysts for Activation of CO_2 in Several Typical Processes, In S. L. Suib (Ed.), New and Future Developments in Catalysis: Activation of Carbon Dioxide, Elsevier, Amsterdam (2013a) 189–222.

Liu, P., Yang, Y., White, M. G., Theoretical perspective of alcohol decomposition and synthesis from CO_2 hydrogenation, Surf. Sci. Rep., 68 (2013b) 233–272.

Liu, C., Guo, X., Guo, Q., Mao, D., Yu, J., Lu, G., Methanol synthesis from CO_2 hydrogenation over copper catalysts supported on MgO-modified TiO_2, J. Mol. Catal. A. Chem., 425 (2016) 86–93.

Liu, L., Fan, F., Bai, M., Xue, F., Ma, X., Jiang, Z., Fang, T., Mechanistic study of methanol synthesis from CO_2 hydrogenation on Rh-doped Cu(1 1 1) surface. Mol. Catal., 466 (2019) 26–36.

Loricera, C. V., Alvarez-Galvan, M. C., Guil-Lopez, R., Ismail, A. A., Al-Sayari, S. A., Fierro, J. L. G., Structure and reactivity of sol-gel V/SiO_2 catalysts for the direct conversion of methane to formaldehyde, Top Catal., 60 (2017) 1129–1139.

Lorkovic, I., Noy, M., Weiss, M., Sherman, J., McFarland, E., Stucky, G. D., Ford, P. C., C₁ coupling via bromine activation and tandem catalytic condensation and neutralization over CaO/zeolite composites, Chem. Commun., (2004) 566–567.

Lu R., K. Kawamoto, Preparation of mesoporous CeO_2 and monodispersed NiO particles in CeO_2, and enhanced selectivity of NiO/CeO_2 for reverse water gas shift reaction, Mater. Res. Bull., 53 (2014) 70–78.

Lu, W., Zhou, L., Oxidation of C-H Bonds, John Wiley & Sons; Hoboken, New Jersey (2017).

Mahyuddin, M. H., Shiota, Y., Yoshizawa, K., Methane selective oxidation to methanol by metal-exchanged zeolites: a review of active sites and their reactivity, Catal. Sci. Technol., 9 (2019) 1744–1768.

Majewski, A. J., Wood, J., Tri-reforming of methane over $Ni@SiO_2$ catalyst, Int. J. Hydrog. Energy, 39 (2014) 12578–12585.

Marlin, D. S., Sarron, E., Sigurbjörnsson, O., Process advantages of direct CO_2 to methanol synthesis, Front. Chem., 6 (2018) 446.

Maxwell, G. R., Synthetic Nitrogen Products: A Practical Guide to the Products and Processes, Plenum Publishers, New York (2004) 347–360.

McCormick, R. L., Al-Sahali, M. B., Alptekin, G. O., Partial oxidation of methane, methanol, formaldehyde, and carbon monoxide over silica: global reaction kinetics, Appl. Catal. A. G., 226 (2002) 129–138.

McFarland, E., Unconventional chemistry for unconventional natural gas, Science, 338 (2012) 340–342.

Mesters, C., A selection of recent advances in C₁ chemistry, Ann. Rev. Chem. Biomol. Eng., 7 (2016) 223–238.

Mota, N., Guil-Lopez, R., Pawelec, B. G., Fierro, J. L. G., Navarro, R. M., Highly active Cu/ZnO-Al catalyst for methanol synthesis: effect of aging on its structure and activity, RSC Adv., 8 (2018) 20619–20629.

Mureddu, M., Ferrara, F., Pettinau, A., Highly effcient $CuO/ZnO/ZrO_2@SBA$-15 nanocatalysts for methanol synthesis from the catalytic hydrogenation of CO_2, Appl. Catal. B. Env., 258 (2019) 117941.

Nakagawa, K., Ikenaga, N., Teng, Y., Kobayashi, T., Suzuki, T., Partial oxidation of methane to synthesis gas over iridium–nickel bimetallic catalysts, Appl. Catal. A. G., 180 (1999) 183–193.

Nguyen, L. D., Loridant, S., Launay, H., Pigamo, A., Dubois, J. L., Millet, J. M. M., Study of new catalysts based on vanadium oxide supported on mesoporous silica for the partial oxidation of methane to formaldehyde: catalytic properties and reaction mechanism J. Catal., 237 (2006) 38–48.

Olah, G. A., Gupta, B., Farina, M., Feldberg, J. D., Ip, W. M., husain, A., Karpeles, R., Lammertsma, K., Melhotra, A. K., Trivedi, N. J., Electrophilic reactions at single bonds. Selective monohalogenation of methane over supported acid or platinum metal-catalysts and hydrolysis of methyl halides over gamma-alumina-supported metal-oxide hydroxide catalysts-a feasible path for the oxidative conversion of methane into methyl alcohol/dimethylether, J. Am. Chem. Soc., 107 (1985) 7097–7105.

Olah, G. A., Goeppert, A., Czaun, M., Prakash, G. K. S., Bi-reforming of methane from any source with steam and carbon dioxide exclusively to metgas (CO$_2$–H$_2$) for methanol and hydrocarbon synthesis, J. Am. Chem. Soc., 135 (2013) 648-650.

Olah, G. A., Molnár, Á., Prakash, G. K. S., Hydrocarbon Chemistry, 3rd Ed., Wiley, New York (2018) 123–325.

Osterwalder, N., Stark, W. J., Direct coupling of bromine-mediated methane activation and carbon-deposit gasification, ChemPhysChem, 8 (2007) 297–303.

Pakhare, D., Spivey, J., A review of dry (CO$_2$) reforming of methane over noble metal catalysts, Chem. Soc. Rev., 43 (2014) 7813–7837.

Parmaliana, A., Arena, F., Frusteri, F., Martınez-Arias, A., López Granados, M., Fierro, J. L. G., Effect of Fe-addition on the catalytic activity of silicas in the partial oxidation of methane to formaldehyde, Appl. Catal. A. G., 226 (2002) 163–174.

Pasupulety, N., Driss, H., Alhamed, Y. A., Alzahrani, A. A., Daous, M. A., Petrov, L., Influence of preparation method on the catalytic activity of AU/CU-ZN-AL catalysts for CO$_2$ hydrogenation to methanol, Comptes Rendus Acad. Bulg. Sci., 68 (2015) 1511–1518.

Peringer E., Podkolzin S.G., Jones M.E., Olindo R., Lercher J.A., LaCl$_3$-based catalysts for oxidative chlorination of CH4, Top Catal, 38(1–3) (2006), 211–220.

Pierella LB, Saux C, Caglieri SC, Bertorello HR, Bercoff PG, Catalytic activity and magnetic properties of Co–ZSM-5 zeolites prepared by different methods, Appl Catal. A. Gen., 347 (2008) 5–61.

Podkolzin, S. G., Stangland, E. E., Jones, M. E., Peringer, E., Lercher, J. A., Methyl chloride production from methane over lanthanum-based catalysts, J. Am. Chem. Soc., 129 (2007) 2569–2576.

Porosoff, M. D., Baldwin, J. W., Peng, X., Mpourmpakis, G., Willauer, H. D., Potassium promoted molybdenum carbide as a highly active and selective catalyst for CO$_2$ conversion to CO, ChemSusChem, 10 (2017) 2408–2415.

Posada-Perez, S., Ramirez, P. J., Gutierrez, R. A., Stacchiola, D. J., Vines, F., Liu, P., Illas, F., Rodriguez, J. A., The conversion of CO$_2$ to methanol on orthorhombic β-Mo$_2$C and Cu/β-Mo$_2$C catalysts: mechanism for admetal induced change in the selectivity and activity, Catal. Sci. Technol., 6 (2016) 6766–6777.

Quiroga, M. M. B., Luna, A. E. C., Kinetic analysis of rate data for dry reforming of methane. Ind. Eng. Chem. Res., 46 (2007), 5265–5270.

Rasmussen, P. B., Holmblad, P. M., Askgaard, T., Ovesen, C. V., Stoltze, P., Norskov, J. K., Chorkendorff, I., Catal. Lett., 26 (1994) 373–381.

Ravi, M., Ranocchiari, M., van Bokhoven, J. A., The direct catalytic oxidation of methane to methanol – a critical assessment, Angew. Chem. Int. Ed., 56 (2017) 16464–16483.

Reddy, G. K., Loridant, S., Takahashi, A., Delichère, P., Reddy, B. M., Reforming of methane with carbon dioxide over Pt/ZrO$_2$/SiO$_2$ catalysts – effect of zirconia to silica ratio, Appl. Catal. A. G., 389 (2010) 92–100.

Rezaei, M., Alavi, S. M., Sahebdelfar, S., Yan, Z. F., Syngas production by methane reforming with carbon dioxide on noble metal catalysts, J. Nat. Gas Chem., 15 (2006) 327–334.

Rostrup-Nielsen, J.R., Sehested, J., and Norskov, J.K. Hydrogen and synthesis gas by steam- and CO$_2$ reforming. Adv. Catal., 47 (2002) 65–139.

Saeidi, S., Amin, N. A. S., Rahimpour, M. R., Hydrogenation of CO$_2$ to value-added products – a review and potential future developments, J. CO$_2$ Util., 5 (2014) 66–81.

Sahebdelfar, S., Takht Ravanchi, M., Heterogeneous Catalytic Hydrogenation of CO₂ to Basic Chemicals and Fuels, In Kumar, A., Sharma, S. (Eds.), Chemo-Biological Systems for CO₂ Utilization, CRC Press (2021).

Sanchez-Sanchez, M., Lercher, J. A., Oxidative Functionalization of Methane on Heterogeneous Catalysts, In Pombeiro, A. J. L., Fátima, M., da Silva, C. G. (Eds.), Alkane Functionalization, Wiley, Boca Raton (2019).

Schwarz, H., Chemistry with methane: concepts rather than recipes, Angew. Chem. Int. Ed., 50 (2011) 10096–10115.

Shimura, K., Fujitani, T., Effects of promoters on the performance of a VO_x/SiO_2 catalyst for the oxidation of methane to formaldehyde, Appl. Catal. A. G., 577 (2019) 44–51.

Smeets, P. J., Groothaert, M. H., Schoonheydt, R. A., Cu based zeolites: a UV-vis study of the active site in selective methanol oxidation at low temperatures, Catal. Today, 110 (2005) 303–309.

Song, C. S., Pan, W. Tri-reforming of methane: a novel concept for catalytic production of industrially useful synthesis gas with desired H_2/CO ratios, Catal. Today, 98 (2004) 463–484.

Stroebe, M., Hoffmann, V. H., Zogg, A., Scheringer, M., Hungerbuhler, K., Environmentally oriented design and assessment of chemical products and processes, Chimia, 55 (2001) 887–891.

Su, X., Yang, X., Zhao, B., Huang, Y., Designing of highly selective and high-temperature endurable RWGS heterogeneous catalysts: recent advances and the future directions, J. Energy Chem., 26 (2017) 854–867.

Sun, Q., Ye, J., Liu, C., Ge, Q., In_2O_3 as a promising catalyst for CO_2 utilization: a case study with reverse water gas shift over In_2O_3. Greenhouse Gas Sci. Tech., 4 (2014) 140–144.

Sushkevich, V. L., van Bokhoven, J. A., Effect of Brønsted acid sites on the direct conversion of methane into methanol over copper-exchanged mordenite, Catal. Sci. Tech., 8 (2018) 1414–4150.

Sushkevich, V. L., van Bokhoven, J. A., Methane-to-methanol: activity descriptors in copper-exchanged zeolites for the rational design of materials, ACS Catal., 9 (2019) 6293–6304.

Tada, S., Watanabe, F., Kiyota, K., Shimoda, N., Hayashi, R., Takahashi, M., Nariyuki, A., Igarashi, A., Satokawa, S., Ag addition to $CuO-ZrO_2$ catalysts promotes methanol synthesis via CO_2 hydrogenation, J. Catal., 351 (2017) 107–118.

Taifan, W., Baltrusaitis, J., CH_4 conversion to value added products: potential, limitations and extensions of a single step heterogeneous catalysis, Appl. Catal. B. Env., 198 (2016) 525–547.

Taifan, W., Arvidsson, A. A., Nelson, E., Hellman, A., Baltrusaitis, J., CH_4 and H_2S reforming to CH_3SH and H_2 catalyzed by metal promoted Mo_6S_8 cluster: a first-principles micro-kinetic study, Catal. Sci. Technol., 7 (2017a) 3546–3554.

Taifan, W., Baltrusaitis, J., Mini review: direct catalytic conversion of sour natural gas (CH_4 + H_2S + CO_2) components to high value chemicals and fuels, Catal. Sci. Technol., 7 (2017b) 2919–2929.

Takht Ravanchi, M., Sahebdelfar, S., Catalytic conversions of CO_2 to help mitigate climate change: recent process developments, Proc. Safety Env. Prot., 145 (2021) 172–194.

Thacker, C. M., Miller, E., Carbon disulfide production: effect of catalysts on reaction of methane with sulfur, Ind. Eng. Chem., 36 (1944) 182–184.

Tijm, P. J. A., Waller, F. J., Brown, D. M., Methanol technology developments for the new millennium, Appl. Catal. A. G., 221 (2001) 275–282.

Tisseraud, C., Comminges, C., Belin, T., Ahouari, H., Soualah, A., Pouilloux, Y., LeValant, A., The Cu-ZnO synergy in methanol synthesis from CO_2, Part 2: origin of the methanol and CO selectivities explained by experimental studies and a sphere contact quantification model in randomly packed binary mixtures on Cu-ZnO coprecipitate catalysts, J. Catal., 330 (2015) 533–544.

Tisseraud, C., Comminges, C., Pronier, S., Pouilloux, Y., LeValant, A., The Cu-ZnO synergy in methanol synthesis Part 3: impact of the composition of a selective Cu-ZnO$_x$ core-shell catalyst on methanol rate explained by experimental studies and a concentric spheres model, J. Catal., 343 (2016) 106–114.

Tomkins, P., Ranocchiari, M., van Bokhoven, J. A., Direct conversion of methane to methanol under mild conditions over Cu-zeolites and beyond, Acc. Chem. Res., 50 (2017) 418–425.

Tosoni, S., Pacchioni, G., Oxide-supported gold clusters and nanoparticles in catalysis: a computational chemistry perspective, ChemCatChem, 11 (2019) 73–89.

Toyir, J., Ramirez de la Piscina, P., Fierro, J. L. G., Homs, N., Catalytic performance for CO$_2$ conversion to methanol of gallium-promoted copper-based catalysts: influence of metallic precursors, Appl. Catal. B. Env., 34 (2001) 255–266.

Tsang, S.C., Claridge, J.B. and Green, M.L.H. Recent advances in the conversion of methane to synthesis gas. Catal. Today, 23 (1995) 3–15.

Tuomi, S., Guil-Lopez, R., Kallio, T., Molybdenum carbide nanoparticles as a catalyst for the hydrogen evolution reaction and the effect of pH, J. Catal., 334 (2016) 102–109.

Ud Din, I., Shaharun, M. S., Alotaibi, M. A., Alharthi, A. I., Naeem, A., Recent developments on heterogeneous catalytic CO$_2$ reduction to methanol, J. CO$_2$ Util., 34 (2019) 20–33.

Usman, M., Wan Daud, W. M. A., Recent advances in the methanol synthesis via methane reforming processes, RSC Adv., 5 (2015) 21945–21972.

van der Oosterkamp, P. F., Synthesis Gas Generation – Industrial, In Horváth, I. T. (Ed.), Encyclopedia of Catalysis, Wiley, Germany (2003), 457–482.

Verykios, X. E., Catalytic dry reforming of natural gas for the production of chemicals and hydrogen, Int. J. Hyd. Energy, 28 (2003) 1045–1063.

Vogt, C., Kranenborg, J., Monai, M., Weckhuysen, B. M., Structure sensitivity in steam and dry methane reforming over nickel: activity and carbon formation, ACS Catal., 10 (2020) 1428–1438.

Vourros, A., Garagounis, I., Kyriakou, V., Carabineiro, S. A. C., Maldonado-Hodar, F. J., Marnellos, G. E., Konsolakis, M., Carbon dioxide hydrogenation over supported Au nanoparticles: effect of the support, J. CO$_2$ Util., 19 (2017) 247–256.

Wang, X., Chen, H., Sachtler, W. M. H., Catalytic reduction of NO$_x$ by hydrocarbons over Co/ZSM-5 catalysts prepared with different methods, Appl. Catal. B. Env., 26 (2000) L227–L239.

Wang, W., Zhang, Y., Wang, Z., Yan, J.M., Ge, Q., Liu, C.J., Reverse water gas shift over In2O3–CeO2 catalysts, Catal. Today 259 (2016) 402–408.

Wang, Y. H., Kattel, S., Gao, W. G., Li, K. Z., Liu, P., Chen, J. G., Wang, H., Exploring the ternary interactions in Cu-ZnO-ZrO$_2$ catalysts for efficient CO$_2$ hydrogenation to methanol, Nat. Commun., 10 (2019) 1166.

Wang, W., Qu, Z., Song, L., Fu, Q., CO$_2$ hydrogenation to methanol over Cu/CeO$_2$ and Cu/ZrO$_2$ catalysts: tuning methanol selectivity via metal-support interaction, J. Energy Chem., 40 (2020) 22–30.

Weissermel, K., Arpe, H. J., Industrial Organic Chemistry, 3rd Ed., Wiley, Germany (1997).

Woertink, J. S., Smeets, P. J., Groothaert, M. H., Vance, M. A., Sels, B. F., Schoonheydt, R. A., Solomon, E. I., Proc. Natl. Acad. Sci. USA, 106 (2009) 18908–18913.

Xu, J., Froment, G. F., Methane steam reforming, methanation and water-gas shift: 1. Intrinsic kinetics, AIChE J., 35 (1989) 88–96.

Yang, Y., Mims, C. A., Disselkamp, R. S., Kwak, J. H., Peden, C. H. F., Campbell, C. T., (Non) formation of methanol by direct hydrogenation of formate on copper catalysts, J. Phys. Chem. C, 114 (2010) 17205–17211.

Yang, Y., Mims, C. A., Mei, D. H., Peden, C. H. F., Campbell, C. T., Mechanistic studies of methanol synthesis over Cu from CO/CO$_2$/H$_2$/H$_2$O mixtures: the source of C in methanol and the role of water, J. Catal., 298 (2013) 10–17.

Yang, Y., Mei, D. H., Peden, C. H. F., Campbell, C. T., Mims, C. A., Surface-bound intermediates in low-temperature methanol synthesis on copper: participants and spectators,
ACS Catal., 5 (2015) 7328–7337.

Zhang, Q., He, D., Han, Z., Zhang, X., Controlled partial oxidation of methane to methanol/
formaldehyde over Mo-V-Cr-Bi-Si oxide catalyst, Fuel, 81 (2002) 1599–1603.

Zhang, Q., He, D., Zhu, Q., Direct partial oxidation of methane to methanol: reaction zones
and role of catalyst location, J. Nat. Gas Chem., 17 (2008) 24–28.

Zhang, A., Sun, S., Komon, Z. J. A., Osterwalder, N., Gadewar, S., Stoimenov, P., Auerbach,
D. J., Stucky, G. D., McFarland, E. W., Improved light olefin yield from methyl bromide coupling over modified SAPO-34 molecular sieves, Phys. Chem. Chem. Phys.,
13 (2011) 2550–2555.

Zhao B., Pan, Y. X., Liu, C., The promotion effect of CeO₂ on CO₂ adsorption and hydrogenation over Ga₂O₃, Catal. Today 194 (2012) 60–64.

Zhong, J., Yang, X., Wu, Z., Liang, B., Huang, Y., Zhang, T., State of the art and perspectives in heterogeneous catalysis of CO₂ hydrogenation to methanol, Chem. Soc. Rev.,
49 (2020) 1385–1413.

Zichittella, G., Paunovic, V., Amrute, A. P., Perez-Ramirez, J., Catalytic oxychlorination versus oxybromination for methane functionalization, ACS Catal., 7 (2017) 1805–1817.

4 Methane Conversions

4.1 INTRODUCTION

Methane is an attractive feedstock in C_1 chemistry due to its large resources (natural gas, shale gas, methane hydrates, etc.). Other advantages as a feedstock are source-independent composition and ease of purification form heteroatoms. It has a high hydrogen content which can be used in product treatment or upgrading. Another incentive for methane conversion is low density of methane which impacts its transportation cost. As most methane natural gas fields are located in remote areas, the direct conversion of methane into transportable liquid fuels and chemicals would be highly desirable (Lunsford, 2000).

Currently indirect routes via syngas are used for methane conversion in which the syngas section accounts for the majority of the cost (Figure 4.1). From a conceptual point of view, the direct routes should have notable economic advantage over indirect routes, but so far very few direct routes proceeded to a commercial scale. The direct conversion of methane into useful chemicals is still an important issue in heterogeneous catalysis (Schwach et al., 2017). The product yield per single pass through the reactors is usually small, while the separation of products is difficult and costly.

Nevertheless, there are few commercialized methane conversion reactions which produce higher hydrocarbon product without an oxidant such as high-temperature self-coupling to acetylene. The theoretical yield of these reactions, however, is limed by thermodynamic constraints. Unfortunately, apart from methane conversion into synthesis gas, hydrogen cyanide, acetylene and in minor amounts to chlorinated methane, industrial methane conversion pathways are not yet competitive to oil-based production of chemicals and fuels (Horn and Schlögl, 2015).

There are other opportunities for direct conversion of methane into valuable higher hydrocarbons such as light olefins and aromatics (benzene, toluene, etc.). These reactions have received much attention.

4.2 CHEMISTRY OF METHANE

Methane is the lightest alkane. It is highly stable with apolar molecules (Table 4.1). Although the individual C–H bonds are polar, the molecule in non-polar due to its symmetry.

The high stability of methane molecule (highest among hydrocarbons) and its symmetrical structure makes its effective activation and catalytic conversion as a challenge in C_1 chemistry. Thus, it is difficult to activate one C–H bond without affecting the other bonds.

Many reactions of methane proceed via free radical intermediates. The homolytic dissociation strength of the C–H bond (439 kJ/mol) is ca. 46 kJ/mol higher than that in methanol and 170 kJ/mol higher than that of the π-bond of ethylene (as potential methane conversion products), meaning that the products are more

DOI: 10.1201/9781003279280-4

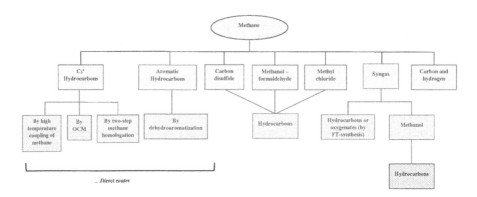

FIGURE 4.1 Possible pathways for conversion methane into higher hydrocarbons and oxygenates (modified from Choudhary et al., 2003).

susceptible than methane to consecutive reactions such as oxidation (Gunsalus et al., 2017). Consequently, functionalization of methane by free radical mechanisms may be inherently problematic.

The high stability of methane molecules renders its conversion into higher hydrocarbons thermodynamically highly unfavorable (Figure 4.2). However, at high temperatures (typically above 700°C), some of the transformations (e.g., to acetylene and aromatics) become favorable. Nevertheless, the decomposition of both reactant and products becomes favorable, which increases the potential for reduction of product yield and coke deposition.

The use of an oxidant, most commonly oxygen, can remove the thermodynamic barrier. Therefore, according to the absence or presence of oxygen, the methane conversions are examined as nonoxidative and oxidative conversions, respectively.

4.3   NON-OXIDATIVE CONVERSIONS

Oligocondensation is referred to a series of reactions by which the C–C bond is formed by hydrogen loss resulting in heavier hydrocarbon backbones. According to operating conditions, acetylene, olefins or aromatics may be produced. Acetylene was a basic chemical till 1960s when it was replaced by ethylene for most applications because of the lower cost of ethylene and its easier and safer handling.

TABLE 4.1

Thermochemical Properties of Methane

Compound	Normal Boiling Point (°C)	Dipole Moment (D)	$\Delta H°_f$ (kJ/mol)	$\Delta G°_f$ (kJ/mol)
CH₄	−161.50	0	−74.87	−50.84

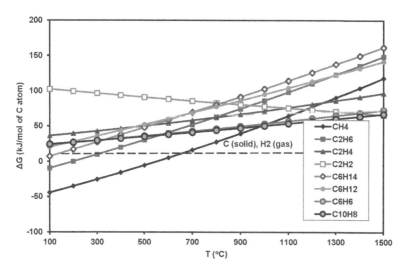

FIGURE 4.2 Gibbs free energy of formation per carbon atom for methane and selected hydrocarbons as a function of temperature.

4.3.1 HIGH-TEMPERATURE SELF-COUPLING

The conversion of methane and other hydrocarbons into acetylene (e.g., Eq. 4.1) is highly endothermic:

$$2CH_4 \rightarrow C_2H_2 + 3H_2 \quad \Delta H^0_{298} = 376 \, kJ/mol \quad (4.1)$$

It requires high reaction temperatures (>1,200°C) and the product should be quenched (in a few milliseconds) to stabilize the acetylene product by avoiding its dissociation. The reason for these extreme conditions can be inferred from Figure 4.2. It shows that acetylene is highly unstable under normal temperatures. At high temperatures, acetylene becomes more stable than other hydrocarbons but it is still unstable and can decompose to its constituent elements (below 4,000°C). This is also the case for the hydrocarbon feeds.

Figure 4.2 also shows that methane requires higher reaction temperatures compared to higher hydrocarbons. Equilibrium calculations show that acetylene formation from methane becomes apparent only above 700°C and is appreciable above 1,100°C (Figure 4.3).

Under high reaction temperatures, the rapid formation of acetylene and its subsequent decomposition competes. Side reactions are also promoted at higher temperatures. Thus, to achieve reasonable yields, short residence times (typically 0.1–10 ms) should be employed.

4.3.1.1 Mechanism

The reactions go through a free radical mechanism initiate by formation of CH_3 and H radicals which may recombine and end or arrive in a series of chain reactions giving acetylene and other products (Choudhary et al., 2003).

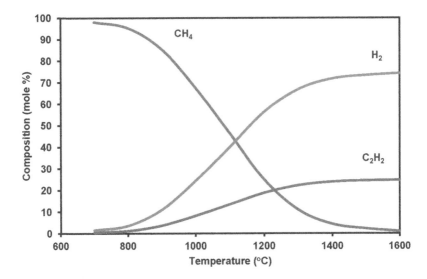

FIGURE 4.3 Equilibrium composition of methane coupling to acetylene versus temperature (modified after Pässler et al., 2007).

It is generally believed that the overall reaction is thermal coupling of methane occurring via high-temperature stepwise dehydrogenation (Holmen et al., 1995):

$$2CH_4 \rightarrow C_2H_6 + H_2 \rightarrow C_2H_4 + H_2 \rightarrow C_2H_2 + H_2 \rightarrow C + H_2 \qquad (4.2)$$

At high temperatures (>1,100°C), the main reaction chain goes via a free radical mechanism (Holmen et al., 1995):

$$CH_4 \rightarrow CH_3^\bullet + H_2 \qquad (4.3)$$

$$H^\bullet + CH_4 \rightarrow CH_3^\bullet + H_2 \qquad (4.4)$$

$$2CH_3^\bullet \rightarrow C_2H_6 \qquad (4.5)$$

Reaction (4.3) is the primary source of free radicals and the rate determining step. The ethane can involve in secondary reactions such as unimolecular:

$$C_2H_6 \rightarrow 2CH_3^\bullet \qquad (4.6)$$

Which is the reverse of reaction (4.5), and also

$$C_2H_6 + H^\bullet \rightarrow C_2H_5^\bullet + H^\bullet \qquad (4.7)$$

$$C_2H_6 + CH_3^\bullet \rightarrow C_2H_5^\bullet + CH_4 \qquad (4.8)$$

$$C_2H_5^\bullet \rightarrow C_2H_4 + H^\bullet \qquad (4.9)$$

And, subsequently, the secondary reactions of ethylene produce acetylene according to the below sequence:

$$C_2H_4 + H^\bullet \rightarrow C_2H_3^\bullet + H_2 \tag{4.10}$$

$$C_2H_4 + CH_3^\bullet \rightarrow C_2H_3^\bullet + CH_4 \tag{4.11}$$

$$C_2H_3^\bullet \rightarrow C_2H_2 + H^\bullet \tag{4.12}$$

At lower temperatures or higher residence times, methylation of ethylene should also be considered.

The supply of a large energy input at high temperatures in a short time is an important issue. This heat can be supplied by exothermic partial oxidation of the hydrocarbon feed in an autothermal process:

$$CH_4 + O_2 \rightarrow CO + H_2 + H_2O \quad \Delta H^0_{298} = -278 \, \text{kJ/mol} \tag{4.13}$$

$$CO + H_2O \rightarrow CO_2 + H_2 \quad \Delta H^0_{298} = -41 \, \text{kJ/mol} \tag{4.14}$$

The reaction temperature being determined by the CH_4/O_2 ratio, in addition to short residence time, is essential to obtain good acetylene yields.

4.3.1.2   Industrial Processes

Acetylene is commercially produced from a variety of sources such as coal (calcium carbide, CaC_2 process) and hydrocarbons (thermal or cracking processes). It is also a by-product of steam crackers. There are several uncatalyzed thermal processes for converting hydrocarbons ranging from methane to light naphtha to crude oil into acetylene. The trend in process development has been towards using heavier oil fractions, residual oils and even coal (Schobert, 2014). The common feature of the processes is rapid heat transfer rate to high (>1,400°C) temperatures, very short contact times (1–10 ms), low acetylene partial pressures and rapid quenching of pyrolysis gases. Acetylene is extracted from the cracked gas containing 5–20 vol.% C_2H_2 using selective solvents such as dimethylformamide (DMF), N-methylpyrrolidone (NMP), kerosene, methanol or acetone and is finally purified (Weissermel and Arpe, 1997).

The thermal processes can be classified as allothermal and autothermal processes. The allothermal processes could be further classified by heat transfer method which could be direct, e.g., by electric heating or indirect by a heat carrier. Autothermal processes need O_2 separation plants (increasing complexity and plant size) and yield a product containing H_2O and carbon oxides. Allothermal processes do not need O_2 and produces potentially less contaminated C_2H_2/H_2 but their design is more difficult compared to the burner of autothermal plants.

The BASF (Sachsse–Bartholomé) process is autothermal based on partial combustion. A premixed flame is used. The products are quenched by water or oil. Water is easier to use but oil is more effective.

Montecatini and Société Beige dAzote (SBA)/Kellogg developed two commercial autothermal processes which are principally similar to the Sachsse–Bartholomé process, but there is a difference in the design of the burners and the Montecatini process uses excess pressures (for several bars) (Weissermel and Arpe, 1997).

The Hüls electric arc process is an electrothermic process and an example of direct heat transfer allothermal processes capable of using natural gas, refinery gases and LPG as feed. An electric arc between water-cooled carbon steel supplies the reaction heat. It reduces hydrocarbon consumption by about 50% by eliminating combustion, but is economic in locations where low-cost electric power is available. The HEAP (Hydrogen Electric Arc Pyrolysis) process of Hoechst–Hüls uses plasma consisting of 30%–65% H atoms which means a high enthalpy density (Weissermel and Arpe, 1997).

The *regenerative furnace pyrolysis* or *Wulff process* is an example of indirectly heated process being more flexible considering operating conditions. It uses alternate heating in ovens with brick lattice acting as heat carrier. They are heated by burning the fuel and by alternating the input to the feed when cracking occurs. Soot formation is a problem and ethane and propane are the preferred feeds (Pässler et al., 2012). In the Hoeescht process, hot combustion gases act as the heat carrier.

The modern processes are dominated by autothermal or partial combustion technologies (Olah et al., 2018). The overall yield of acetylene and ethylene can be up to 50% with ethylene to acetylene ratio of 0.1–3. The BASF processes belong to this group of processes. The Sachsse–Bartholomé process is the first one with the greatest applications.

Microwave plasma reactors were used for selective acetylene synthesis (>90%) from methane. Under ambient temperature and atmospheric pressure, methane conversion into acetylene was done by using direct current pulse discharge (Kado et al., 1999). At methane conversions of 16%–52%, acetylene selectivity was >95%. In this case, effective removal of deposited carbon and stabilizing the state of discharge was accomplished by using oxygen. High power pulsed radio frequency and microwave carbon catalyzed processes can also result in high selectivities (Ioffe et al., 1995).

Acetylene is a versatile raw material for a number of important compounds, including vinyl chloride, vinyl acetate, acrylonitrile, and chloroprene, all of which are monomer intermediates for polymer production. However, as a consequence of the development of petrochemical industry, nearly all of these processes became obsolete. Furthermore, for the production of C_2 chemicals (specifically vinyl monomers), ethylene was a more economical feedstock in large scale; despite higher activities and selectivities with typically fewer reaction steps in acetylene chemistry. Due to the dangerous nature of acetylene, its handling costs are high, but it would remain an important precursor for producing chemicals (Trotuş et al., 2014).

4.3.2 Two-Step Methane Homologation

The two-step homologation (also called low-temperature coupling) of methane is a non-oxidative route to convert methane into higher hydrocarbons. In its simplest form:

$$2CH_4 \rightarrow C_2H_6 + H_2 \quad \Delta H^0_{500K} = +67.8 \, \text{kJ/mol} \, C_2H_6, \quad \Delta G^0_{500K} = +70.5 \, \text{kJ/mol} \, C_2H_6 \quad (4.15)$$

The reaction is endothermic and strongly thermodynamic limited.

The process involves two-step synthesis for direct coupling of methane. In the first step, methane is dissociatively adsorbed on a metal surface with some C–C bond formation at high temperature (~450°C) and low hydrogen pressure. Subsequently the resulted surface carbonaceous species are hydrogenated at low temperature (<200°C) and high hydrogen pressure to form methane and higher hydrocarbons. Some hydrogenolysis can also occur (Bradford, 2000). The interest is rooted from lower operating temperature and CO_x-free products when compared to oxidative routes such as OCM (oxidative coupling of methane) and PO (partial oxidation). The hydrogenation products obey Schulz-Flory distribution (Holmen, 2009).

It was independently discovered by two groups of researchers (Belgued et al., 1991; Koerts and Vansanten, 1991). At 250°C and over a EUROPT-1 platinum catalyst, an isothermal methane conversion process was reported by Belgued et al., (1991). In the presence of pure methane (as feed), C_2H_6 and H_2 formation was observed, and the amount of C_2H_6 production was an order of magnitude less than the H_2. However, these two products disappeared when time-on-stream exceeds 8 min, which was caused by the accumulation of surface carbonaceous residue. Then flow was switched from CH_4 to H_2 and formation of saturated hydrocarbons ranging from C_1 to C_7 was observed, with 19.3% conversion of the deposited methane to C_{2+} hydrocarbons. Later, comparative studies of Co, Ru, and Pt catalysts indicated that Ru is the most active catalyst for the homologation of methane (Belgued 1992).

4.3.2.1 Catalytic Effects

The homologation of methane is based on the capability of transition metals in the dissociative adsorption of methane and producing CH_x (with $x = 0, 1, 2$ or 3) surface species and hydrogen. In the next step, the produced carbonaceous surface intermediates are hydrogenated to form higher hydrocarbons (Figure 4.4). In the temperature range of 175–527°C, the surface residues were formed and hydrogenated to produce hydrocarbons up to C_5 in a temperature range of 27–127°C (Olah et al., 2018).

Although the original works were based on Pt as the active metal, it appeared that better results could be obtained with metals having reputation of forming an extensive C–C bond from C_1 species such as FT catalysts, thus Co and Ru received attention (Bond, 2006).

4.3.2.1.1 Metal Selection

Platinum, Co, Ru, and Rh were the metals used in the original studies. For methane decomposition, cobalt, platinum and ruthenium were the main metals used; as they exhibit good methane activation and carbon homologation and are frequently used in Fischer–Tropsch or methane activation processes. For two-step homologation process, these metals are also suitable; ruthenium and platinum have excellent properties for this reaction and cobalt is available (with lower costs) with a reasonable activity.

Koerts et al. (1992a) compared various transition metals and the order of activity was Co, Ru, Ni, Rh > Pt, Re, Ir > Pd, Cu, W, Fe and Mo. The order was similar to that for hydrogen-deuterium exchange. For 5% Ru/SiO_2 catalyst and at 100% methane conversion, a total C_{2+} yield of 13% was obtained (Koerts et al., 1992a). For

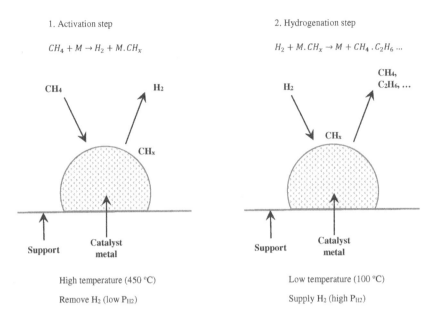

1. Activation step

$$CH_4 + M \rightarrow H_2 + M.CH_x$$

2. Hydrogenation step

$$H_2 + M.CH_x \rightarrow M + CH_4 . C_2H_6 \dots$$

High temperature (450 °C)

Remove H₂ (low P$_{H2}$)

Low temperature (100 °C)

Supply H₂ (high P$_{H2}$)

FIGURE 4.4 Schematic representation of two-step cycle for homologation of methane (M designated the reduced metal). Equations are not stoichiometrically balanced.

improving ethane yields, other precious metals supported on silica were evaluated but they had low efficiency. Co/SiO_2, Ru/SiO_2, $Rh/CeO_2/SiO_2$ and EUROPT-1 (6.3% Pt on SiO_2) are examples of the studied catalysts.

Over SiO_2-supported transition metals (W, Mo, Fe, Cu, Au, Ru, Rh, Pd, Ni, Co, Re, Pt and Ir), volcano plots were constructed in terms of metal-carbon bond strength (as measured by heats of adsorption for atomic carbon on reduced metal surfaces, Q_{M-C}) which, in accordance with the Sabatier principle, was considered as empirical evidence for the important role of metal-carbon bond strength on catalyst efficiency, selectivity to C_{2+} hydrocarbons and chain-growth probability (Bradford, 2000). According to experimental data of Koerts et al. (1992a), it was concluded that metals exhibiting a bond strength, Q_{M-C}, in the range of 140–169 kcal/mol (that is, Ru, Co, Fe and Ni) are the most effective metals for CH_4 homologation, at least under the experimental conditions.

4.3.2.1.2 Particle Size Effects

The CH_x species resulting from dissociative adsorption of methane depends on metal type, metal structure, support type, metal-support interactions and operating parameters (temperature, pressure and contact time) (Bradford, 2000). Strong metal-support bonding results in metal carbide formation. During dissociative adsorption of methane on transition metal surfaces, a concomitant of electron donation from σ bond to the surface and back donation from metal to methane occurs (Bradford 1999, 2000). Thus, surface CH_x fragments mostly occupy the locations where their tetera-valence is completed. As a consequence of geometric and electronic constraints, the formation of CH_x species from CH_4 dissociation is structure-sensitive in many cases

(Bradford and Vannice, 1999). It has been reported that dehydrogenation of CH_4 over Pt nanoparticle is more than that over larger particle being interpreted by more low-coordinated Pt site (corners and edges) on nanoparticles which decrease the energy barrier (Vines et al., 2010). A large ensemble of Pt atoms contains a larger portion of low-coordination number sites (steps, kinks).

4.3.2.1.3 Metal Support Effects

Regarding support effects, Soltan Mohammad Zadeh and Smith (1998, 1999) evaluated methane homologation on supported Co catalysts. They reported that a Co/Al_2O_3 catalyst was more active in methane decomposition compared to Co/SiO_2 that showed a higher activity in the hydrogenation step and a greater C_{2+} selectivity. Moreover, higher migratory of carbonaceous species on alumina supported catalyst was also reported. On the other hand, the reducibility of Co ions was increased by Pt addition which consequently enhanced methane adsorption (Lu et al., 1992). It was further reported that the Co-Pt/NaY system had almost complete methane conversion and a selectivity of 83.6% for C_{2+} hydrocarbons in the hydrogenation step, which is due to the synergistic effect of Co-Pt on the C–C bond formation.

Solymosi et al. (1994) showed that for methane decomposition over supported Pd, the TON was in the order $TiO_2 > Al_2O_3 > SiO_2 > MgO$. The difference was attributed to different particle sizes and easier mobility of Pd to the support.

A study on promotion effect of K on Co-based catalysts illustrated that it strongly depends on the support type (Boskovic et al., 1996). For SiO_2 support, the addition of K promoted the dissociation of methane, but decreased the activity of hydrogenation of carbonaceous species with an increase of C_{2+} C-based selectivity from 14% to 36%. With Al_2O_3 support, the addition of K increased the decomposition of methane with a marginal increase in C_{2+} selectivity. The C_{2+} yield decreased in both cases.

4.3.2.2 Mechanism

Much effort has been put into the understanding of the carbon surface species associated with methane decomposition. Kinetic and high-resolution electron energy-loss spectroscopy (HREELS) studies on Ru single crystals revealed vinylidene species (CCH_2) as the possible key intermediate in C_2 production (Wu et al., 1994). The methylidyne (CH) formed in methane adsorption step can be hydrogenated back to methane or dimerized to vinylidene which subsequently forms ethane upon hydrogenation.

Temperature-programmed hydrogenation of surface carbon resulted from methane decomposition over Rh/SiO_2 catalyst at 400°C after 3 min revealed three distinct peaks at <50, 100–200 and >400°C designated as C_α, C_β and C_γ, respectively, with a decreasing order of reactivity (Koerts et al., 1992a). These species were considered to be the surface intermediates in homologation reaction (Bezemer, 2017). C_α is highly active and mobile chemisorbed carbidic phase, C_β is amorphous surface carbon and C_γ is inactive, strongly bonded graphitic carbon. While the first two hydrogenate to methane and C_{2+} hydrocarbons, C_γ produces only methane. The latter has been reported to be graphitic in nature and located on the support (Koerts et al., 1992a).

An "aging" phenomenon occurred by exposing the carbonaceous species to high temperatures, due to which less active species were formed from active carbonaceous species (Koerts and Vansanten 1991; Koerts et al., 1992b). Hence, the C_{2+} yield was influenced by the amount, type, age and thermal history of the carbonaceous deposit.

A dual temperature sequence (to favor thermodynamics for each step) can be used for two-step homologation of methane; a fast methane adsorption at about 400°C followed by hydrogenation at lower temperature (approximately 100°C) (Solymosi et al., 1992). Alternatively, an isothermal condition can also be used, that is, both steps performing at the same temperature in the range of 200–320°C. In a Ru-catalyzed process using a Pd-Ag membrane reactor, both methane conversion and higher hydrocarbons yields could be significantly enhanced at a lower temperature (Olah et al., 2018).

Soltan Mohammad Zadeh and Smith (1998) developed a kinetic model for the decomposition of CH_4 over supported Co catalyst under conditions relevant to methane homologation (450°C, 101 kPa).

Neglecting the combination reaction of surface carbon species, the surface reaction was approximated with following steps:

$$CH_4(g) + 2S \rightarrow CH_3S + HS \tag{4.16}$$

$$CH_nS + S \rightarrow CH_{n-1}S + HS \quad n = 1-3 \tag{4.17}$$

$$2HS \rightarrow H_2 + 2S, \tag{4.18}$$

where S is a Co site. The model included the migration of CH_x fragments to the support. The migration step was shown to be essential to explain the kinetics for longer times (2–7 min) or high methane concentrations, that is, a rapid decrease in the CH_4 decomposition rate which continued despite a nominal coverage of the surface Co by CH_x that was >1.

There are two major problems with commercialization of methane homologation: thermodynamic limitation that results in very low yields and two-step process which is difficult for industrial operation requiring further engineering work. Development of better catalyst and using novel reactors such as membrane reactor could improve the process performance (Tang et al., 2014).

4.3.3 METHANE DEHYDROAROMATIZATION

For the first time, Wang et al. (1993) reported the possibility of catalytic transformation of methane to benzene under nonoxidative conditions over catalysts containing transition metal ions (Zn and Mo) on HZSM-5 zeolite at 700°C in a fixed-bed reactor. Thereafter, this catalytic process received much attention and many research efforts were paid for its improvement.

4.3.3.1 Chemistry and Thermodynamic of Reactions

The main reactions in catalytic nonoxidative conversion of methane into aromatics are as follows:

$$6CH_4 \Leftrightarrow C_6H_6 + 9H_2 \quad \Delta H_{298}^0 = 532 \, kJ/mol \quad \Delta G_{400}^0 = 398 \, kJ/mol \tag{4.19}$$

$$10CH_4 \Leftrightarrow C_{10}H_8 + 16H_2 \quad \Delta H^0_{298} = 889\,\text{kJ/mol} \quad \Delta G^0_{400} = 669\,\text{kJ/mol} \quad (4.20)$$

$$8CH_4 \Leftrightarrow C_8H_{10} + 11H_2 \quad \Delta H^0_{298} = 617\,\text{kJ/mol} \quad \Delta G^0_{400} = 490\,\text{kJ/mol} \quad (4.21)$$

$$7CH_4 \Leftrightarrow C_7H_8 + 10H_2 \quad \Delta H^0_{298} = 574\,\text{kJ/mol} \quad \Delta G^0_{400} = 442\,\text{kJ/mol} \quad (4.22)$$

The other reaction product is CO_x-free hydrogen which is also valuable. The reactions are endothermic and accompanied by increase in total moles; therefore, they are thermodynamically favored at higher temperatures and lower pressures.

Thermodynamic equilibrium calculations show that coke and hydrogen are the main products and practically no benzene is formed (Moghimpour Bijani et al., 2012). Consequently, the side reactions should be kinetically controlled by using an appropriate catalyst. Under coke free conditions, for instance, the equilibrium conversion of methane is only about 12% with 50% selectivity to benzene and 50% selectivity to naphthalene at 700°C (Wang et al., 1997). By using the appropriate (shape selective) catalyst, the selectivity of benzene can be increased at the expense of that of naphthalene and other more bulky aromatics.

Two approaches have been proposed to relieve thermodynamic limitation, that is, oxidative aromatization, e.g.:

$$6CH_4 + 4.5O_2 \rightarrow C_6H_6 + 9H_2O \quad \Delta H^0_{298} = -1846\,\text{kJ/mol} \quad \Delta G^0_{298} = -1624\,\text{kJ/mol} \tag{4.23}$$

and co-aromatization with an alkane or alkene (Spivey and Hutchings 2014):

$$2CH_4 + C_4H_{10} \rightarrow C_6H_6 + 6H_2 \quad \Delta H^0_{298} = 396\,\text{kJ/mol} \quad \Delta G^0_{298} = -36.2\,\text{kJ/mol} \tag{4.24}$$

4.3.3.2 Catalytic Systems

Many catalysts with different active species and supports have been used for methane dehydroaromatization (CH_4 DHA). Under non-oxidative conditions, methane has been converted only over bifunctional catalyst containing metallic and Brønsted acid sites at about 700°C and atmospheric pressures.

4.3.3.2.1 Active Metals

According to the open literature, the active metals used in catalyst formulation are transition metals including Mo, Zn, Cu, Pt, Ni, W, Fe, Re, Rh, Mn or their combinations such as Pt-Mo, Mo-Ir, W-Zn, W-La, Mo-Zn, W-Mg, W-Li, W-Mg-Zn, Mo-Fe, Mo-Ni, Co-Zn, Mo-Ru, Mo-Pd, Mo-Pd-Ru, Co-Ga, Mo-W and Mo-Co.

In most research works, molybdenum followed by tungsten have been used as active metals and in practice they showed highest performance compared to other transition metals. Most of the catalysts have been prepared by incipient wetness impregnating of the active metal, often with the ammonium salt of Mo on the support (Solymosi and Szoke, 1998; Osawa et al., 2003; Matus et al., 2009). Nevertheless, solid-state ion exchange (Ding et al., 2001; Petkovic and Ginosar, 2014) and physical mixing have also been used.

Upon addition of molybdenum to the HZSM-5 support, the catalytic activity shows a significant increase. Therefore, the presence of the metallic phase on HZSM-5 seems to be essential. The performance of the metal phase increase in the order:

$$Ni < Pt < Cu < Zn < Mn < Re < W < Mo$$

The catalyst activity in CH_4 DHA reaction is affected by metal loading level. Xu et al. (1995) studied the effect of Mo loading on catalytic performance at 500°C and 1 bar. The maximum methane conversion was obtained for 2–3 wt.% of Mo. By further increase in Mo loading, methane conversion decreased by passing a minimum at about 6wt.% of Mo. For higher loadings, the conversion does not change appreciably. In contrast, the selectivity increased with Mo loading and passed a maximum at about 6 wt.%. Upon further increase of Mo, the trend of aromatic selectivity was opposite of that for C_2 hydrocarbons. This implies that ethylene is a primary product.

XRD and BET measurements showed that the Mo species are well dispersed in zeolite channels which, due to their interaction with zeolites, decrease their crystallinity. For Mo loading larger than 4.5 wt.%, the Mo species might migrate into zeolite channels and form MoO_3 crystals upon calcination. If these crystals grow sufficiently, they cause partial blockage of zeolite channels thereby decreasing their surface area and increasing their average pore diameter (Xu et al., 1995). Similar experiments in bench scale showed the best catalytic performance for 4 wt.% Mo although change of performance in the range of 2–4 wt.% was not significant (Lu et al., 1999). The small difference in the optimum loading could be due to different activation conditions of the catalysts. The catalysts reported by Xu et al., (1995) were calcined in air whereas those reported in Lu et al. (1999) were activated in helium. In other references for the Mo/HZSM-5 catalysts prepared by impregnation method, the optimum Mo loading was also reported about 4 wt.% (Skutil and Taniewski, 2006; Vosmerikov et al., 2006). Excessive loading of Mo causes extraction of Al from zeolite framework and formation of $Al_2(MoO_4)$ phase which is not active in CH_4 DHA. Increasing Mo loading also causes reduced Brønsted acid sites in zeolite channels (Hongmei and Yide, 2006).

4.3.3.2.2 Catalyst Supports for CH_4 DHA

In most studies, zeolites have been used as catalyst support for CH_4 DHA reaction. The results show that the acidity and structure of the support are two main factors influencing the conversion of methane into aromatics. Thus, catalyst activity increases with acid strength of the supports and benzene forms only over Mo/HZSM-5 and Mo/γ-Al_2O_3 catalysts (Pinglian et al., 1997). It appears that coke easily forms on Mo/γ-Al_2O_3 catalysts. This shows that the effect of acidity is very important. On the other hand, superacid solid catalysts such as Mo/ZrO_2 with Hammett acidity (a measure of acidity) $H_0 \leq -11.35$ have been tested under similar conditions. Unexpectedly, except for a small amount of CO, no aromatic product was observed over these catalysts. Therefore, acidity is also an important factor in CH_4 DHA. Upon increasing SiO_2/Al_2O_3 ratio, the acidity of the support and aromatization activity decreases (Pinglian et al., 1997).

Zhang et al. (1998) evaluated different zeolite supports for aromatization of methane. The catalysts were prepared by crushing appropriate amounts of MoO_3 with the zeolite support followed by calcination at 500°C under flow of air for 3 h. The H-type silica-alumina zeolites such as ZSM-5, ZSM-8, ZSM-11 and β have two-dimensional structures with pore diameters close to dynamic diameter of benzene molecule (6 Å) and, therefore, are suitable supports for CH_4 DHA. Among the catalysts, MoO_3/HZSM-11 showed the best activity and stability. The performance of MoO_3/HZSM-8 was lower than MoO_3/HZSM-5 and its activity was better than MoO_3/Hβ. The order of methane aromatization activity was as follows:

MoO_3/H-ZSM-11 > MoO_3/H-ZSM-5 > MoO_3/HZSM-8 > MoO_3 / H-β > MoO_3/H-MCM-41 > MoO_3/H-SAPO-3 > MoO_3/H-MOR ≈ MoO_3/H-X ≈ MoO_3 / H-Y > MoO_3/H-SAPO-5 ≈ MoO_3/H-SAPO-11

Generally, zeolites with channel and pore sizes appropriate for diffusion of aromatic products without caged structures (which are suitable for coke formation) show relatively stable activity in CH4 DHA reaction. The optimum amount of MoO_3 on HZSM-5 is 3–5 wt.%, on HZSM-8, 5–7 wt.% and on H-β is 7–8 wt.%. The optimum SiO_2/Al_2O_3 ratio for HZSM-11 in CH_4 DHA reaction has been found as 25–50. On 3 wt% Mo/HZSM-5 catalysts with various SiO_2/Al_2O_3 ratios tested 700°C and 1 bar, the highest Brønsted acidity and best performance (10.2% CH_4 conversion and 62.8 hydrocarbon selectivity) was obtained for an Si/Al ratio of about 40 (Liu et al., 1999).

Mo/H-ZSM-5 catalysts showed the best performance in DHA reaction. They are commonly synthesized by impregnation using oxide precursors (e.g., ammonium heptamulibdate and molybdenum trioxide). Other synthesis methods such as solid-phase synthesis and chemical vapor deposition (CVD) are also employed. Most of the Mo-oxo precursors are too large to fit 10-M pores of the zeolites. Hoverer, upon heating mobile suboxide are formed which diffuse into pores and produce high dispersion on Brønsted acid sites (Kosinov and Hensen, 2017).

The major aspect of Mo/H-ZSM-5 catalysts is Brønsted acid sites, state of Mo (Mo_2C or MoC_xO_y) and channel structure of the zeolite (Spivey and Hutchings, 2014).

The best Mo-loading is within 2%–6% with 10% methane conversion and 50%–60% benzene selectivity (Olah et al., 2018).

From Fourier Transform Infrared Spectroscopy (FTIR) studies along with catalyst performance evaluations it was concluded that Mo/HZSM-5 containing 60% of remaining number of original Brønsted acid sites exhibited good performance in CH_4 DHA reaction (Kusmiyati and Amin, 2005).

Ding et al (2002) investigated the structure of 1–3.6%Mo/H-ZSM-5 catalysts and showed that the dispersed MoO_3 migrates into the channels to form MoO_x species. After treating at 700°C, the MoO_x species were exchanged with acid sites to form the dimer $[Mo_2O_5]^{2+}$ bridging to framework oxygens (Eq. 4.25 and Scheme 4.1). Higher Mo loadings enhanced dimer formation. Borry et al. (1999) showed that MoO_3 vapors reacts with acid sites to form bridging $[Mo_2O_5]^{2+}$ clusters:

$$MoO_x + H^+ \rightarrow (MoO_2(OH))^+ \rightarrow Mo_2O_5^{+} + H_2O \qquad (4.25)$$

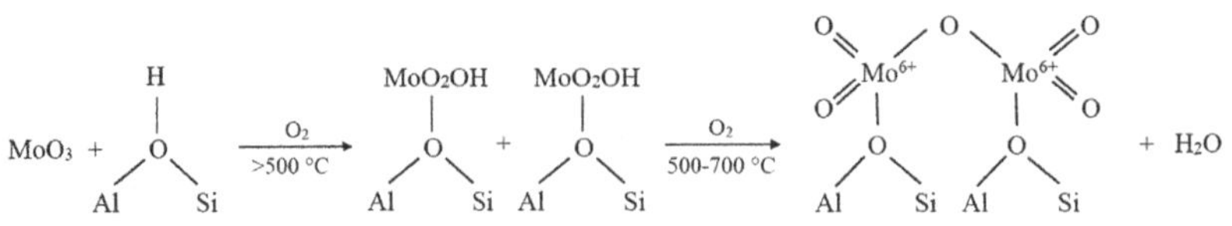

SCHEME 4.1   Formation of bridging Mo clusters.

In order to improve the stability and activity and reduce coke formation, different "promoters" have been used including Cu, Ru and K, among which Cu showed positive effects (Spivey and Hutchings, 2014).

4.3.3.2.3   Structural Modifications

Microporous zeolites impose substantial diffusion limitation in DHA reaction because of the much larger crystal size than that of micropore diameters (Kosinov and Hensen, 2017). Large pore zeolites are not suitable for DHA because the shape selectivity of 10-membered-ring (10MR) are lost. Preparation of nanosized zeolites is another solution, but technical problems such as separation are important challenges. Hierarchical or mesoporous zeolites with both micropores and mesopores are promising alternatives. The general procedures for preparation of these zeolites are top-down (e.g., desilication and dealumination) and bottom up (hard and soft mesogen templates).

Hu et al. (2014) showed that mesoporous TNU-9 prepared hydrothermally with phenyltriethoxysilane as mesoporgen had a better performance (higher benzene productivity and higher stability) than the conventional one.

Tempelman et al. (2016) used hierarchical MCM-22 and illustrated that reducing the domain of crystal sizes increases benzene selectivity. Liu et al. (2012) synthesized H-ZSM-5 by templating with dimethyloctadecyl[3-(trimethoxysilyl)propyl] ammoniumchloride.

Despite its benefits by reducing mass transfer limitation, a major drawback of this approach is that it can have detrimental effect on catalyst performance by increasing external acid sites where shape selectivity is absent and leading to heavy aromatics. Therefore, for an optimal catalyst, a compromise between the micropore volume and mesopore surface is necessary. This is characterized by the so called hierarchical factor (HF) which is the product of fractional micropore volume and mesopore surface area (Kosinov and Hensen, 2017):

$$HF = \left(\frac{V_{micro}}{V_{total}} \right) \times \left(\frac{S_{meso}}{S_{total}} \right) \tag{4.26}$$

Wu et al. (2015) found a volcano type dependence of benzene production and HF for a number of MFI zeolites, illustrating that an accurate design for an optimum catalyst is necessary.

A common technique for deactivating surface acid sites is silylation, for example, by tetraethoxysilane (TEOS). The selection of an appropriate silylating agent is important because its diffusion into micropores will deactivate internal acid sites. This approach has been useful for improving benzene selectivity and catalyst stability of hierarchical zeolits (Tempelman et al., 2016; Kikuchi et al., 2006).

Yet another approach is the synthesis of core-shell H-ZSM-5-silicalte 1 catalyst in which pure silica zeolites is grown on alumina silicate. The resulting acid-free surface improves dehydroraromatization performance of the catalyst, that is, high selectivity of aromatics and catalytic stability (Jin et al., 2013).

In summary, the main features of the bifunctional catalysts are the presence of Brønsted acid sites, state of Mo species (Mo_2C or MoO_xC_y) and zeolite channel structure which strongly influence activation and aromatization of methane. The physical properties of the zeolite strongly influence the selectivity of zeolite supported catalysts.

4.3.3.3 Mechanism and Kinetics

4.3.3.3.1 Reaction Mechanism

Mo/H-ZSM-5 is generally believed to act as a bifunctional catalyst in methane aromatization.

The activation of methane has been the subject of many research works and various active sites and mechanisms have been proposed for CH_4 DHA reaction. Xu et al. (1995) proposed that the intermediate of the reaction is ethylene which is subsequently converted into benzene by polymerization. In the first paper of Wang et al. (1993), the activation of methane over HZSM-5 and Mo/HZSM-5 catalysts were considered to initiate via a carbenium ion mechanism:

$$CH_4 + Mo^{6+} \rightarrow CH_3^+ + [Mo\text{-}H]^{5+} \tag{4.27}$$

$$CH_4 + H^+(s) \rightarrow CH_3^+(s) + H_2 \tag{4.28}$$

According to their research and according to the mechanism proposed for aromatization of light paraffins (C_2–C_4), Xu et al. (1995) concluded that methane is activated by the formation of CH_3^+ by Brønsted acid sites and Mo in zeolite channels. In other words, heterolytic cleavage of methane occurs in solid acid medium in the presence of Mo species. They considered the following mechanistic steps for ethylene formation:

1. The C–H bonds of methane are polarized by interaction with MnO_3 in HZSM-5 zeolite channels:

$$\mathbf{CH_4} \qquad \mathbf{H^- \text{--} CH_3^+}$$
$$\overline{|MoO_3|} \rightarrow \overline{\lceil MoO_3 \rceil}$$

2. The polarized methane molecule reacts with Brønsted acid sites of HZSM-5 (H_z^+) to form carbine intermediate by a catalytic cycle:

$$\begin{array}{ccc} & & H_z^+ \\ H_z^+ & & \\ / & & \\ H^- \text{--} CH_3^+ & H_2 + CH_3^+ & CH_2 \\ & & \| \\ \overline{|MoO_3|} \rightarrow & \overline{\lceil MoO_3 \rceil} \rightarrow & \overline{|MoO_3|} \end{array}$$

3. The pseudo-carbine molybdenum results in the formation of ethylene as the primary product of the reaction via dimerization of $CH_2 = MoO_3$.
4. Aromatization of ethylene occurs on HZSM-5 zeolite to produce benzene and methylbenzene.

4.3.3.3.2 Reaction Kinetics

Iliuta et al. (2003) studied the kinetics of CH_4 DHA over Ru-Mo/HZSM-5 catalysts in a differential fixed-bed reactor under atmospheric pressure at 600–700°C using methane and argon mixture feed with methane volume fractions of 50%–90%. Different models based on Langmuir–Hinshelwood–Huogen–Watson mechanisms were tested for fitting kinetic data. The activation energies, adsorption enthalpies and entropies were calculated based on Arrhenius and vant Hoff equations.

The final reaction rate was obtained from a series of elementary steps shown below:

$$CH_4 + S \xrightarrow{k_1} CH_4S \tag{4.29}$$

$$CH_4S \xrightarrow{k_2} CH_2S + H_2 \tag{4.30}$$

$$CH_2S \xrightarrow{k_3} \frac{1}{2}C_2H_4 + S \tag{4.31}$$

$$\frac{1}{2}C_2H_4 \xrightarrow{k_4} \frac{1}{6}C_6H_6 + \frac{1}{2}H_2 \tag{4.32}$$

In that research, three alternative models were considered:

Model 1: the adsorption of methane (reaction 4.29) step was assumed as the rate determining step (RDS)

Model 2: the surface reaction (reaction 4.30) step was assumed as RDS

Model 3: the dimerization of CH_2 and desorption (reaction 4.31) step were assumed as RDS

Among these models, model 2 assuming dehydrogenation of adsorbed methane is RDS showed better agreement with experimental data. The rate expression obtained from this model is shown in Eq. (4.33).

$$r_2 = k_2 \frac{P_{CH4} - \dfrac{1}{K_P} P_{C2H6}^{1/6} P_{H2}^{3/2}}{1 + K_1 P_{CH4} + \dfrac{K_3}{K_4} P_{C6H6}^{1/6} P_{H2}^{1/2}} \tag{4.33}$$

In which

$$K_4 = \exp\left(-\frac{\Delta G_4}{RT}\right) \tag{4.34}$$

$$K_P = \frac{P_{C6H6e}^{1/6} P_{H2e}^{3/2}}{P_{CH4e}} = \exp\left(-\frac{\Delta G}{RT}\right) \tag{4.35}$$

According to these results, the activation energy of reaction was obtained as 88.2 kJ/mol, the enthalpy of adsorption of methane over active sites −73.4 kJ/mol and adsorption entropy as −75.3 kJ/mol.K. The model showed favorable fit with experimental data.

Yao et al. (2008) investigated the intrinsic kinetics of CH_4 DHA under nonoxidative conditions over Mo/HZSM-5 catalyst in a tubular quartz reactor under atmospheric pressure and 640–700°C and space velocities 700–2,100 ml/g.h.

Three main reactions were considered for CH_4 DHA under nonoxidative conditions:

$$6CH_4 \leftrightarrow C_6H_6 + 9H_2 \tag{4.19}$$

$$CH_4 \leftrightarrow C + 2H_2 \tag{4.36}$$

$$C_6H_6 \leftrightarrow 6C + 3H_2 \tag{4.37}$$

Two of the above reactions (i.e., Eqs. 4.19 and 4.36) were considered as independent reactions. The Langmuir–Hinshelwood patterns were considered for intrinsic kinetics. The reaction rates were calculated as a function of fugacity of the gaseous species. According to the Langmuir-Hinshelwood mechanism, the reaction rates may be obtained as Eqs. 4.38 and 4.39

$$r_1 = \frac{k_1 \cdot f_{CH4}^6 \cdot (1 - \beta_1)}{\left(1 + K_{CH4} \cdot f_{CH4} + K_{H2} \cdot f_{H2} + K_{C6H6} \cdot f_{C6H6}\right)^6} \tag{4.38}$$

$$r_2 = \frac{k_2 \cdot f_{CH4} \cdot (1 - \beta_2)}{\left(1 + K_{CH4} \cdot f_{CH4} + K_{H2} \cdot f_{H2} + K_{C6H6} \cdot f_{C6H6}\right)} \tag{4.39}$$

$$k_i = A_i \cdot \exp\left(-\frac{E_i}{R}\left(\frac{1}{T} - \frac{1}{\bar{T}}\right)\right) \tag{4.40}$$

$$K_i = \exp\left(a_i - b_i\left(\frac{1}{T} - \frac{1}{\bar{T}}\right)\right) \tag{4.41}$$

$$\beta_1 = \frac{f_{C6H6} \cdot f_{H2}^9}{K_{f1} \cdot f_{CH4}^6} \tag{4.42}$$

$$\beta_2 = \frac{f_{H2}^2}{K_{f2} \cdot f_{CH4}} \tag{4.43}$$

in which the subscripts 1 and 2 refer to Eqs. (4.19) and (4.36), respectively. K_f is the equilibrium constant, k_1 and k_2 are rate constants of reactions, respectively, K_i is the adsorption constant of species i, β is the approach to equilibrium and $\bar{T}$ is the temperature which is 670°C. The fugacity of species i, f_i, under ambient pressures and temperature range of 640–700°C could be approximated and replaced by its partial pressure.

The Levenberg–Marquardt algorithm was used for kinetic parameters estimation. Statistical test and residual error distribution curves showed good agreement of experimental data with calculated values and also that the Langmuir-Hinshelwood model was appropriate for describing intrinsic kinetics of CH_4 DHA under the used operating conditions.

4.3.3.4   Reactor Types and Operating Conditions

4.3.3.4.1   Reactors for CH_4 DHA

In most references, fixed-bed reactors were used for CH_4 DHA. Cook et al. (2009) studied CH_4 DHA over 4 wt.% Mo_2C/ZSM-5 catalyst (0.16–0.4 mm in diameter) in fixed-bed and fluidized-bed reactors. The fluidized-bed reactor exhibited better yields but catalyst deactivation was much more rapid. In the fluidized-bed reactor, the rate of aromatic formation after 60 h on stream was about 15% higher than that in fixed-bed reactor (6.3 mmol C/g_{cat}.h in fluidized-bed reactor, 5.5 mmol C/g_{cat}.h in fixed-bed reactor). The rate of naphthalene formation increased with time in the fluidized-bed reactor. These results could be due to longer residence time of reaction intermediates in the zeolite resulting in the conversion of C_2 intermediates into heavier products such as naphthalene. In fluidized-bed reactors, higher mechanical strengths are required due to crushing of catalyst particle as a result of collision and friction with each other and reactor walls.

4.3.3.4.2   Effect of Reaction Temperature and Space Velocity

In temperature range of 620–740°C, methane conversion and aromatic selectivity increased with temperature over a 2wt% Mo/HZSM-5 catalyst, while C_2 selectivity decreased (Xu et al., 1995). Thermodynamically, higher temperatures favor the endothermic CH_4 DHA reactions.

It is noteworthy that too high temperatures cause rapid catalyst deactivation due to extensive coke formation. Therefore, optimization of the temperature seems to be essential. In different references, CH_4 DHA reaction mostly operated at about 700°C.

In space velocity range of 1,000–1,900 cm³ g⁻¹ h⁻¹, methane conversion and aromatic selectivity decreased with methane feed space velocity while selectivity to ethane and ethylene as well as ethylene to ethane ratio increased (5%Mo/HZSM-5 at 725°C) (Skutil and Taniewski, 2006). This implies that ethylene is the primary product of CH_4 DHA reaction and that aromatics are obtained by reaction progress from ethylene.

CH_4 DHA tests have been often performed at space velocities around 1,500 ml/g.h.

4.3.3.5   Surface Carbon Species

The main problem for a better understanding of CH_4 DHA reaction for its development and improvement is the deposition of heavy carbonaceous species during reaction.

Jiang et al. (1999) studied the carbonaceous species deposited on Mo/HZSM catalyst using the Nuclear Magnetic Resonance (NMR) technique. They showed that two carbon types deposit on the catalyst. One type deposits occur on partially reduced Mo species and the other type occurs on acid sites of the zeolite. The former is responsible for the activation of methane and for conversion of carbine-like or carbidic species into ethylene. The latter is active in the formation of aromatics from ethylene.

Yuan et al. (1997) characterized the deposited carbonaceous species by UV-Ramman Spectroscopy and concluded that the carbonaceous species deposited on Mo/HZSM-5 are essentially polyaromatic in nature.

The results of temperature-programmed oxidation (TPO) of spent Mo/HZSM-5 catalysts prepared by impregnation showed three temperature peaks at 459, 511 and 558°C (Liu et al., 2006). The carbonaceous species that burn at higher temperatures are mainly those formed on Brønsted acid sites of the zeolites while those burning at lower temperatures are those formed on Mo carbide or Mo oxycarbides. The coke which oxidizes at early stages at about 459°C is over Mo carbide species located on the external surface of zeolite while TPO peak at about 511°C is related to coke over Mo species located within zeolite channels. In the early stages of DHA reaction, the molybdenum oxide species are reduced and carbided with methane and the active Mo species (molybdenum carbide or oxycarbide) are formed (induction period). The activated methane species may be re-adsorbed on active Mo species and form coke. After an induction period, benzene is formed.

Dehydrogenation and oligomerization of monocyclic aromatic products also results in deposition of aromatic type carbon species on Brønsted acid sites. Therefore, a high-temperature peak appears in TPO of the spent catalysts. It has been shown that the deposited coke associated with Mo species is reactive and its formation is reversible; however, the coke formed on Brønsted acid sites is neutral and its formation is irreversible. This type of coke is the main cause of the deactivation of Mo/HZSM-5 catalysts (Liu et al., 2006).

Liu et al. (2007) investigated the carbonaceous species deposited on Mo/HZSM-5 and Ga-Mo/HZSM-5 catalysts in DHA reaction using X-ray Photoelectron Spectroscopy (XPS), X-ray Auger Electron Spectroscopy (XAES), TG/DTG, HRTEM and Differential Thermal Analysis (DTA). In early stages of reaction, MoO_3 is converted into Mo_2C (or MoC_x) and $Mo_2O_xC_y$ and a large amount of hydrogenated carbonaceous species such as CH_x ($x = 1-3$) over Mo species. These carbonaceous species are easily removed by burning with air below 500°C. After an induction period of 80 min, dehydrogenation of methane over Mo sites and aromatization on Brønsted acid sites occur. The presence of alkalene species on catalyst promotes the formation of aromatic compounds. In the last stage of reactions, carbonaceous species deposited on the catalyst comprised 79.4% hydrogenated carbon and 9.1% carbon comprising higher aromatic compounds. The hydrogenated carbon species such as CH_x ($x = 1-3$) over the catalyst gradually lose hydrogen and transform to amorphous carbon or graphite which are difficult to remove. Nevertheless, before the formation of aromatic carbon, its removal is possible. Deposition of heavy aromatic-type carbon is the main cause of catalyst deactivation.

4.3.3.6   Catalyst Deactivation

The selectivity of coke is significant that is the main cause of rapid catalyst deactivation. Over 6% Mo/HZSM-5, for example, the coke selectivity was 26%–29% for CH_4 conversion in the range of 6.5–8.1% at 725°C and 3 bar (Chu and Qiu, 2003).

Dealumination of zeolite by steam has increased aromatic selectivity and reduced coke deposition. The loss of activity is initially very rapid but slows down with time. The selectivity of benzene, however, is rather constant and high with time on stream but that of naphthalene passes a maximum (Olah et al., 2018; Spivey and Hutchings 2014).

Although highly dispersed MoC_x species are necessary for Mo-support interaction and optimum performance, those formed in the micropores of the zeolite contribute to deactivation by pore blocking making Brønsted acid sites inaccessible (Tempelman et al., 2015).

Coke deposition is inevitable in DHA of methane. The addition of various additives to feed such as CO, CO_2, H_2, NO and O_2 improved the catalyst life.

A reaction-regeneration cyclic operation with 1.5-h reaction and 0.5-h regeneration increased the benzene production three-folds from 33 to 97 $g_{benzene}/kg_{cat}.h$ (Portilla et al., 2015).

4.3.3.7   Coaromatization of Methane

The main obstacles for commercial realization of DHA of methane are high operating temperatures and low aromatic yields. In early studies, Choudhary et al. (1997) reported that the reaction condition and aromatic yield can be improved by addition of light alkanes or alkanes to methane feed. Their study showed that Ga/H-ZSM-5 can activate methane in the presence of lower alkanes or alkenes to form higher hydrocarbons (10%–45% at 400–600°C). The results were interpreted by activation of methane in the presence of hydrocarbons. Co-aromatization of methane by modified H-ZSM-5 catalysts has been reviewed (Guo et al., 2009).

2% Zn/ZSM-5 cannot promote CH_4 DHA at atmospheric pressure. However, the synergy between ZnO nanoclusters and Brønsted acid sites enables formation of surface methoxy or methyl species (Olah et al., 2018). Coaromatization of CH_4 with light alkanes over Zn- and Ga- modified H-Beta with 13C CP/MAS NMR spectroscopy showed that in coaromatization of $^{13}CH_4$ and propane over Ga/H-Beta, the labeled carbons enter both in aromatic ring and methyl substitutes (Luzgin et al., 2010). In similar tests with methane as methylating agent on Zn/H-Beta, methane entered aromatic ring formed form propane (Luzgin et al., 2008).

Alcohols have been also used as feed additives. 5%Mo/HZSM5 (Si/Al = 55) doped with 2% Ga resulted in 13.7% aromatic yield ($X = 15.5\%$, $S = 88.7\%$ with CH_4/ CH_3OH = 6.5 mol/mol in feed at 670°C) (Majhi and Pant, 2014).

The addition of ^{13}CO in methane feed also showed the involvement of ^{13}C in the formation of aromatics. Over an Mo/HMCM-49 catalyst, the methane conversion did not show significant increase, but the catalyst lifetime showed slight increase (Yao et al., 2010).

There is no general agreement on the positive role of addition of hydrocarbons into methane DHA feed. While it is favored by some authors (e.g., Chu and Qiu, 2003; Echevsky et al., 2004; Baba, 2005), it has been criticized by other researchers (e.g., Naccache et al., 2002; Bradford et al., 2004; Luzgin et al., 2009, Moghimpour Bijani et al., 2013).

4.3.3.8 Processes

Dehydroaromatization of light alkanes has been commercialized as the Cyclar process. Unlike methane and ethane, aromatization of propane and higher alkanes is much easier to achieve. The UOP/BP Cyclar process for the manufacture of petrochemical-grade benzene, toluene, and xylenes (BTX) by catalytic aromatization of propane and butanes uses Ga-doped zeolites in stacked radial-flow reactors combined with a continuous catalyst regeneration (CCR) unit. A further development (BP/Amoco-UOP Cyclar process) uses liquefied petroleum gases (propane/butanes) over Al-modified zeolites at > 425°C and 0.2–0.4 MPa (Cornils et al., 2013).

4.4 OXIDATIVE CONVERSIONS

4.4.1 OXIDATIVE COUPLING OF METHANE

The direct conversion of methane into lower olefins is highly desirable to supply ever-increasing demand for these important petrochemical and refining feedstocks more economically form low-cost and abundant natural gas sources. The direct coupling of methane to C_{2+} hydrocarbons (e.g., Eqs. 4.44 and 45) is strongly endothermic and equilibrium-limited:

$$2CH_4 \rightarrow C_2H_6 + H_2 \quad \Delta H^0_{298K} = +64.4 \, \text{kJ/mol} \, C_2H_6 \tag{4.44}$$

$$2CH_4 \rightarrow C_2H_4 + 2H_2 \quad \Delta H^0_{298K} = +201 \, \text{kJ/mol} \, C_2H_6 \tag{4.45}$$

To overcome the thermodynamic limitation, oxidative coupling of methane (OCM) has been explored. Oxidative coupling of methane, also termed oxidative oligomerization of methane, is thermodynamically favorable.

The pioneering work of Keller and Bhasin (1982) illustrated that methane can form C_{2+} hydrocarbons in the presence of oxygen as oxidant. Since then, all around the world, academia and chemical companies have planned to develop a process being economical and can upgrade methane. Due to the shortage in fossil fuel and increase in crude oil price, oxidative coupling of methane is a new practical route. In this reaction, biogas or natural gas can be used as feedstock for the production of different chemicals. OCM technology can reduce costs, energy and environmental emissions for ethylene production.

Methane oxidative activation is an energy-effective route for its activation that needs a small external energy input. All available processes for oxidative conversion of methane are catalytic. Due to very high energy of the C–H bond of methane molecule (439 kJ/mol), high temperatures (>400°C) are needed for its catalytic oxidation. At a high temperature, less stable intermediates are formed by catalytic activation and oxidation of methane (Arutyunov and Strekova, 2017).

Despite the benefits of the OCM reaction, a viable catalyst for its commercialization has not been developed yet. The main problem for its commercialization is maintaining high performance and long lifetime at high operating temperatures.

Another drawback of the OCM reaction is that C$_2$ yield cannot exceed 17%–20%. This restriction limits its large-scale application (Arutyunov and Strekova, 2017).

4.4.1.1 Chemistry of OCM

Oxidative coupling of methane is a catalytic process in which oxygen and methane are reacted at high temperatures and ethane and ethylene are produced. The desired reactions leading to ethylene formation are

$$2CH_4 + 0.5O_2 \rightarrow C_2H_6 + H_2O \quad \Delta H^0_{298K} = -177\,kJ/mol\,C_2H_6 \qquad (4.46)$$

$$C_2H_6 + 0.5O_2 \rightarrow C_2H_4 + H_2O \quad \Delta H^0_{298K} = -105\,kJ/mol \qquad (4.47)$$

The reactions are exothermic and thermodynamically favorable. The competitive side reactions are deep oxidation of methane which are more strongly exothermic:

$$CH_4 + 2O_2 \rightarrow CO_2 + 2H_2O \quad \Delta H^0_{298K} = -802\,kJ/mol \qquad (4.48)$$

$$CH_4 + 1.5O2 \rightarrow CO + 2H_2O \quad \Delta H^0_{298K} = -519\,kJ/mol \qquad (4.49)$$

Similarly, the desired C$_{2+}$ products are more reactive than methane and can participate in cascade deep oxidation reactions. Thus, the products are C$_2$H$_6$, C$_2$H$_4$, H$_2$O, H$_2$, CO, CO$_2$ and trace amounts of higher hydrocarbons and C$_1$ oxygenate (CH$_3$OH and CH$_2$O). This limits the single-pass maximum C$_2$ yield. Generally, oxidative coupling reactions occur at 650–880°C, which is in the limit of catalytic combustion where C$_{2+}$ selectivity is dictated by the availability of O$_2$ (Lunsford, 1995).

The main research topic of OCM process is improving selectivity and stability of the catalyst. As OCM is a highly exothermic reaction, reactor operation at lower temperature (<800°C) is highly desirable. Moreover, the separation of unconverted methane and by-products (such as hydrogen, carbon oxides, ethane, water and C$_3$–C$_4$ hydrocarbons) is costly and highly energy-intensive that should be minimized. Hence, in an efficient OCM process design, OCM catalysts and reactor operation must be improved, and separation and purification units must have cost-effective performance (Godini et al., 2019).

Researchers reported that for the OCM reaction, best catalysts have about 20% methane conversion with 80% ethane and ethylene selectivity (Khojasteh Salkuyeh and Adams, 2015).

The use of alterative and soft oxidants such as N$_2$O and CO$_2$ has been proposed to avoid undesirable deep oxidation reactions. The use of CO$_2$ is more desirable because of its environmental implications.

Thermodynamic equilibrium calculations show that the yields of ethane and ethylene are sufficiently high (13% and 57%, respectively) in the CO$_2$-OCM at 800°C, (Krylov and Mamedov, 1995). However, the experimental yields of C$_2$ hydrocarbons over metal oxide catalysts has been lower than 9% at 850°C and CO$_2$/CH$_4$ = 2 due to kinetic limitations (Nishiyama and Aika, 1990).

4.4.1.2 Catalysts

Since the discovery of OCM reaction, a large number of metal oxides (simple and mixed oxides) have been tested for the reaction. A variety of metal oxides are effective in C–H bond activation. Non-reducible group IIA oxides showed the best catalytic performance in OCM. Alkali metal oxides, carbonates and some lanthanides forming sesquioxide (M_2O_3, such as La_2O_3) but not higher oxides (e.g., MO_2) also showed good performances (Elkins and Hagelin-Weaver, 2013).

The oxide catalysts are generally promoted or mixed oxides of group IA and IIA elements or modified or pure oxides of transition metals. Impregnation, sol-gel, flame spray pyrolysis and precipitation are different methods of catalyst synthesis. In order to improve methane conversion and ethylene selectivity and decrease catalyst deactivation, the catalyst can be modified with other oxides, chloride sites or metals and the reaction conditions (such as space velocity and temperature) can be changed (Kalyani et al., 2014; Lei et al., 2014).

Ito et al. (1985) reported that by oxidative coupling of methane, reasonable C_2 yields could be obtained. According to their research, on PbO/Al_2O_3, 58% C_2 selectivity with 5% CH_4 conversion was obtained, and according to Ito et al. (1985) research, on the Li/MgO catalyst, 50% C_2 selectivity with 28% CH_4 conversion was obtained.

In a recent statistical analysis of 1870 data sets on catalyst compositions and their performances in OCM reaction, Sr, Ba, Mg, Ca, La, Nd, Sm, Ga, Bi, Mo, W, Mn and Re were identified as the key elements used as oxides with Li, Na and Cs as cationic dopants (Zavyalova et al., 2011).

Effective catalysts can be categorized to five groups according to their components (Lunsford, 1995):

*   Monophasic oxides
*   Ions of group IA or IIA supported on basic oxides (such as Sr/La_2O_3, Ba/MgO and Li/MgO)
*   Transition metal oxides containing group IA ions
*   Highly basic pure oxides among which lanthanide oxide series (except CeO_2) are the best
*   Any of the above-mentioned materials promoted with chloride ions

4.4.1.2.1 Monophasic Oxides

Different mixed-metal monophasic oxides were studied in the coupling reaction but their major drawback is that under severe operating conditions, they are subject to phase separation. Dissanayake et al. (1993) evaluated Ba-Sn, Ba-Pb and Ba-Bi perovskites among which only Ba-Sn perovskite resisted phase separation and was not selective. For the other two perovskites, barium segregated to the surface. In all catalysts, surface transformation is a common problem that adversely affects the performance and is a serious problem in commercial operation.

4.4.1.2.2 Group IA and IIA Catalysts

In the 1990s, OCM reaction was evaluated on group IA and IIA compounds for pure or mixed oxide systems and the use of MgO, BaO, Li_2O and CaO as support.

The main role of the catalyst is providing favorable surface sites for methane conversion into radical species and subsequent reaction for hydrocarbon production (Kus et al., 2002).

4.4.1.2.3 Pure or Modified Transition Metal Oxides

Low catalytic activity, fast deactivation and low ethylene selectivity are major drawbacks of pure transition metal oxide catalysts. Hence, in order to improve their catalytic properties, they must be modified. Recently, it was reported that by changing calcination temperature during catalyst synthesis, for pure oxides or their mixed counterparts, an improvement in catalyst performance can be obtained (Kus et al., 2003).

The Na_2WO_4 gained attention in OCM reaction. In order to optimize activity/selectivity properties, these catalysts are promoted with Mn or supported on different oxide supports.

4.4.1.2.4 Chloride-Containing Catalysts

Chloride ion addition to OCM catalyst has valuable effect on its properties, specifically on the C_2H_4/C_2H_6 product ratio. Chlorine can be used as part of the catalyst or as organo-chlorine compounds to be added to reagents. Typically, for chlorided catalysts, C_2H_4/C_2H_6 ratio of 3–5 was obtained which is an attractive finding as ethylene is the favorable product. Lunsford (1995) reported alkali-metal bismuth oxy-chloride catalysts (such as $LiCa_2Bi_3O_4Cl_6$) by which $C_2H_4/C_2H_6 > 35$ was obtained. The main drawback of these chlorinated catalysts is that they lost their activity within few hours.

In the chlorine-containing systems, chlorine can dehydrogenate C_2H_6 in the gas phase and this homogeneous reaction is responsible for obtaining large C_2H_4/C_2H_6 ratios. In the presence of water, most of the chlorine is lost as HCl. Operating the reaction at lower temperatures can minimize rate of chlorine loss. As oxidative dehydrogenation of ethane, in comparison to oxidative coupling of methane, is faster on chlorinated catalysts, the favorable C_2H_4/C_2H_6 ratio can be obtained by these catalysts. The role of Cl^- ions is similar to CO_2 for selectivity (Ahmed and Moffat, 1990a).

It has been demonstrated that alkali chloride-Mn-Na_2WO_4/SiO_2 catalysts exhibit high ethylene yields in the OCM reaction. However, these catalysts are rapidly deactivated through the evaporation of the alkali chloride. To maintain constant high yields of ethylene, a continuous supply of alkali chloride vapor to a Mn-Na_2WO_4/SiO_2 catalyst is necessary (Hiyoshi and Sato, 2016). In addition, chlorine-containing organic compounds were used as a modifier of the OCM reaction. The gas-phase doping of tetrachloromethane (CCl_4) to alkali metal-Mn/SiO_2 catalysts was reported to be effective for ethylene formation (Ahmed and Moffat, 1990b).

4.4.1.3 Mechanism

OCM can be regarded as oxidative dehydrodimerization of methane. Many of the metal oxide catalysts show similar conversion-selectivity results, that is, nearly linear decrease of selectivity with conversion, e.g., by increasing oxygen concentration in the feed. This has been formulated as rule of 100% which states that the sum of conversion and selectivity of the catalysts is about 100 (Sahebdelfar et al., 2012). This similarity implies common active sites and mechanisms. Based on a large number of experimental evidences (electron paramagnetic resonance (EPR), mass spectroscopy, isotope labeling studies) it is generally accepted that the primary role of the catalyst is producing methyl free radicals which couple in subsequent steps (Olah et al. 2018).

As different catalysts are used for the OCM reaction, there is not a single type of center being responsible for CH_4 activation. Hence, according to the used catalyst, potential active centers will be evaluated. Li/MgO is a common catalyst for the OCM reaction. There are different oxygen ions in the low coordination state; namely O^-, O_2^-, O_2^{2-} and O^{2-}. As in the coupling reaction, only metal oxides are effective and in the presence of a gas-phase oxidant (N_2O or O_2), methane reactivity is greater, oxygen ions must be paid attention to. For Li/MgO catalysts, $[Li^+O^-]$ centers are possible sites for methane activation. These centers are present at high temperatures and form only in an O_2 atmosphere (Lunsford, 1995).

For the OCM reaction over oxide catalysts, the first and rate-determining step is methane interaction with catalyst by which $CH_3^{\cdot}$ radicals are formed (Eq. 4.50) and escaped to reactor volume. This is a multi-stage heterogeneous-homogeneous process occurring on the catalyst surface and in gas-phase. In this reaction, the role of catalyst is limited to the generation of methyl radicals:

$$CH_4 + O_S^- \rightarrow CH_3^{\bullet} + OH_S^-$$ (4.50)

where O_S and OH_S are surface oxygen and hydroxyl groups, respectively.

In the next step, the surface OH groups interact with O_2 and the surface is re-oxidized in a catalytic cycle (Eq. 4.51).

$$4OH_S + O_2 \rightarrow 2H_2O + 4O_S$$ (4.51)

According to the data of product distribution in the reactor outlet, ethane is the primary obtained hydrocarbon and it is formed by the coupling of $CH_3^{\cdot}$ radicals. In the oxidative coupling reaction, gas-phase reactions are of high importance.

Different factors influence production rate of $CH_3^{\cdot}$ radicals, and, consequently, CH_4 conversion into C_2 products. CO_2 has a negative effect on the production of $CH_3^{\cdot}$ radicals that enter gas-phase, where they may couple to each other or enter into chain-branching reactions and finally CO_x formation.

Different researchers tried to model these homogeneous-heterogeneous reactions. The difficult step is modeling reactions of radicals or intermediates with the surface (Lunsford, 1995).

Ito et al. (1985) proposed a model for heterogeneous reactions (Eqs. 4.52–4.54) over Li/MgO catalyst:

$$CH_4 + O_S^- \rightarrow CH_3^{\bullet} + OH_S^-$$ (4.52)

$$2OH_S^- \rightarrow H_2O + O_S^{2-} + \square$$ (4.53)

$$O_S^{2-} + \square + 1/2O_2 \Leftrightarrow 2O_S^-$$ (4.54)

in which □ stands for an oxide ion vacancy and S is for an adsorbed species.

Regarding the C–H bond activation mode, according to Eq. (4.52), the C–H bond is homolytically broken by abstraction of a hydrogen atom. Other researchers (Maitra et al., 1992) focused on catalyst basicity and proposed that the C–H bond is heterolytically broken (Eq. 4.55) as CH_4 is an acid:

$$CH_4 + M^{n+} - O^{2-} \longrightarrow \overset{\underset{|}{CH_3^-}\ \overset{|}{H^+}}{M^{n+} - O^{2-}} \overset{O_2}{\longrightarrow} \overset{\underset{|}{O_2^-}\ \overset{|}{H^+}}{M^{n+} - O^{2-}} + CH_3^{\cdot} \quad (4.55)$$

The oxidative coupling of methane consists of sequential partial oxidation of CH_4 into C_2H_6 and finally C_2H_4 (Eqs. 4.46–4.49). At first, oxygen and methane reacts and water and ethane are produced. By *in-situ* conversion steps, the produced C_2H_6 is changed to C_2H_4. Of course, higher hydrocarbons are produced as well in trace amounts. In the OCM reaction, ethylene is the favorable product; but according to most mechanistic studies, ethane is the first product. The drawback of this process is that any slight increase in O_2 concentration may shift the reaction to the production of carbon oxides. In the large-scale application of this process, the type of oxygen feed (i.e., air or pure oxygen) used is important. If the main process is based on pure oxygen, an air separation unit is needed that increase the process cost. If air is used in the process, its nitrogen can handle the exothermic temperature rise. If a selective membrane reactor is used in the process, without the need to an ASU (air separation unit), almost pure oxygen can be obtained. By this approach, technological difficulties and cost implications of the direct air usage can be overcome (Lunsford, 1995; Galadima and Muraza, 2016).

If N_2O is used as an oxidant, Eq. (4.54) is replaced by Eq. (4.56) and heterogeneous reactions (4.57)–(4.59) are considered as well:

$$O_S^{2-} + \square + N_2O \rightarrow 2O_S^- + N_2 \quad (4.56)$$

$$N_2O + O_S^- \rightarrow O_{2S}^- + N_2 \quad (4.57)$$

$$N_2O + O_{2S}^- \rightarrow O_{3S}^- + N_2 \quad (4.58)$$

$$O_{3S}^- \Leftrightarrow O_2 + O_S^- \quad (4.59)$$

According to Eqs. (4.52) and (4.57), CH_4 has an inhibiting effect on N_2O decomposition. For a definite conversion, higher C_{2+} selectivity was obtained by N_2O than O_2. The ozonide ion (O_3^-) is responsible for nonselective C_2H_4 oxidation. This ion is formed via Eq. (4.58) or reverse reaction of Eq. (4.59). Generally, at $CH_4/O_2 > 2$ and $T < 700°C$, most of the reaction occurs on the catalytic surface except for coupling of $CH_3^{\cdot}$ radicals (Lunsford, 1995).

Al-Zahrani et al. (1994) reported that the catalytic activity can be influenced by CO_2 presence in the feed mixture. It decreases production rate of C_2H_4 and C_2H_6. As depicted in Eqs. (4.60–4.67), on the catalyst surface, CO_2 can have competitive adsorption with O_2 and CH_4. Due to the interaction between O_2-ads and CH_4-ads, radical species and, consequently, C_2H_4 and C_2H_6 are produced. Moreover, this interaction can change reaction pathway to CO and H_2O formation.

Competitive adsorption:

$$CH_4 + S \rightarrow CH_4S \tag{4.60}$$

$$CO_2 + S \rightarrow CO_2S \tag{4.61}$$

$$O_2 + S \rightarrow O_2S \tag{4.62}$$

Interaction of adsorbed species and radicals:

$$CH_4S + O_2S \rightarrow CH_3^{\bullet} + OH_2^{\bullet} \tag{4.63}$$

$$CH_3^{\bullet} + CH_3^{\bullet} \rightarrow CH_3CH_3 \tag{4.64}$$

$$CH_3^{\bullet} + O_2 \rightarrow CO, CO_2, H_2O \tag{4.65}$$

$$CH_3CH_3 + 0.5O_2 \rightarrow CH_2CH_2 + H_2O \tag{4.66}$$

$$CH_4S + CO_2S \rightarrow CO, CO_2, H_2O \tag{4.67}$$

The presence of CO_2 has an improving effect on C_2 selectivity, increasing the C_2H_4/C_2H_6 product ratio. For 5% methane conversion, due to the addition of 0.031 bar CO_2, C_2 selectivity increased from 45% to 64%. On increasing the conversion level by the increasing catalyst amount, more CO_2 was derived from C_2H_4 and the ethylene selectivity was enhanced. The reason is that CO_2 poison those sites which are capable of excess oxidation of C_2H_4 to CO_2 (Lunsford, 1995).

During CH_4 coupling, CO and CO_2 are produced as by-products but over most catalysts, CO is converted into CO_2. As depicted in Scheme 4.2, these carbon oxides are derived from all three hydrocarbons. These hydrocarbons compete for common sites and their oxidation cannot be considered separately (Lunsford, 1995).

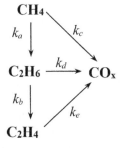

SCHEME 4.2 Reaction network in OCM (Redrawn from Lunsford, 1995).

It was reported that when OCM reaction occurred at low CH$_4$ conversions to have high C$_2$ selectivities, most of the CO$_x$ was obtained directly from CH$_4$. A mechanism should be considered for this direct oxidation. As CH$_3$· radical collides with the catalyst surface 10^5 times faster before it reacts with another CH$_3$· radical in the gas-phase, it might react with surface species (Lunsford, 1995).

For a non-selective oxide (such as CeO$_2$), the catalytic behavior may completely alter by addition of an alkali metal ion that forms a surface carbonate/oxide phase. CeO$_2$ extensively reacts with CH$_3$· radicals and during oxidation CO$_2$ is produced. However by addition of an Na$_2$CO$_3$ monolayer, an active and selective coupling catalyst was obtained. In the presence of the Na$_2$O/Na$_2$CO$_3$ layer, any transfer of electron from CH$_3$· to CeO$_2$ is inhibited and CH$_3$· radicals are prevented from reacting with the surface (Tong et al., 1990).

Tiemersma et al. (2012) obtained power-law rate expressions for the reactions shown in the scheme based on simplified mechanism over a Mn/Na$_2$WO$_4$/SiO$_2$ catalyst prepared by incipient wetness impregnation. For the formation of C$_2$H$_6$ from Eq. (4.46), orders 1 and 0.36 were obtained in partial pressure of methane and oxygen, respectively.

Several kinetic models have been reported for the OCM reaction (Otsuka et al., 1986; Ding et al., 1991, Santamaria et al., 1991, Olsbye et al., 1992). The model of Stansch et al. (1997) for La$_2$O$_3$/CaO catalyst is an adequate one as it considers the main products in a ten-step reaction network. All reactions except for the homogenous dehydrogenation of ethane to ethylene are catalytic. The inhibiting effect of oxygen and carbon dioxide on ethane formation rate is described by Hougen–Watson type rate equation.

4.4.1.4 Reactor Options

For the OCM reaction, there are some typical selectivity-conversion problems due to which higher hydrocarbon yield (> 30%) is not achieved; as below (Tiemersma et al., 2012):

- High methane conversions (that occurred by feeding a large amount of oxygen) are accompanied by poor selectivity of products and large production of unwanted combustion products like CO$_x$
- Ethylene as a highly reactive intermediate product can be changed to undesirable oxidation products via combustion with the excess oxygen

In order to increase the overall yield of products, two concepts can be used: recycling the reactants to the reactor and separating the desirable products from the product mixture (Makri and Vayenas, 2003).

It is worth mentioning that in the large-scale production of ethylene, a large recycle of methane is inevitable and due to the production of large amounts of CO$_x$, there is a significant decrease in carbon efficiency.

Membrane reactor is an interesting choice for the OCM reaction to have an industrially feasible yield. In these reactors, usually a perovskite type membrane is used for oxygen separation by which low concentration of oxygen is kept along the reactor which is favored for C$_{2+}$ production reactions. Due to the high exothermicity of OCM reactions, an optimized cooling system is required for the membrane reactor that complicates its design (Cruellas et al., 2017).

There are some points that must be taken into account for design and construction of an OCM reactor, as below (Cruellas et al., 2017):

- When oxygen concentration along the reactor is kept low, C_{2+} production is favorable
- Due to the high exothermicity of the OCM reaction, a lot of heat must be removed from the system
- Air separation cost is the determining step in feasibility of the OCM process
- As the OCM reaction is a catalytic one, a multiphase reactor must be handled
- C_{2+} products, after formation, can be oxidized; hence a separation unit is needed to improve the reaction yield

In the OCM reactor, the feed stream specification that directly affects ethylene selectivity and yield, has impact on thermally-stable reactor operation and downstream units performance and, overall, on the whole economy of the OCM process. The quantity and source of feed gas components (i.e., oxygen, methane and inert components) are important parameters in determining economy and operation of OCM process (Godini et al., 2019).

4.4.1.4.1   Distributed Oxygen Feeding

Generally, OCM reaction has two challenges; developing a stable and industrially reliable catalyst and performing the reaction at low temperatures. In order to prevent the occurrence of side reaction of methane conversion into CO_x, the selection of a suitable catalyst and reaction conditions and methane to oxygen ratio is of great importance (Galadima and Muraza, 2016).

Oxygen distribution along the reactor is a concept by which selectivity can be improved. According to the published kinetic data (Vatani et al., 2014), reaction order in oxygen for C_2 formation is lower than that of combustion reactions. Hence, oxygen distribution can improve C_2 selectivity. Another advantage of this concept is controlling the hot-spot in fixed-bed reactors.

Reactor with multiple inlets for oxygen feeding is the simplest way for distributed feeding. Another option is the application of the multi-stage reactor that has intermediate gas mixer and un-reacted methane, C_2 and by-products leaving the previous stage mixed with air/oxygen (Schammel et al., 2013). Due to external cooling between stages, this configuration has an advantage for heat management.

The application of porous or dense membranes and electrolytic cells are other techniques to distributed feeding. Interested reader is referred to Cruellas et al. (2017) for their detailed description.

4.4.1.5   Reactor Configuration

Generally, for the OCM reaction, three multi-phase reactor types have been studied; namely fixed-bed, fluidized-bed and bubble column reactors. A solid catalyst is used for fixed-bed and fluidized-bed reactors and a liquid catalyst in the form of a molten salt is used in bubble column reactors. The molten salt catalysts such as lithium carbonate have a melting point lower than OCM reaction temperature and can be used in bubble column reactors (Conway et al., 1989).

As OCM is a highly exothermic reaction, heat management is an important issue in reactor design and operation of OCM reactors. The use of an inert component such as steam or nitrogen in the OCM feed has economic and operational implications. The use of an inert component for diluting is highly recommended in controlling the thermal performance (Godini et al., 2019).

In packed beds, shell and tube heat exchanger is used for external cooling. Microchannel configuration is another alternative for easier temperature control. Hot-spot formation is an issue occurring in all fixed-beds that needs careful consideration (Mazanec et al., 2009).

Regarding heat management, fluidized-beds are the best choice in which, due to gas-solid mixing, a pseudo-isothermal behavior is achieved. When feed gas is at a low temperature, it immediately heats up by the exothermic heat produced in the OCM reaction. In a fluidized-bed configuration, internal heat exchangers can also be used (Cruellas et al., 2017).

Another alternative for heat management is splitting the rector to multiple stages and cooling the intermediate gas streams. In this configuration, a single packed-bed is used in series and each stage has a low CH_4 conversion (Schammel et al., 2013).

For the industrial-scale fixed-bed reactors, the available heat-exchange surface area is low and due to very exothermic reactions, there is a temperature profile in the reactor; even in the presence of external cooling. In order to have enough heat-exchange, a multi-tubular configuration is necessary (Cruellas et al., 2017).

Kaminsky et al. (1997) proposed a shell and tube configuration in which catalyst was coated on tubes. Compared to packed-bed reactors, the main advantage of this configuration is that reaction directly occurs in the tubes where heat exchange with coolant medium is efficient and consequently radial heat-transfer resistance and radial heat transfer temperature gradient is minimized. Another advantage of this configuration is the lack of gas-phase pressure drops.

Besides tubular configurations, micro-channel reactors and honeycomb structures are also proposed for the OCM reaction. The interested reader is refereed to Cruellas et al. (2017) for detailed results.

For fluidized-bed reactors used in OCM reaction, different regimes can be used (Cruellas et al., 2017):

- *Bubbling fluidized-bed*: In this regime, moderate gas velocities are used.
- *Turbulent fluidized-bed*: In this regime, due to high gas velocity, particles and gas are well mixed. In this case, feed gas can have a low temperature. The main drawback of this case is higher separation cost of solid-gas mixture.
- *Re-circulating fluidized-bed*: In this regime, still higher gas velocities are used and large number of solids leaves the reactor with the gas-phase that needs separation outside the reactor and solid catalyst will be recycled back to the reactor.

C_2 selectivity is influenced by mass-transfer resistance, as it can decrease local O_2 concentration at catalyst surface and increase C_2 concentration near the catalyst surface.

4.4.1.6 Process

Conceptually, the OCM process could be divided into three sections: reaction, purification and separation units (Figure 4.5). Natural gas and oxygen are pre-heated to 700°C and continuously fed to the OCM reactor which is at 115 kPa and reaction is performed at 850°C. As this is an exothermic one, its heat is used for the vaporization of HP-BFW (high pressure boiler feed water). The products are compressed to 1,090 kPa and cooled down to 40°C. In the next step, the effluent gases from reaction section are purified in an absorption column by MEA (monoethanolamine) solution. In the next step, in a stripper column, MEA solvent is regenerated and the captured CO_2 is released in a dilute stream. In the separation section, two cryogenic distillation columns are used for C_2H_4 separation. In the first one (demethanizer), unreacted methane is separated from ethane and ethylene stream. In the second one, C_{2+} components are separated; ethylene from the top and ethane from the bottom (Stünkel et al., 2011; Salerno et al., 2012).

4.4.1.6.1 Oxygen–Nitrogen Separation

The integration of air separation inside the reactor can greatly improve overall process economy. For this goal, solid electrolyte cells, perovskite membranes and chemical looping concept are three routes that can be used for O_2–N_2 separation inside the reactor (Cruellas et al., 2017).

In the chemical looping system, without any contact between methane and molecular oxygen, a reaction occurs between methane and O_2-containing solid particles; in which methane is converted into C_2 and gaseous by-products and the solid is reduced. In the next step, by means of air, particles are regenerated separately. After oxidizing back to the original state, particles are used for the OCM reaction. Generally, in chemical looping, two reactors (fluidized- or packed-bed) are needed; one for reaction and one for regeneration (Jubin, 1988).

4.4.1.6.2 Product Separation

Immediate separation of the C_2 products is another idea for improving the selectivity. Counter-current moving-bed chromatographic reactor is an appropriate tool in this regard; in which by the application of solid adsorbents, gas products (before being oxidized to CO_x) are separated. As the catalyst and adsorbent are active at different temperatures (800 and 1,000°C, respectively), reaction and adsorption cannot occur in one reactor (Lee and Hibino, 2011).

4.4.1.7 Recent Developments

For the OCM reaction, a suitable catalyst is the one by which reaction temperature is decreased, ethylene yield is increased and a long lifetime is achieved. Recently, Siluria Technologies announced that nanowire-based biocatalysts can be used for the OCM reaction. Siluria used different innovative technologies for the production

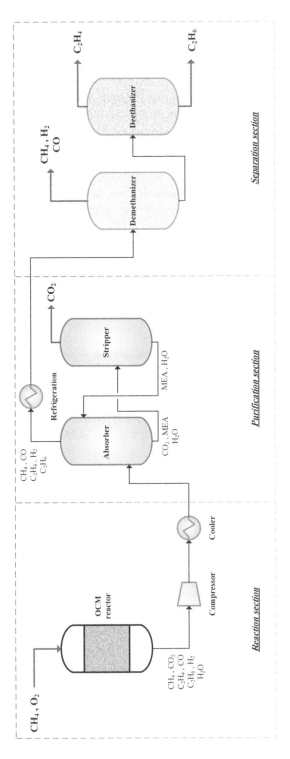

FIGURE 4.5 Flow diagram proposed for OCM (Redrawn from Stünkel et al., 2011 and Salerno et al., 2012).

of OCM catalysts to be commercially viable. The important ones are (http://siluria. com/technology) as follows:

- Nanowire catalyst synthesis technology by which large number of unique, novel inorganic nanowire structures can be created
- Technology for the production of unique templates
- The implication of high throughput screening tools for evaluation of a large number of catalysts at a time

The synthesis procedure of the nanowire catalysts is as follows. (Non-)transition metals or their oxide crystals are doped and grown on biological substrates. In the Siluria Technology for the OCM reaction, with their nanowire catalyst, the OCM reaction occurs in the temperature range of 200–300°C and practical pressures (5–10 atm), and the heat of reaction is used to drive the whole process. A methane conversion of 90% is obtained by this catalyst. The Siluria technology for the OCM reaction is depicted in Figure 4.6. In the first step, natural gas (as the feedstock) is pretreated and unwanted impurities are removed. The produced ethane is transferred into ethylene and C_{3+} hydrocarbons which are separated later. In Siluria's approach, methanation and compression units are also included to ensure increased ethylene yield and methane recyclization (Scher et al., 2013).

The Siluria process is flexible in oxygen and methane sources. For instance, oxygen can be supplied from different sources; such as pure oxygen from a pipeline or an ASU, compressed air from a compressor or enriched air from a VPSA (vapor pressure swing adsorption) unit. Based upon the oxygen source used, the purification process used for the downstream part varies but this oxygen source does not

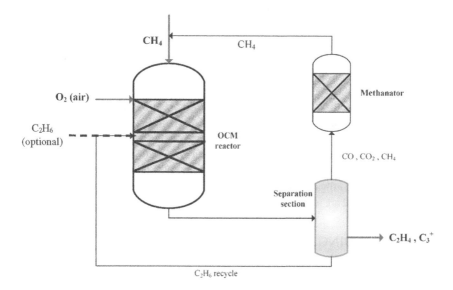

FIGURE 4.6 Simplified flow diagram of the Siluria process (based on Galadima and Muraza, 2016 and http://siluria.com/technology).

have influence on the used catalyst. Propane, CO_2, nitrogen and H_2O content can be controlled to a certain limit by the OCM catalyst. Different sources can be used for methane; such as pipeline natural gas, landfill gas, wellhead natural gas and bio-gas. In this technology, ethane can also be used as a co-feed to be converted into ethylene. This product as the largest petrochemical building block is used in the production of plastics, detergents, coatings, engine coolants, adhesives and other similar products with everyday use (http://siluria.com/technology).

Another advantage of this process is that unreacted methane is recycled back to the reactor for oxidation to ethylene. Moreover, by the hydro-isomerization process, higher *n*-alkanes can be changed into high octane number gasoline (Galadima et al., 2014).

It was reported that nanowires and nanoscale materials are good candidates for the catalyst of OCM reaction, as they can give improved ethylene yield at low temperatures (Aman et al., 2014).

4.4.1.8   Integrating OCM Process with Other Ones

The main features of the OCM process are its large heat of reaction and large amount of unreacted methane. These can be used by other processes. Thus, by integrating the OCM process with another one, an economically attractive process can be obtained.

The commercial methanol plant is based on natural gas ATR (autothermal reforming) process; while in the desired process the unreacted methane from the OCM reactor (a gas-phase stream containing CO, CO_2, CH_4, H_2 and H_2O) is used. This gaseous stream is a by-product with low economic value in ethylene production process, but its CO content is attractive for syngas production. At first, trace amount of impurities (such as ethylene) must be removed, as they may affect syngas production.

In the formaldehyde production, the unreacted methane is used as raw material. The selective oxidation of methane is used in the reactor and CH_2O, CO_2, CO and H_2O are produced. For the production of 117.1 ton $_{formaldehyde}$/year, additional oxygen (as oxidant) is needed. By two distillation columns, all the produced CH_2O in this reactor is purified. After the formaldehyde product is separated, syngas is produced by stream coming from top of the demethanizer column in the formaldehyde purification section. In order to adjust the carbon/hydrogen ratio, CO_2 is added from the OCM purification section (with amines). This ratio depends on CO and CO_2 concentrations from the non-reacted methane. Water is used for cooling the gas stream leaving the reformer and steam is generated for the reformer, before entering the reactor. The syngas is compressed to 30 bar. The operating temperature of reactor is 250°C.

An economic evaluation was performed to compare OCM process with and without recycling this unreacted methane. In both processes, 2,593 t/d pure methane is used as feedstock and 396 t/d ethylene is produced with 99% purity. 117.1 t/d of formaldehyde and 204.2 t/d of methanol are also produced in the alternative process (Salerno et al., 2012).

4.5   CONCLUDING REMARKS

The direct conversion of methane into higher hydrocarbons is still far from wide commercial utilization despite great economic potentials. Low product yields and the need for supplying or removing large amount of heat of reaction for nonoxidative and oxidative route, respectively, are the main technical challenges.

Non-oxidative catalytic conversion of methane into aromatics is performed near atmospheric pressures and high temperatures (>700°C). As production of aromatics with high selectivities is possible, methane aromatization is a potential route for utilization of natural gas. Considerable efforts have been devoted to development of stable and selective catalysts and understanding bifunctional role of the catalyst and the nature of the species deposit during reaction, although they were not fully successful. Currently, overall conversions of about 14% and benzene selectivity of 80% has been achieved over Mo/HZSM-5 catalysts. The synthesis condition and composition of Mo/HZSM-5 have important effect on the state and position of molybdenum on zeolite and thus catalyst performance.

Zeolite pore structure with windows entrance diameters close to the diameter of benzene molecule provides conversion and high selectivity to benzene. Incorporating molybdenum into zeolite effectively modifies the structural properties and acidity of the zeolite. The subsequent changes in physico-chemical properties during CH_4 DHA reaction are often related to formation and deposition of carbonaceous species. Although the bifunctional role of Mo/HZSM-5 catalysts has been accepted, there is not still information regarding the detailed reaction kinetics. There are still questions regarding the nature and distribution of molybdenum sites in zeolite framework. On the other hand, commercialization of nonoxidative conversion of methane into aromatics needs improvements in technology and engineering of the process. An important issue in commercialization of the process is rapid catalyst deactivation due to coke deposition. Therefore, further research on mechanism and kinetic of coke deposition, its suppression or increasing the resistance of the catalyst towards coke deposition is necessary.

Recent researches showed that the application of powdered or bulk modified or pure transition metal compounds and elements of groups IA and IIA have methane conversions of up to 50% with ethylene yields of 10%–25% in the OCM reaction. Hence, an improvement in catalyst formulation is inevitable. Catalyst textural composition, its basicity, its preparation procedure and avoiding complete hydrocarbon oxidation to CO_x species are important catalyst characteristics. In order to improve catalyst performance, its modification with alkali metals (such as Li and Na) and co-doping of oxide catalysts are two approaches, but reaching ethylene yield of 40% is still difficult. Moreover, in order to reach optimum yields of these catalysts, a high temperature of 750–900°C is needed. A major drawback of these catalysts is stability. At the operating temperature, they decompose easily and because of decay in active sites, their handling is difficult. Hence, research studies must focus on solving these problems. It is announced that incorporating dopants (such as WO_2 and CeO_2) can improve catalyst stability at high reaction temperatures.

Recently, Siluria Technologies discovered bio-based nanowires that can present ethylene yield of 90% at temperature lower than 600°C. The critical point is developing a catalyst that inhibits complete hydrocarbon oxidation and improves catalyst lifetime. Different methods are reported for the synthesis of nanowires such as template-free methods, vapor-phase growth and template-assisted synthesis.

Many conversion processes (direct or indirect) need oxygen which has an equivalent or higher price than methane itself.

Despite the above-mentioned difficulties, direct conversion of methane into value-added products is quite active field both in industrial and academic communities. The increased availability of large gas resources (e.g., shale gas) and rising crude oil prices might make these processes economically and technically feasible in the future.

NOMENCLATURE

ASU	air separation unit
ATR	autothermal reforming
BTX	benzene toluene xylenes
CCR	continuous catalyst regeneration
CVD	chemical vapor deposition
DHA	dehydroaromatiztion
DMF	dimethylformamide
DTA	differential thermal analysis
FTIR	Fourier Transform Infra-Red
HEAP	hydrogen electric arc pyrolysis
HF	hierarchical factor
HP-BFW	high pressure boiler feed water
MEA	monoethanolamine
NMP	N-methylpyrrolidone
NMR	nuclear magnetic resonance
OCM	oxidative coupling of methane
PO	partial oxidation
RDS	rate determining step
TEOS	tetraethoxysilane
VPSA	vapor pressure swing adsorption
XAES	X-ray auger electron spectroscopy
XPS	X-ray photoelectron spectroscopy

REFERENCES

Ahmed, S., Moffat, J. B., Role of carbon tetrachloride in the conversion of methane on silica-supported alkali metal added alkaline earth oxide catalysts, Appl. Catal., 58 (1990a) 83–103.

Ahmed, S., Moffat, J. B., The oxidative coupling of methane on Mn/SiO_2 and effect of solid- and gas-phase doping, J. Catal., 125 (1990b) 54–66.

Al-Zahrani, S., Song, Q., Lobban, L. L., Effects of carbon dioxide during oxidative coupling of methane over lithium/magnesia: mechanisms and models, Ind. Eng. Chem. Res., 33 (1994) 251–258.

Aman, D., Zaki, T., Mikhail, S., Ali, M., A simple simulation model for oxidative coupling of methane over $La_{0.6}Sr_{0.4}NiO_3$ nanocatalyst, Chem. Process Eng. Res., 23 (2014) 29–37.

Arutyunov, V. S., Strekova, L. N., The interplay of catalytic and gas-phase stages at oxidative conversion of methane: a review, J. Mol. Catal. A. Chem., 426 (2017) 326–342.

Baba, T., Conversion of methane over Ag^+-exchanged zeolite in the presence of ethane, Catal. Surv. Asia, 9 (2005) 147–154.

Belgued, M., Amariglio, A., Amariglio, H., Conversion of methane into higher hydrocarbons on platinum, Nature, 352 (1991) 789–790.

Belgued, M., Amariglio, H., Pareja, P., Amariglio, A., Saintjust, J., Low temperature catalytic homologation of methane on platinum, ruthenium, and cobalt, Catal. Today, 13 (1992) 437–445.

Bezemer, N. D., Two-step methane homologation in pulsed operation (2017). Mater thesis, Eindhoven University of Technology.

Bond, G. C., Metal Catalyzed Reactions of Hydrocarbons, Boston, MA: Springer (2006).

Borry, R. W., Kim, Y. H., Huffsmith, A., Reimer, J. A., Iglesia, E., Structure and density of Mo and acid sites in Mo-exchanged H-ZSM5 catalysts for nonoxidative methane conversion, J. Phys. Chem. B., 103 (1999) 5787–5796.

Boskovic, G., Soltan Mohammad Zadeh, J., Smith, K. J., K promotion of Co catalysts for the two-step methane homologation reaction, Catal. Lett., 39 (1996) 163–168.

Bradford, M. C. J., Vannice, M. A., CO_2 reforming of CH_4, Catal. Rev. Sci. Eng., 41 (1999) 1–42.

Bradford, M. C. J., Two-step methane conversion to higher hydrocarbons: comment on the relevance of metal-carbon bond strength, J. Catal., 189 (2000) 238–243.

Bradford, M. C. J., Te, M., Konduru, M., Fuentes, D. X., CH_4-C_2H_6-CO_2 conversion to aromatics over Mo/SiO_2/H-ZSM-5, Appl. Catal. A., 266 (2004) 55–66.

Choudhary, V. R., Kinage, A. K., Choudhary, T. V., Low-temperature nonoxidative activation of methane over H-galloaluminosilicate (MFI) zeolite, Science, 275 (1997) 1286–1288.

Choudhary, T. V., Aksoylu, E., Wayne Goodman, D., Nonoxidative activation of methane, Catal. Rev. Sci. Eng., 45 (2003) 151–203.

Chu, W., Qiu, F., Remarkable promotion of benzene formation in methane aromatization with ethane addition, Top. Catal., 22 (2003) 131–134.

Conway, S. J., Szanyi, J., Lunsford, J. H., Catalytic properties of lithium carbonate melts and related slurries for the oxidative dimerization of methane, Appl. Catal., 56 (1989) 149–161.

Cook, B., Mousko, D., Hoelderich, W., Zennaro, R., Conversion of methane to aromatics over Mo_2C/ZSM-5 catalyst in different reactor types, Appl. Catal. A. G., 365 (2009) 34–41.

Cornils, B., Herrmann,W.A., Wong C.-H., Zanthoff, H.-W. (eds.) Catalysis from A to Z, Weinheim, Germany: Wiley-VCH (2013).

Cruellas, A., Melchiori, T., Gallucci, F., van Sint Annaland, M., Advanced reactor concepts for oxidative coupling of methane, Catal. Rev., 59 (2017) 234–294.

Ding, X., Yu, Z., Wang, X., Shen, S., The pathway of oxidative coupling of methane on a La_2O_3/$BaCO_3$ Catalyst, Stud. Surf. Sci. Catal., 61 (1991) 65–71.

Ding, W., Li, S., Meitzner, G. D., Iglesia, E., Methane conversion to aromatics on Mo/HZSM5: structure of molybdenum species in working catalysts, J. Phys. Chem. B., 105 (2001) 506–513.

Ding, W., Meitzner, G. D., Iglesia, E., The effects of silanation of external acid sites on the structure and catalytic behavior of Mo/H-ZSM5, J. Catal., 206 (2002) 14–22.

Dissanayake, D., Kharis, K. C. C., Lunsford, J. H., Rosynek. M. P., Catalytic partial oxidation of methane over Ba-Pb, Ba-Bi, and Ba-Sn perovskites, J. Catal., 139 (1993) 652–663.

Echevsky, G. V., Kodenev, E. G., Kikhtyanin, O. V., Parmon, V. N., Direct insertion of methane into C_3-C_4 paraffins over zeolite catalysts: a start to the development of new one-step catalytic processes for the gas-to-liquid transformation. Appl. Catal. A., 258 (2004) 159–171.

Elkins, T. W., Hagelin-Weaver, H. E., Oxidative coupling of methane over unsupported and alumina-supported samaria catalysts, Appl. Catal. A. G., 454 (2013) 100–114.

Galadima, A., Leke, L., Ismaila, A., Sani, S., Abdullahi, I., Upgrading n-Heptane via catalytic hydroisomerisation, Chem. Mater. Res., 6 (2014) 63–68.

Galadima, A., Muraza, O., Revisiting the oxidative coupling of methane to ethylene in the golden period of shale gas: a review, J. Ind. Eng. Chem., 37 (2016) 1–13.

Godini, H. R., Azadi, M., Penteado, A., Khadivi, M., Wozny, G., Repke, J. U., A multi-perspectives analysis of methane oxidative coupling process based on miniplant-scale experimental data, Chem. Eng. Res. Des., 151 (2019) 56–69.

Gunsalus, N. J., Koppaka, A., Park, S. H., Bischof, S. M., Hashiguchi, B. G., Periana, R. A., Homogeneous functionalization of methane, Chem. Rev., 117 (2017) 8521–8573.

Guo, J., Lou, H., Zheng, X., Energy-efficient coaromatization of methane and propane, J. Nat. Gas Chem., 18 (2009) 260–272.

Hiyoshi, N., Sato, K., Oxidative coupling of methane over $Mn-Na_2WO_4/SiO_2$ catalyst with continuous supply of alkali chloride vapor, Fuel Proc. Tech., 151 (2016) 148–154.

Holmen, A., Olsvik, O., Rokstad, O. A., Pyrolysis of natural-gas-chemistry and process concepts. Fuel Proc. Tech., 42 (1995) 249–267.

Holmen, A., Direct conversion of methane to fuels and chemicals, Catal. Today, 142 (2009) 2–8.

Hongmei, L., Yide, X., H_2-TPR study on Mo/HZSM-5 catalyst for CH_4 dehydroaromatization, Chin. J. Catal., 27 (2006) 319–323.

Horn, R., Schlögl, R., Methane activation by heterogeneous catalysis, Catal. Lett., 145 (2015) 23–39.

Hu, J., Wu, S., Liu, H. Ding, H., Li, Z., Guan, J., Kan, Q., Effect of mesopore structure of TNU-9 on methane dehydroaromatization, RSC Adv., 4 (2014) 26577–26584.

Iliuta, M. C., Iliuta, I., Grandjean, B. P. A., Larachi, F., Kinetics of methane nonoxidative aromatization over Ru-Mo/HZSM-5 catalyst, Ind. Eng. Chem. Res., 42 (2003) 3203–3209.

Ioffe, M. S., Pollington, S. D., Wan, J. K. S., High-power pulsed radio-frequency and microwave catalytic processes: selective production of acetylene from the reaction of methane over carbon, J. Catal., 151 (1995) 349–355.

Ito, T., Wang, J. X., Lin, C. H., Lunsford, J. H., Oxidative dimerization of methane over a lithium-promoted magnesium oxide catalyst, J. Am. Chem. Soc., 107 (1985) 5062–5068.

Jiang, H., Wang, L., Cui, W., Xu, Y., Study on the induction period of methane aromatization over Mo/HZSM-5: partial reduction of Mo species and formation of carbonaceous deposit, Catal. Lett., 57 (1999) 95–102.

Jin, Z., Liu, S., Qin, L., Liu, Z, Wang, Y., Xie, Z., Wang, X. Methane dehydroaromatization by Mo-supported MFI-type zeolite with core–shell structure (2013) Appl. Catal. A., 453, 295–301.

Jubin, J. C., Fixed bed reactor system, US Patent 4751055 A (1988).

Kado, S., Sekine, Y., Fujimoto, K., Direct synthesis of acetylene from methane by direct current pulse discharge, Chem. Commun., (1999) 2485–2486.

Kalyani, M., Kumar, V. R., Satyanarayana, S. V., Effect of various parameters on oxidative coupling of methane, J. Chem. Eng. Res., 2 (2014) 223–230.

Kaminsky, M. P., Huff, G. A., Calamur, N., Spangler, M. J., Catalytic Wall Reactors and Use of Catalytic Wall Reactors for Methane Coupling and Hydrocarbon Cracking Reactions, US Patent 5599510 A (1997).

Keller, G. E., Bhasin, M. M., Synthesis of ethylene via oxidative coupling of methane, methane: I. Determination of active catalysts, J. Catal., 73 (1982) 9–19.

Khojasteh Salkuyeh, Y., Adams, T. A., A novel polygeneration process to co-produce ethylene and electricity from shale gas with zero CO_2 emissions via methane oxidative coupling, Energy Conv. Man., 92 (2015) 406–420.

Kikuchi, S., Kojima, R., Ma, H., Bai, J., Ichikawa, M. Study on Mo/HZSM-5 catalysts modified by bulky aminoalkyl-substituted silyl compounds for the selective methane-to-benzene (MTB) reaction, J. Catal., 242 (2006) 349–356.

Koerts, T., Vansanten, R.A., A low-temperature reaction sequence for methane conversion. J. Chem. Soc. Chem. Commun., (1991) 1281–1283.

Koerts, T., Deelen, M. J. A. G., van Santen, R. A., Hydrocarbon formation from methane by a low-temperature two-step reaction sequence, J. Catal., 138 (1992a) 101–114.

Koerts, T., Leclercq, P. A., van Santen, R. A., Homologation of olefins with methane on transition metals, J. Am. Chem. Soc., 114 (1992b) 7272–7278.

Kosinov, N., Hensen, E. J. M., Nonoxidative Dehydroaromatization of Methane, In Sels, B., Voorde, M. V. (Eds.) Nanotechnology in Catalysis: Applications in the Chemical Industry, Energy Development, and Environment Protection, 1st Ed., Germany: Wiley (2017).

Krylov, O. V., Mamedov, A. K., Heterogeneous catalytic reactions of carbon dioxide. Russ. Chem. Rev., 64 (1995) 877– 900.

Kus, S., Otremba, M., Torz, A., Taniewski, M., The effect of gas atmosphere used in the calcination of MgO on its basicity and catalytic performance in oxidative coupling of methane, Appl. Catal. A., 230 (2002) 263–270.

Kus, S., Otremba, M., Taniewski, M., The catalytic performance in oxidative coupling of methane and the surface basicity of La_2O_3, Nd_2O_3, ZrO_2 and Nb_2O_5, Fuel, 82 (2003) 1331–1338.

Kusmiyati, K., Amin, N. A. S., Dual effects of supported W catalysts for dehydroaromatization of methane in the absence of oxygen, Catal. Lett., 102 (2005) 69–77.

Lee, B., Hibino, T., Efficient and selective formation of methanol from methane in a fuel cell-type reactor, J. Catal., 279 (2011) 233–240.

Lei, Y., Chu, C., Li, S., Sun, Y., Methane activations by lanthanum oxide clusters, J. Phys. Chem. C., 118 (2014) 7932–7945.

Liu, S., Wang, L., Ohnishi, R., Ichikawa, M., Bifunctional catalysis of Mo/HZSM-5 in the dehydroaromatization of methane to benzene and naphthalene XAFS/TG/DTA/MASS/FTIR characterization and supporting effects, J. Catal., 181 (1999) 175–188.

Liu, H., Bao, X., Xu, Y., Methane dehydroaromatization under nonoxidative conditions over Mo/HZSM-5 catalysts: identification and preparation of the Mo active species, J. Catal., 239 (2006) 441–450.

Liu, B. S., Jiang, L., Sun, H., Au, C. T., XPS, XAES, and TG/DTA characterization of deposited carbon in methane dehydroaromatization over Ga-Mo/ZSM-5 catalyst, Appl. Surf. Sci., 253 (2007) 5092–5100.

Liu, H., Yang, S., Hu, J., Shang, F., Li, Z., Xu, C., Guan, J., Kan, Q., A comparison study of mesoporous Mo/H-ZSM-5 and conventional Mo/H-ZSM-5 catalysts in methane non-oxidative aromatization, Fuel Proc. Tech., 96 (2012) 195–202.

Lu, G. M., Hoffer, T., Guczi, L., Reducibility and co-hydrogenation over Pt and Pt-Co bimetallic catalysts encaged in NaY-zeolite, Catal. Lett., 14 (1992) 207–220.

Lu, Y., Xu, Z., Tian, Z., Zhang, T., Lin, L., Methane aromatization in the absence of an added oxidant and the bench scale reaction test, Catal. Lett., 62 (1999) 215–220.

Lunsford, J. H., The catalytic oxidative coupling of methane, Angew. Chem. Int. Ed. Eng., 34 (1995) 970–980.

Lunsford. J. H, Catalytic conversion of methane to more useful chemicals and fuels: a challenge for the 21st century, Catal. Today, 63 (2000) 165–174.

Luzgin, M. V., Rogov, V. A., Arzumanov, S. S., Toktarev, A. V., Stepanov, A. G., Parmon, V. N., Understanding methane aromatization on a Zn-modified high-silica zeolite, Angew. Chem. Int. Ed., 47 (2008) 4559–4562.

Luzgin, M. V., Rogov, V. A., Arzumanov, S. S., Toktarev, A. V., Stepanov, A. G., Parmon, V. N., Methane aromatization on Zn-modified zeolite in the presence of a co-reactant higher alkane: how does it occur? Catal. Today, 144 (2009) 265–272.

Luzgin, M. V., Gabrienko, A. A., Rogov, V. A., Toktarev, A. V., Parmon, V. N., Stepanov, A. G., The 'Alkyl' and 'Carbenium' pathways of methane activation on Ga-modified zeolite BEA:^{13}C solid-state NMR and GC-MS study of methane aromatization in the presence of higher alkane, J. Phys. Chem. C., 114 (2010) 21555–21561.

Maitra, A. M., Campbell, I., Tyler, R. J., Influence of basicity on the catalytic activity for oxidative coupling of methane, Appl. Catal. A., 85 1992) 27–46.

Makri, M., Vayenas, C. G., Successful scale up of gas recycle reactor-separators for the production of C_2H_4 from CH_4, Appl. Catal. A. G., 244 (2003) 301–310.

Majhi, S., Pant, K. K., Direct conversion of methane with methanol toward higher hydrocarbon over Ga modified Mo/HZSM-5 catalyst, J. Ind. Eng. Chem., 20 (2014) 2364–2369.

Matus, E. V., Sukhova, O. B., Ismagilov, I. Z., Tsikoza, L. T., Ismagilov, Z. R., Peculiarities of dehydroaromatization of CH_4-C_2H_6 and CH_4 over Mo/ZSM-5 catalysts, React. Kin. Cat. Lett., 98 (2009) 59–67.

Mazanec, T., Simmons, W., Brophy, J., Pesa, F., Tonkovich, A. L., Litt, R., Qiu, D., Silva, L., Lamont, M., Fanelli, M., Oxidative Coupling of Methane, US Patent 20090043141 A1 (2009).

Moghimpour Bijani, P., Sohrabi, M., Sahebdelfar, S., Thermodynamic analysis of nonoxidative dehydroaromatization of methane, Chem. Eng. Tech., 35 (2012) 1825–1832.

Moghimpour Bijani, P., Sohrabi, M., Sahebdelfar, S., Nonoxidative aromatization of CH_4 using C_3H_8 as a coreactant: thermodynamic and experimental analysis, Ind. Eng. Chem. Res., 53 (2013) 572–581.

Naccache, C. M., Mériaudeau, P., Sapaly, G., Tiep, L. V., Taârit, Y. B., Assessment of the low-temperature nonoxidative activation of methane over H-Galloaluminosilicate (MFI) zeolite: A C-13 labelling investigation, J. Catal., 205 (2002) 217–220.

Nishiyama, T., Aika, K.-I., Mechanism of the oxidative coupling of methaneusing carbon dioxide as an oxidant over PbO-MgO. J. Catal., 122 (1990) 346–351.

Olah, G.A., Molnár, A., Prakash, G.K.S., 2018. Hydrocarbon Chemistry, 3rd Ed., Wiley, New York, 125–236.

Olsbye, U., Desgrandchamps, G., Jens, K. J., Kolbe, S., A kinetic study of the oxidative coupling of methane over a BaCO$_3$/La$_2$O$_n$(CO$_3$)$_{3-n}$ catalyst: I. Determination of a global reaction scheme and the influence of heterogeneous and homogeneous reactions, Catal. Today, 13 (1992) 209–218.

Osawa, T., Nakano, I., Takayasu, O., Dehydrogenation of methane over Mo/ZSM-5: Effects of additives in the methane stream, Catal. Lett., 86 (2003) 57–62.

Otsuka, K., Jinno, K., Morikawa A., Active and selective catalysis for the synthesis of C_2H_4 and C_2H_6 via oxidative coupling of methane, J. Catal., 100 (1986) 353–359.

Pässler, P., Hefner, W., Buckl, K., Meinass, H., Meiswinkel, A., Wernicke, H. J., Ebersberg, G., Müller, R., Bässler, J., Behringer, H., Mayer, D., Acetylene, In Ullmann's Encyclopedia of Industrial Chemistry, Wiley-VCH, Weinheim (2007).

Petkovic, L. M., Ginosar, D. M., Comparison of two preparation methods on catalytic activity and selectivity of Ru-Mo/HZSM5 for methane dehydroaromatization, J. Fuels, 2014 (2014) 7–13.

Pinglian, T., Zhusheng, X., Tao, Z., Laiyuan, C., Liwu, L., Aromatization of methane over different Mo-supported catalysts in the absence of oxygen, Reac. Kin. Cat. Lett., 61 (1997) 391–396.

Portilla, M. T., Llopis, F. J., Martínez, C., Non-oxidative dehydroaromatization of methane: an effective reaction-regeneration cyclic operation for catalyst life extension, Catal. Sci. Technol., 5 (2015) 3806–3821.

Sahebdelfar, S., Takht Ravanchi, M., Gharibi, M., Hamidzadeh, M., Rule of 100: an inherent limitation or performance measure in oxidative coupling of methane, J. Nat. Gas Chem., 21 (2012) 308–313.

Salerno, D., Arellano-Garcia, H., Wozny, G., Techno-Economic Analysis for the Synthesis of Downstream Processes from the Oxidative Coupling of Methane Reaction, In Karimi, I. A., Srinivasan, R., (Eds), Proceedings of the 11th International Symposium on Process Systems Engineering, Singapore (2012).

Santamaria, J. M., Miro, E. E., Wolf, E. E., Reactor simulation studies of methane oxidative coupling on a Na/NiTiO$_3$ Catalyst, Ind. Eng. Chem. Res., 30 (1991) 1157–1165.

Schammel, W. P., Wolfenbarger, J., Ajinkya, M., Ciczeron, J. M., Mccarty, J., Weinberger, S., Edwards, J. D., Sheridan, D., Scher, E. C., McCormick, J., Oxidative coupling of methane systems and methods, WO Patent 2013177433 A2 (2013).

Scher, E. C., Zurcher, F. R., Cizeron, J. M., Schammel, W. P., Tkachenko, A., Gamoras, J., Karshtedt, D., Nyce, G., Nanowire catalysts: A patent, EP2576046 A2 (2013).

Schobert, H., Production of acetylene and acetylene-based chemicals from coal, Chem. Rev., 114 (2014) 1743–1760.

Schwach, P., Pan, X., Bao, X., Direct conversion of methane to value-added chemicals over heterogeneous catalysts: Challenges and prospects, Chem. Rev., 117 (2017) 8497–8520.

Skutil, K., Taniewski, M., Some technological aspects of methane aromatization (direct and via oxidative coupling), Fuel Proc. Tech., 87 (2006) 511–521.

Soltan Mohammad Zadeh, J., Smith, K. J., Kinetics of CH$_4$ decomposition on supported cobalt catalysts, J. Catal., 176 (1998) 115–124.

Soltan Mohammad Zadeh, J., Smith, K. J., Two-step homologation of methane on supported Co catalysts, J. Catal., 183 (1999) 232–239.

Solymosi, F., Erdohelyi, A., Cserenyi, J., A comparative study on the activation and reactions of CH$_4$ on supported metals, Catal. Lett., 16 (1992) 399–405.

Solymosi, F., Erdohelyi, A., Cserenyi, J., Felvegi, A., Decomposition of CH$_4$ over supported Pd catalysts, J. Catal., 147 (1994) 272–278.

Solymosi, F., Szoke, A., Conversion of ethane into benzene on Mo$_2$C/ZSM-5 catalyst, Appl. Catal. A. G., 166 (1998) 225–235.

Spivey, J. J., Hutchings, G., Catalytic aromatization of methane, Chem. Soc. Rev., 43 (2014) 792–803.

Stünkel, S., Drescher, A., Wind, J., Brinkmann, T., Repke, J. U., Wozny, G., Carbon dioxide capture for the oxidative coupling of methane process – a case study in mini-plant scale, Chem. Eng. Res. Des. 89 (2011) 1261–1270.

Stansch, Z., Mleczko, L., Baerns, M., Comprehensive kinetics of oxidative coupling of methane over the La$_2$O$_3$/CaO catalyst, Ind. Eng. Chem. Res., 36 (1997) 2568–2579.

Tang, P., Zhu, Q., Wu, Z., Ma, D., Methane activation: the past and future, Energy Env. Sci., 7 (2014) 2580–2591.

Tempelman, C. H. L., Zhu, X., Hensen, E. J. M., Activation of Mo/HZSM-5 for methane aromatization, Chinese J. Catal., 36 (2015) 829–837.

Tempelman, C. H. L., Portilla, M. T., Martinez-Armero, M. E., Mezari, B., de Caluwe, N. G. R., Martinez, C., Hensen, E. J. M., One-pot synthesis of nano-crystalline MCM-22, Micro. Meso. Mater., 220 (2016) 28–38.

Tiemersma, T. P., Tuinier, M. J., Gallucci, F., Kuipers, J. A. M., van Sint Annaland, M., A kinetics study for the oxidative coupling of methane on a Mn/Na$_2$WO$_4$/SiO$_2$ catalyst, Appl. Catal. A. G., 433–434(2012) 96–108.

Tong, Y., Rosynek, M. P., Lunsford, J. H., The role of sodium carbonate and oxides supported on lanthanide oxides in the oxidative dimerization of methane, J. Catal., 126 (1990) 291–298.

Trotuş, I. T., Zimmermann, T., Schüth, F., Catalytic reactions of acetylene: a feedstock for the chemical industry revisited, Chem. Rev., 114 (2014) 1761–1782.

Vatani, A., Jabbari, E., Askarieh, M., Torangi, M. A., Kinetic modeling of oxidative coupling of methane over Li/MgO catalyst by genetic algorithm, J. Nat. Gas Sci. Eng., 20 (2014) 347–356.

Vines, F., Lykhach, Y., Staudt, T., Lorenz, M. P. A., Papp, C., Steinruck, H. P., Libuda, J., Neyman, K. M., Gorling, A., Methane activation by platinum: critical role of edge and corner sites of metal nanoparticles, Chem. Eur. J., 16 (2010) 6530–6539.

Vosmerikov, A. V., Zaykovskii, V. I., Korobitsyna, L. L., Kodenev, E. G., Kozlov, V. V., Echevskii, G. V., Catalysts for non-oxidative methane conversion, Studies Surface Sci. Catal., 162 (2006) 913–920.

Wang, L., Tao, L., Xie, M., Xu, G., Huang, J., Xu, Y., Dehydrogenation and aromatization of methane under non-oxidizing conditions, Catal. Lett., 21 (1993) 35–41.

Wang, D., Lunsford, J. H., Rosynek, M. P., Characterization of a Mo/ZSM-5 catalyst for the conversion of methane to benzene. J. Catal., 169 (1997) 347–358.

Weissermel, K., Arpe, H.-J., Industrial Organic Chemistry, Weinheim, Germany: Wiley-VCH (1997).

Wu, M.C., Lenzsolomun, P., Goodman, D.W., Two-step, oxygen-free route to higher hydrocarbons from methane over ruthenium catalysts, J. Vac. Sci. Technol., 12 (1994) 2205–2209.

Wu, Y., Emdadi, L., Oh, S.C., Sakbodin, M., Liu, D. Spatial distribution and catalytic performance of metal-acid sites in Mo/MFI catalysts with tunable meso-/microporous lamellar zeolite structures, J. Catal., 323 (2015) 100–111.

Xu, Y., Liu, S., Wang, L., Xie, M., Gue, X., Methane activation without using oxidants over Mo/HZSM-5 zeolite catalysts, Cat. Lett., 30 (1995) 135–149.

Yao, B., Chen, J., Liu, D., Fang, D., Intrinsic kinetics of methane aromatization under non-oxidative conditions over modified Mo/HZSM-5 catalysts, J. Nat. Gas Chem., 17 (2008) 64–68.

Yao, S., Sun, C., Li, J., Huang, X., Shen, W., A ^{13}CO isotopic study on the CO promotion effect in methane dehydroaromatization reaction over a Mo/HMCM-49 catalyst, J. Nat. Gas Chem., 19 (2010) 1–5.

Yuan, S., Li, J., Hao, Z., Feng, Z., Xin, Q., Ying, P., Li, C., The effect of oxygen on the aromatization of methane over the Mo/HZSM- 5 catalyst, Catal. Lett., 63 (1997) 73–77.

Zavyalova, U., Holena, M., Schlögl, R., Baerns, M., Statistical Analysis of past catalytic data on oxidative methane coupling for new insights into the composition of high-performance catalysts, ChemCatChem, 3 (2011) 1935–1947.

Zhang, C. L., Li, Sh., Yuan, Y., Zhang, W. X., Wu, T. H., Lin, L. W., Aromatization of methane in the absence of oxygen over Mo-based catalysts supported on different types of zeolites, Catal. Lett., 56 (1998) 207–213.

5 Synthesis Gas Chemistry

5.1 INTRODUCTION

Synthesis gas (a mixture of H_2, CO and CO_2 with various proportions) can be produced from many carbonaceous sources including fossil fuels (natural gas, coal), renewables (biomass and its derivatives), carbon dioxide and organic wastes using reforming or gasification technologies. The availability and flexibility of syngas is the key for its wide application in chemical synthesis or using it as a source of its individual components (H_2 and CO). Consequently, it can play a role in sustainable production of valuable chemicals and fuels.

The commercially important reactions in syngas chemistry include Fischer–Tropsch synthesis (FTS), hydroformylation (oxo synthesis) and methanol synthesis by which hydrocarbons (methane and higher hydrocarbons) and oxygenates (e.g., methanol and higher alcohols) can be produces.

FTS is becoming a core technology in syngas chemistry for the conversion of natural gas, coal and biomass into clean transportable liquid fuels such as diesel and gasoline. Therefore, synthesis gas is a platform feedstock in C_1 chemistry for producing a wide range of valuable chemicals. Unlike other C_1 feedstocks, CO-based processes (such as methanol synthesis, FTS, Oxo synthesis, etc.) have long been commercialized and are in operation.

This chapter deals with reactions of carbon monoxide and hydrogen of syngas for the production of bulk chemicals being relevant to C_1 chemistry.

5.2 CHEMICAL PROPERTIES OF CARBON MONOXIDE

Carbon monoxide is a nonpolar molecule (Table 5.1). It is an unsaturated compound; therefore, this dominates in its chemical properties. From the viewpoint of the oxidation state, it lies been carbon and CO_2, and, thus, it is a metastable molecule being relatively inert under low temperature and pressures but highly active at elevated temperatures and pressures.

CO is a ligand with many applications in coordination chemistry. Coordination with metals may be linear, bridging or capping, among others (Figure 5.1).

The carbonyl complexes of low-valent transition metals are stable due to high π acceptor capacity of CO. CO is the ligand of many important homogeneous catalysts such as those of methanol carbonylation (Chapter 7) and olefin hydroformylation (Cornils et al., 2013).

The low boiling point and toxicity of CO makes its transportation hazardous. Therefore, it is mostly produced and utilized in site.

TABLE 5.1

Thermochemical Properties of Carbon Monoxide

Compound	Normal BP (°C)	Dipole Moment (D)	ΔH°_{f} (kJ/mol)	ΔG°_{f} (kJ/mol)
CO	−191.5	0.122	−110.5	−137.28

5.3 FISCHER–TROPSCH SYNTHESIS

FTS is catalytic hydrogenation of carbon monoxide to hydrocarbons (and oxygenates). It can be described by the following set of reactions:

$$CO + 2H_2 \rightarrow -(-CH_2-)- + H_2O \quad \Delta H^0_{298} = -204\,kJ/mol \tag{5.1}$$

$$2CO + H_2 \rightarrow -(-CH_2-)- + CO_2 \quad \Delta H^0_{298} = -165\,kJ/mol \tag{5.2}$$

Side reactions under operating conditions include

$$CO + H_2O \rightarrow H_2 + CO_2 \quad \Delta H^0_{298} = -39.8\,kJ/mol \tag{5.3}$$

$$CO + 3H_2 \rightarrow CH_4 + H_2O \quad \Delta H^0_{298} = -247\,kJ/mol \tag{5.4}$$

$$2CO \rightarrow C + CO_2 \quad \Delta H^0_{298} = -172\,kJ/mol \tag{5.5}$$

$$xCO + yM \rightarrow M_xC_y \tag{5.6}$$

FTS is a highly exothermic reaction. Thermodynamically, the formation of methane (Eq. 5.4) and higher hydrocarbons (Eqs. 5.1 and 5.2) is energetically favorable. Equilibrium calculations show that methanation reaction is thermodynamically more favorable than formation of C_{2+} hydrocarbons (Anderson, 1984). Consequently, an effective catalyst should kinetically control the former in favor of formation of higher molecular weight products. In addition to the catalyst type, operating conditions are effective on product selectivity. As a consequence, according to operating temperature, two general modes of FTS exist. The low-temperature FTS (220–250°C), is

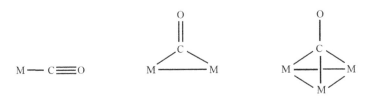

FIGURE 5.1 Some common geometries of CO coordination with metals (M).

used for heavy linear alkanes, while high-temperature FTS (320–350°C) is used for the production of light olefins, gasoline and oxygenates. The operating pressure is typically 20–30 bar in either case.

FTS can be considered as a polymerization reaction in which a C_1 unit is added to the growing chain. The primary products are α-olefins and paraffins with the former can participate in secondary reactions. The main factors affecting CO conversion and product selectivity are the catalyst type, reactor type and operating conditions (Zhang et al., 2010).

5.3.1 CATALYSTS

Most of Group VIII (Groups 8–10 according to IUPAC) metals are active in catalytic hydrogenation of CO giving methanol, methane and higher hydrocarbons as the characteristic products. Vannice (1975) obtained the following order for average molecular weight of products: Ru > Fe > Co > Rh > Ni > Ir > Pt > Pd.

Only Ni, Co, Fe and Ru show sufficiently high activities for commercial uses. Nickel is not a suitable active component due to its high hydrogenation activity which results in high methane selectivity. Furthermore, under typical reaction conditions, Ni forms volatile carbonyls which result in active metal loss and problems in downstream. Ru is highly active, but is not used for large-scale applications because of its high price. Thus, only iron and cobalt are used in commercial catalyst formulations (Dry, 2004). Iron-based catalysts can be used in bulk form. However, cobalt catalysts are used as supported form because of the high price of Co. Both catalysts require chemical promotion to achieve high selectivity and stability. They should be reduced to a metallic form by hydrogen before being used for FTS.

Commercial catalysts can be classified to the following groups according to active metal type and its introduction:

- Fused iron catalysts
- Precipitated iron catalysts
- Supported cobalt catalyst

5.3.1.1 Iron-Based Catalysts

Metallic iron shows very low activity in the FTS reaction. However, under reaction conditions, it gradually transforms into active form. This activation process may take, for example, several weeks to achieve optimum activity in the case of the iron catalyst used in medium-pressure FTS. During this period, called carburization, phase changes along with formation of surface carbon, carbides and oxides takes place. Further structural changes of the catalyst occur during its lifetime, affecting its activity and selectivity (Olah et al., 2018).

It is generally believed that the active phase of iron catalysts in FTS is associated with iron carbides. The formation of these carbides strongly depends on catalyst formulation, activation process and the operating conditions. XRD and Mössbauer absorption spectroscopy have been used to identify iron nanoparticles and several carbide phases associated with FTS which were reported as ε-Fe_2C, ε'-$Fe_{2.2}C$, Fe_7C_3, χ-Fe_5C_2 and θ-Fe_3C (cementite). The predominant carbide phases found, however, are the monoclinic Hägg carbide (χ $-Fe_5C_2$) and the hexagonal ε'–$Fe_{2.2}C$ carbide. Debate emerged regarding the activity and selectivity of individual iron carbides (Wezendonk et al., 2018).

Promoters play a particularly important role in the performance of iron-based FT catalysts (Zhang et al., 2010). For iron catalysts, alkalinity of the sites is important and the order of effectiveness is Li < Na < K < Rb which depends on its amount and support type as well (Dry, 2002). Alkali metal ions, especially K^+, are the most common promoter of Fe-based catalysts acting as electronic promoters by affecting electronic character of Fe. They can modify the activity and selectivity by enhancing CO adsorption and inhibiting that of H_2. This results in a slight increase in catalyst activity, increased 1-alkene selectivity, increased chain growth probability and improved resistance of iron towards oxidation by the resulting water product (Gaube and Klein, 2008).

Copper is another additive in precipitated iron catalysts. It does not affect selectivity, but enhances the reducibility of iron during catalyst activation phase. Thus, lower reduction temperatures become possible thereby causing less sintering (Pendyala et al, 2015).

In precipitated iron catalysts, K and Cu promote activity and selectivity while SiO_2 and Al_2O_3 act as structural stabilizers. The active phase is apparently carbide. However, iron oxides which are active in WGS (water gas shift) are also formed. Consequently, flexibility in feed H_2/CO ratio is achieved. Due to low H_2/CO in coal-based syngas, it is used only in combination with iron.

In preparation of the fused iron catalyst, an alkali promoter K_2O for improving performance and structural promoters (Al_2O_3 and MgO) for increasing surface area are added to a bath of molten magnetite (1,500°C). The melt is cooled rapidly, crushed and reduced (400°C) before being used in fluidized bed reactors.

For a fixed bed application, iron is dissolved in nitric acid and promoters (alkaline and copper) are added into the solution. The precipitate is formed by the addition of ammonia or sodium carbonate which is subsequently filtered, dried and formed (e.g., by extrusion). For slurry reactors where microspherical particles are required, drying and shaping occur simultaneously in a spray dryer.

At $T > 340°C$, iron shows a unique high selectivity to light olefins and low selectivity to methane. Its applicability for heavy wax is limited due to the tendency for forming elemental carbon and catalyst deactivation. The large amount of water formed is also inhibitory to activity and deactivates the catalyst by oxidizing iron species and thus lowers the conversion. Thus, it is necessary to reduce P_{H_2O} in reaction medium e.g., by using series reactors with inter-stage water removal. Therefore, high recycle ratios are necessary after the removal of water.

5.3.1.2 Cobalt Catalysts

Cobalt is the favored active metal for synthesis of long-chain hydrocarbons because of its high activity and high selectivity to heavy linear alkanes. The active phase is metal and cobalt has low tendency to carbidizing. Cobalt shows low WGS activity. Water is the main oxygenated product which affects the activity and selectivity, although some oxygenates are produced as well. N_2, CO_2 and CH_4 when present in the feed are usually regarded as inert (Tsakoumis et al., 2010).

In contrast to iron, olefins tend to re-enter in chain growth process by re-adsorption on Co catalysts which increase the chain length.

Because of its high price, cobalt is usually dispersed on high surface area oxide supports. Cobalt catalysts are typically synthesized by impregnation or precipitation with incipient wetness impregnation of the support with a cobalt salt solution. Cobalt(II) nitrate ($Co(NO_3)_2$) is the most commonly used cobalt salt (Khodakov et al., 2007). The next steps are drying, calcination and reduction with H_2. Typical Co loading is 10–30 wt.%. Cobalt is usually poorly dispersed on oxide supports.

Different materials such as alumina, silica, titania, zeolites and carbon materials have been used as support in preparation of the catalyst. Alumina has been widely used in commercial catalyst due to its desirable and controllable properties. The support strongly influences the performance of the final catalyst in FTS through interacting with metal affecting the structure and electronic density of the metal particles (Gholami et al., 2020).

The reduction of small Co particles over Al_2O_3 and SiO_2 supports is difficult due to strong metal-support interactions (SMSI) leading to the formation of mixed metal-support compounds. The addition of noble metals of group VIII such as Pt, Rh and Ru has beneficial effects on catalyst activity by facilitating reduction of CoO species. Over the Co catalyst, FTS is structure sensitive and the turn-over number (TON) decreases for particle size smaller than ca. 8–10 nm (Melaet et al., 2014). This effect could not be attributed to SMSI as it has been observed over inert supports such as carbon materials (Bezemer et al., 2006).

In general, Co is less strongly affected by chemical and structural promoters (Dry, 2002). Pt, Re and Ru are used to inhibit carbon or oxide formation which lead to deactivation. Structural promotes such as Mn, Zn, B and Si are not uncommon. Other promoters include some rare earth metal oxides (Zhang et al., 2010).

The application of alkali metals as promoters for cobalt-based catalysts is rare as they decrease the FTS reaction rate.

La(Co,Fe)O_3 perovskites have also been investigated as precursors of FTS catalysts (Bedel et al., 2003). Perovskites are mixed oxides of the general formula ABO_3 in which A^{3+} and B^{3+} can be partially substituted without changing the crystal structure. They have been used as catalysts, particularly in oxidation reactions.

In La-based La(Co$_x$Fe$_{1-x}$)O_3 perovskite, the crystal system is orthorhombic for $x < 0.5$ and rhombohedral for $x \geq 0.5$ with the latter cannot be used as FTS catalyst precursor because reduction only results in complete reduction of Co^{3+} into Co^{2+} (Bedel et al., 2003).

Over La$_{(1-y)}$Co$_{0.4}$Fe$_{0.6}$O$_{3-\delta}$ perovskite-based catalyst, a high selectivity of C_2–C_4 olefins was obtained (Bedel et al., 2005). The partial reduction of the perovsike generated small Co^0 particles in interaction with a deficient perovskite.

The effect of Co substitution in LaFe$_{1-x}$Co$_x$O$_3$ ($x = 0.0, 0.1, 0.2, 0.3, 0.4, 0.5$ and 1.0) perovskite prepared by amorphous citrate precursor method and its consequence on FTS with simulated biomass-derived syngas (H_2/CO/CO$_2$/CH$_4$/N$_2$ molar ratio of 32%/32%/12%/18%/6%, respectively) was studied by Escalona et al. (2010). The highest CO conversion was obtained for $x = 0.2$ which was attributed to the formation of cobalt segregates on the surface as revealed by XRD results. In addition to C_6–C_{18+} hydrocarbons, large amount of methane (36.7 mol %) was also formed. On the other hand, pure LaCoO$_3$ perovskite sample showed the highest degree of poorly dispersed Co that favored the formation of short chain (C$_8$–C$_9$) hydrocarbons due to larger Co segregated particle size. Temperature programmed reduction (TPR) results illustrated that the pure LaFeO$_3$ sample, in contrast, was essentially irreducible with no FTS activity.

5.3.1.3 Ruthenium Catalysts

Ruthenium-based catalysts are most active FT catalysts. Ruthenium is capable of formation of methane at temperatures as low as 100°C, whereas high MW (molecular weight) hydrocarbons are formed at higher pressures (*P*>100 bar). At low temperatures (100°C) and very high pressures (1,000–2,000 bar), the product is termed polymethylene (MW>1,000,000 g/mol) which is equivalent to high-density polyethylene (Pichler and Buffleb, 1940). Ru is active as metal and do not need stabilizing promoters. Carbides of Ru are unknown at typical FTS conditions (Adesina 1996). High molecular weight waxes are produced at temperatures as low as 150°C (423 K). An advantage is that it is active under high water and oxygen pressures, which is attractive for converting biomass-derived syngas (Eslava et al., 2017). As for Co, it is not active in RWGS (reverse water gas shift).

The high price of Ru excludes it from commercial applications; nevertheless, it is widely used in academic research because of its high activity, high probability of chain growth and operating under milder conditions.

5.3.1.4 Support Effects

The selection of an appropriate support is an important step in the design of FTS catalysts. Active component dispersion, electronic modification and metal support interaction are important considerations being affected by support texture and acidity (Wang et al., 2013). The desirable properties of the support in CO hydrogenation include high surface area for high dispersion of the active component, proper interaction with active metal for facilitating its reduction and resistance to sintering during preparation and reaction, high porosity for improved mass transfer, high heat conductivity for dissipation of heat of reaction and attrition resistance (Cheng et al., 2017).

Metal oxides such as Al$_2$O$_3$, SiO$_2$, TiO$_2$ and ZrO$_2$ have been conventionally used especially for supported Co catalysts (Cheng et al., 2017). Novel supports such as nanocarbon material, N-doped carbon, SiC and metal organic framework (MOF) have recently received attention as potential supports. Over conventional supports, the strong precursor-support interaction causes difficulty in reduction of active metal oxide. This interaction is much lower with carbon material.

Reuel and Bartholomew (1984) compared the performance of different supports for cobalt catalysts. The order of activity for 3 wt% loaded cobalt catalysts (at 250°C, 1 atm) was MgO < carbon < Al$_2$O$_3$ < SiO$_2$ < TiO$_2$, while the order for interaction was SiO$_2$ < Al$_2$O$_3$ < TiO$_2$.

Acidic zeolites have been studied as support for increasing isoalkane selectivity in FTS. However, these were found to be not suitable due to their low activity and high methane yields (Li et al., 2004).

5.3.1.5 Comparison

Table 5.2 compares Fe and Co catalysts for FTS. Cobalt is more expensive, but more durable. For the conversion of carbon-containing feedstocks into liquid fuels via synthesis gas (XTL) applications, cobalt is best suited for gas to liquid (GTL) applications. Iron is more suited for biomass and coal-derived syngases because of their low H_2/CO_x ratios.

5.3.2 MECHANISM AND KINETICS

FTS can be viewed as a polymerization of CO under reductive conditions in which C_1 units are added to the growing chain (Tsakoumis et al., 2010). 1-Alkenes, the main isomeric olefins obtained, and alcohols are considered to be the primary products (Olah et al., 2018). The former can participate in secondary reactions complicating the reaction path.

The product distribution could be predicted by assuming chain growth via addition of one carbon at each time to the growing chain. Thus, FTS includes the three basic steps initiation, growth and termination (Olah et al., 2018). Accordingly, a mechanism should entail activation of CO, formation of monomer species and addition of the monomer to the growing chain.

There is a considerable controversy about FT synthesis mechanism. Several mechanisms have been proposed for primary reactions differing in chain initiator

TABLE 5.2
A Comparison of Cobalt and Iron Catalysts in FTS

Parameter	Cobalt Catalyst	Iron Catalyst
Feed gas H_2/CO (mol/mol)	Limited (2.0–2.3)	Wider (0.5–2.5), because of WGS activity
Operating temperature (°C)	190–240	200–350
Side reactions	Negligible	WGS
Operating pressure (bar)	5–15	10–60
Activity at high conversions	High (no water product inhibition)	Low (water product inhibition)
Product spectrum	Mostly n-paraffins, also α-olefins	n-paraffins+ α-olefins, also oxygenates
Chain growth probability (α)	>0.94	>0.95
Promoters	Nobel metals, oxides (e.g., La_2O_3, ZrO_2)	Alkali metals
Cost	More expensive (~1,000 times)	Cheap
Lifetime	Durable	Shorter life (coke deposition, attrition)
Sulfur tolerance (ppm)	>0.1	>0.2

Sources: Khodakov et al. (2007); Mehariya et al. (2020).

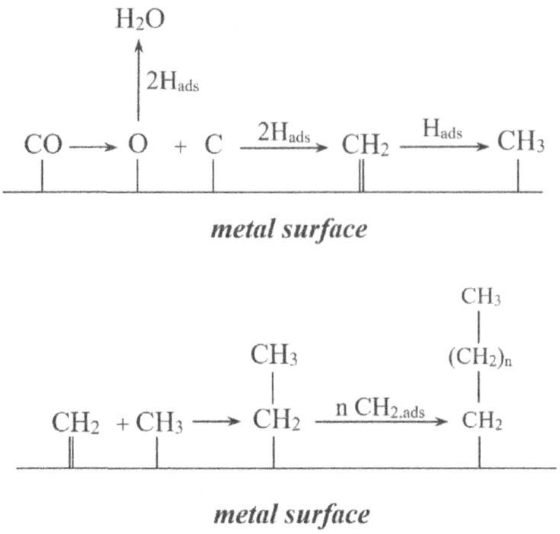

FIGURE 5.2 Schematic representation of carbide mechanism (Redrawn from Franssen et al., 2011).

and monomeric species. Three principal routes are: the carbide (or alkyl) mechanism, the hydroxymethylene (or enolic) mechanism and CO insertion mechanism and their variants (Olah et al, 2018).

The carbide mechanism was first proposed by Fischer and Tropsch (1926a, b) via carbidizing of the metal surface of the catalyst by CO followed by hydrogenating of the C atoms to methylene groups, which then polymerized to hydrocarbons. It is generally accepted that CO is adsorbed dissociatively and forms surface C* and O* species which subsequently hydrogenated to the adsorbed CH_x* and OH* species, respectively. CH_3* is the chain initiator and CH_2* is the surface monomer species. Experimental evidence of stepwise insertion of syngas-derived CH_x units for carbide mechanism was obtained by Biloen et al (1979). By dissociating labeled [13]CO over Ni, Co and Ru catalysts, and then exposing the surface to syngas with [12]CO, they obtained abundant [13]CH_4 product and hydrocarbons with more than one ^{13}C incorporated in their growing chain illustrating that the oxygen free species CH_x ($x = 0$–3) from syngas were the possible intermediate in methanation and they were capable of being incorporated into the growing chain. This mechanism, however, cannot give information about oxygenates formation. Figure 5.2 depicts the steps of carbene mechanism.

To explain oxygenate formation, enolic mechanism encompasses a surface enolic (–CHOH) species resulted from hydrogenation of adsorbed CO as the initiator of the surface reaction. The enolic species also involves in propagation by linking up of neighboring monomers with the elimination of water to form -COH-CH_3 species which can hydrogenate to an alkene and H_2O (Figure 5.3). The primary products are thus oxygenates and α-olefins.

The alternative CO insertion mechanism could also account for the formation of alcohols was proposed by Pichler and Schulz (1970). According to this mechanism, the adsorbed CO* acts as the monomer by continuous addition into metal-alkyl bond of the growing chain initiated as CH_3*. The initiation step is

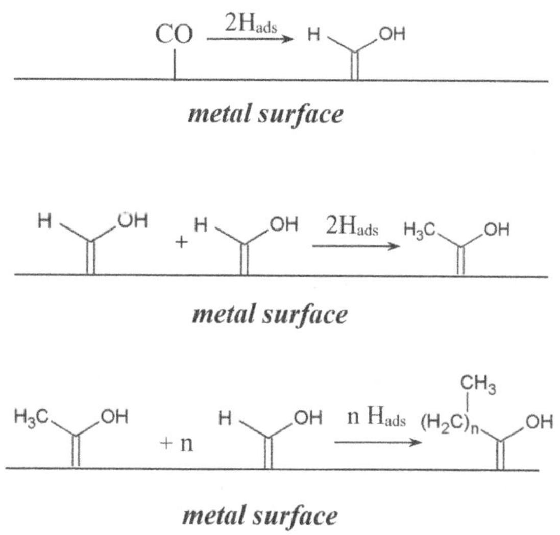

FIGURE 5.3 Schematic representation of hydroxyl-carbene mechanism (Redrawn from Franssen et al., 2011).

insertion of a CO* to an M–H bond. The termination step as α-olefin or paraffin is similar to the carbide mechanism but alcohol formation is via surface acyl (C_nH_zCO*) intermediates (Figure 5.4). This mechanism predicts oxygenate formation better than others.

The carbide mechanism is supported by a large number of studies whereas evidences for hydrocarbon formation from enol and CO insertion mechanisms are

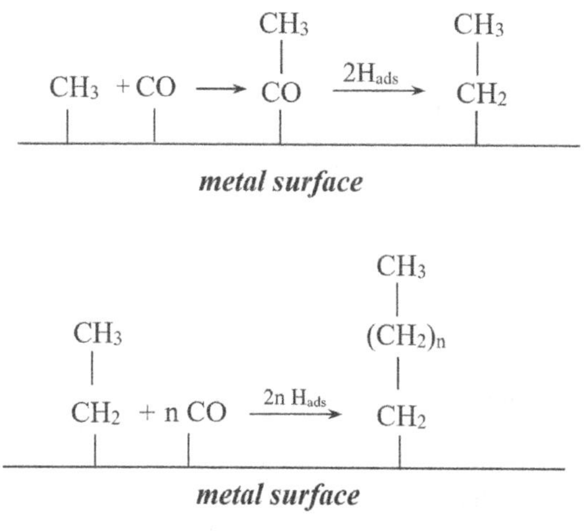

FIGURE 5.4 Schematic representation of CO-insertion mechanism (Redrawn from Franssen et al., 2011).

poor although they may be possible routes for aldehyde and alcohol formation (Bartholomew and Farrauto, 2006). Currently, the carbide-type mechanism is the most widely accepted one. Nevertheless, some authors proposed that more than one mechanism might act to account for the very complex composition of the products (Hindermann et al., 1993).

If the desorbed primary products interact with another active site, secondary reactions occur (van der Laan, 1999). For α-olefins, the probable secondary reactions are (i) n-paraffins production by hydrogenation, (ii) isomerization, (iii) cracking and hydrogenolysis, (iv) growing chain insertion that is mostly effective for C_2H_4 and C_3H_6, and (v) hydrocarbon chains readsorption and initiation (Novak et al., 1982).

The product analysis by Schulz (1939) showed that the primary product distribution is similar to that of polymerization process, that is, ASF (Anderson-Schulz-Flory).

The statistics of chain length shows a linear relation between the mole fraction of the molecules containing n carbon and carbon number n known as ASF distribution, originally developed for free radical polymerization (Olah et al., 2018). Similar treatment was used for FTS with additional assumption that unbranched 1-alkenes are the primary products (Friedel and Anderson, 1950):

$$F_n = \frac{n(1-\alpha)\alpha^{n-1}}{\sum\limits_{i=1}^{\infty}\alpha^i} = n(1-\alpha)^2\alpha^{n-1} \tag{5.7}$$

where F_n is the fraction of hydrocarbons in the product with carbon number n and α is the probability that an intermediate undergoes chain growth. For α independent of chain growth

$$\alpha = r_p/(r_p + r_t) \tag{5.8}$$

In which r_p and r_t are chain growth and chain termination rates, respectively. If waxes are the desired products, the α value should be larger than 0.9.

The product distribution typically shows deviations from ideal ASF distribution with a maximum for methane, a minimum for C_2 products and a monotonous increase in chain growth probability with hydrocarbon chain length for higher products ($n>10$) (Figure 5.5).

The exceptions have been explained by thermodynamic favorability of C_1 formation compared to C–C bond formation, or kinetic favorability of C_1 hydrogenation compared to chain termination for methane (Schulz, 2003) or a different formation path (Schulz, 2013). For C_2 minimum, strong binding of ethylene and easier growth compared to higher hydrocarbons has been proposed. The deviation for C_{10+} compounds has been treated by two types of active sites and thus two α values or higher diffusion limitation for higher products which increase their readsorption.

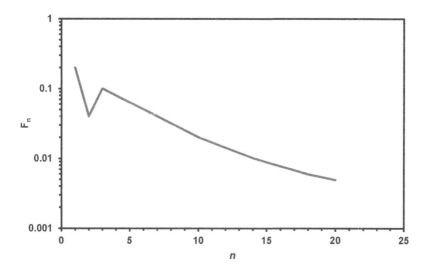

FIGURE 5.5   Classical ASF distribution with anomalies at C_1, C_2 and C_{10+} products.

Kuipers et al. (1995) reported exponential growth of paraffin/1-olefin ratio with the chain length, attributed to chain length-dependent physisorption of olefins (van der Laan and Beenackers, 1999) or heat of dissolution (Schulz et al., 1988; Tau et al., 1991) or diffusion in wax solution. The paraffin-to-olefin ratio depends on hydrogenation propensity of the Fischer–Tropsch catalyst with cobalt exhibiting higher hydrogenating activity than iron (de Klerk, 2008).

Dry (1976) derived the rate of Fischer–Tropsch for iron catalysts by combined enol/carbide mechanisms assuming that the adsorption of CO and H_2O are much more stronger than that of CO_2 and H_2:

$$rate_{FT} = kP_{H2}P_{CO}/\left(P_{CO} + aP_{H2O}\right) \qquad (5.9)$$

This kinetic expression illustrates the inhibitory effect of water in FTS over iron catalysts.

Yates and Satterfield (1991) used a bimolecular surface reaction with the Langmuir–Hinshelwood mechanism and derived the following rate expression for cobalt catalysts in slurry reactors:

$$rate_{FT} = kP_{H2}P_{CO}/\left(1 + K_I P_{CO}\right)^2 \qquad (5.10)$$

Eqs. (5.9 and 5.10) are established kinetic equations for overall syngas consumption rates of FTS over iron and cobalt catalysts, respectively.

For predicting the selectivity, the rate of formation of individual hydrocarbons (and WGS for iron catalysts) is required. The kinetics of FT synthesis reactions over different catalysts have been reviewed by van der Laan and Beenackers (1999) and more recently by Méndez and Ancheyta (2020).

5.3.3 CATALYST DEACTIVATION

The FTS catalysts are prone to deactivation by several mechanisms. These can be summarized as follows (Bartholomew and Farrauto, 2006):

1. Build-up of carbonaceous deposits which can block active sites
2. Oxidation of active phase to a less active or inactive phase by product H_2O and CO_2 (e.g., oxidation of Fe_5C_2 to Fe_3O_4), formation of inactive metal-support compounds
3. Poisoning by sulfur and/or nitrogen compounds in the feed
4. Reduction of active surface area by hydrothermal sintering of active phase (of minor importance for fused bulk iron catalysts)
5. Catalyst attrition

Both iron and cobalt are permanently poisoned by sulfur compounds (>0.02 mg/dm^3) (Dry, 2002).

The primary impact of deactivation on catalyst performance is the loss of activity with time. The effect on selectivity is more subtle some of which can be anticipated. In general, the selectivity shifts towards lighter compounds with increased methane selectivity. The alkene content of the product, however, depends on increasing or decreasing hydrogenation activity with time-on-stream.

In LTFT (low temperature FT) precipitated iron catalysts, sintering and oxidation are more important causes of deactivation (Bukur et al., 2016). XRD analysis of fresh and spent catalysts showed crystal growth with time (surface area reduction from 200 to 50 m^2/g corresponding to 20% activity loss; Dry, 2003). In a series of tests in tubular fixed-bed reactors of Sasol R&D pilots, the deactivation was evaluated along the bed length by careful unloading and analysis of the catalysts. The spent catalyst showed highest deactivation in the entrance and exit of the catalyst bed. The least deactivation was observed in the middle of the bed. The BET surface area of (dewaxed) catalysts showed progressive decline with increasing porosity and Fe_3O_4 level along the bed. The sulfur content of the catalyst decreased along the bed. Thus, it was concluded that the main cause of the deactivation was sulfur-poisoning at the entrance and sintering and oxidation at the exit where the highest H_2O partial pressure occurred.

In HTFT (high temperature FT), carbon deposition is the main cause of catalyst deactivation. It can cause disintegration of the catalyst which increase crushing and attrition of the catalyst particles. The carbonaceous products are predominantly aromatic in nature. Boudoward reaction results in carbon deposition which after a few weeks superpass iron on an atomic basis. Sintering has a minor role in deactivation of fused iron HTFT catalysts.

Because of its high price, deactivation is a major challenge in the case of cobalt-based FTS catalysts. In addition to poisoning, fouling and sintering; oxidation or formation of mixed cobalt-support compounds and formation of cobalt carbide (Co_2C) could play a role (Tsakoumis et al., 2010).

Activity measurement of a Co-based catalyst with time in a demo plant showed two distinct stages of deactivation (Van Berge and Everson, 1997). The initial deactivation rate was rapid but it leveled out with time. The former could be due to accumulation of long-chained waxes inside the pores and decrease of diffusivity within pores which is reversible and lasts for a few days to a week, while the second (possibly due to sulfur-poisoning) is irreversible and therefore is important from an operation point of view. This change in the deactivation rate with time suggests a combination of several phenomena.

5.3.4   Reactor Options

The FTS reaction is highly exothermic and produces a wide spectrum of products ranging from noncondensable gases to liquid and solid hydrocarbons. Furthermore, it should be operated in relatively narrow temperature range. These features should be considered in the development and design of the reactor for FTS. Therefore, heat removal and temperature control are important issues in design of commercial reactors. Otherwise, temperature rise could result in high selectivity of methane and lower hydrocarbons and accelerating catalyst deactivation. According to the operation temperature and process technologies, the operating temperature ranges from ~220–240°C, 270–280°C and ~350°C for low, medium and high temperature operations, respectively, with a pressure ranging from 25 to 45 bar (Chorkendorff and Niemantsverdriet, 2017).

The reactors commercialized for FTS include (van Der Laan and Beenackers, 1999, Figure 5.6):

- Fixed bed (typically multitubular) reactors (e.g., in Arge: Sasol, SMDS: Shell)
- Slurry phase reactors (e.g., in SSPD: Sasol, Gas Cat: Energy International, AGC21: Exxon)
- Circulating fluidized bed reactors (Syntol: Sasol)
- Fluidized bed reactors (SAS: Sasol)

Iron catalysts are employed in the high-temperature mode for the synthesis of gasoline and linear low molecular weight olefins. On the other hand, low-temperature processes utilize either iron or cobalt catalyst for the production of high molecular weight linear waxes (Ail and Dasappa, 2016).

The fixed bed and slurry reactors are used for low-temperature operations for the production of heavy linear alkanes with Fe or Co catalysts, whereas fluid-bed type reactors are used for high- temperature operation with Fe catalyst for the production of α-olefins or gasoline-range hydrocarbons.

5.3.5   Refining and Upgrading FT Products

FTS is non-selective and produces hydrocarbon products with a wide range of molecular weights collusively called synthesis crude oil or syncrude. Therefore, they should be further refined like conventional crude to obtain high-quality liquid fuels such as gasoline, jet fuel and especially diesel. This should be performed in separate facilities in down-stream of the GTL plant. The wax of LTFS is predominantly linear paraffins

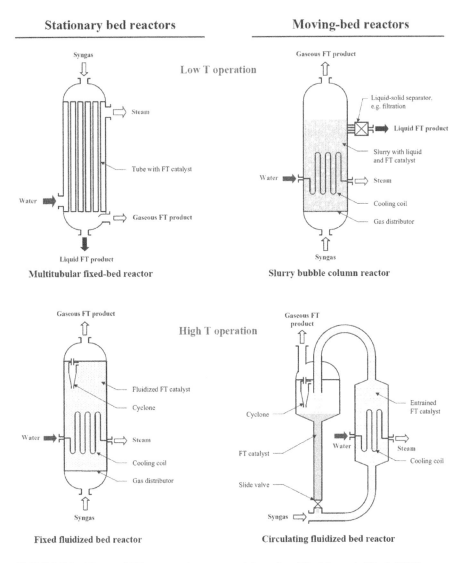

FIGURE 5.6 Types of FT reactors in commercial use (modified from de Klerk 2013).

with small fraction of olefins and oxygenates. The hydrocraking of paraffins and hydrogenation of olefins can be done under relatively mild operating conditions.

For applying the catalyst developed for conventional hydroprocessing to syncrude, one should consider the differences between the two especially in hydrocarbon compositions and heteroatoms. Syncrude virtually lacks sulfur and metallic compounds. Thus, when applying metal sulfide-based hydroprocessing catalysts, sulfur containing compounds should be added to the feed to avoid desulfurization of the catalyst. On the other hand, the carboxylic acid containing streams could cause corrosion problems and metal leaching (de Klerk, 2008).

All around the world, there are few commercial Fischer–Tropsch syncrude-based refineries. Sasol Synfuels East & West (Secunda, South Africa), Oryx GTL (Ras Laffan, Qatar), Shell Middle Distillate Synthesis (Bintulu, Malaysia), Sasol 1 (Sasolburg, South Africa) and PetroSA (Mossel Bay, South Africa) are examples (de Klerk, 2008).

5.3.6 PROCESS TECHNOLOGIES

FTS of hydrocarbons is now a well-established process. This process comprises syngas generation, FTS and product upgrading, with the first step being most costly. The heart of a particular process resides in the FTS reactor.

In 1954, ARGE (consortium) Ruhrchemie/Lurgi developed the ARGE process that has a water-cooled tube bundle reactor (id 46 mm, length 12 m) using precipitated $Fe-K_2O-SiO_2$ catalyst. The process was designed in a way to produce maximum yields of diesel range motor fuels. The operating conditions are 220–260°C temperature, 2–3 MPa pressure, H_2/CO ratio of 1.3–2.0 with GHSV of syngas 500–700 h^{-1}. By this process, syngas conversion of 65%–70% is reported (Cornils et al., 2013).

BP developed a fixed-bed Fischer–Tropsch process with modified and coprecipitated Co catalysts that has in situ catalyst regeneration.

The Synthol process is an HTFT one (~340°C) using fused iron catalyst in a circulating entrained-flow fluidized bed reactor. Despite high operation temperature, the selectivity to methane remains minimal (Schulz, 1999).

The improved Synthol (Kellogg process) technology, called Advanced Synthol process (SAS process), uses a conventional fluidized-bed reactor with fused Fe catalysts. This reactor configuration is less complex in comparison to a circulating fluidized bed.

The SASOL Slurry Phase Distillate (SPD) process is an alternative SASOL LTFT development to ARGE technology. The SPD process uses a slurry bubble column reactor with Co catalyst for the manufacture of middle distillates for subsequent cracking. The SPD process, being scheduled for Qatar, is claimed to be much simpler than the conventional ARGE reactor (Ras Laffan project) (Cornils et al., 2013).

Shell Middle Distillate Synthesis (SMDS) is a two-stage LTFT process that mainly produces middle distillates (with boiling point of 150–400°C, C_9–C_{25}), i.e., kerosene and gas oil. The first stage is heavy paraffin synthesis (HPS) via FTS in a fixed-bed reactor followed by hydroisomerization and hydrocracking in a trickle bed reactor, yielding LPG, naphtha, and predominantly kerosene and gas oil (Dry, 2003). A water-cooled, ARGE-type multi-tubular reactor with Zr-promoted Co catalyst on SiO_2 supports is used for HPS step. This catalyst has been specifically developed for high selectivity to long-chain hydrocarbons with CO conversions of >95% at 230°C temperature and 2.6–3 MPa pressure.

Currently, the SAS and SMDS are the two technologies with highest installed capacity worldwide (de Klerk, 2013).

5.4 MODIFICATIONS OF FTS

A number of FTS variants processes have been developed which differ with conventional FTS in feed and catalysts used and therefore in product composition.

5.4.1 Kölbel–Engelhardt Process

In the Kölbel–Engelhardt (KE) reaction, steam is used as the source of hydrogen for conversion of CO into hydrocarbons:

$$3CO + H_2O \rightarrow -(-CH_2-)- + 2CO_2 \quad \Delta H_{297}^0 = -245\,kJ/mol \quad (5.11)$$

This reaction enables utilizing low-value CO-rich gases such as coke oven gas for hydrocarbon synthesis.

Although there are reports implying direct reaction of CO with the adsorbed H_2O or OH, most of the works indicate that the overall reaction is a combination of WGS followed by FTS (Niwa and Lunsford, 1982). Thus, catalysts having both functionalities can be employed in KE reaction. The product distribution, however, is significantly different from that of FTS (Olah et al., 2018).

Active metals in FTS can be used in catalyst formulation. Fe, Co, Ni and Ru slurried in the long-chain hydrocarbon products are used at 240–280°C and 20 bar. The process has been demonstrated on a pilot plant scale (Cornils et al., 2013).

5.4.2 Isosynthesis

Isosyhtesis process is a variant of FTS which is unique for its selective conversion of syngas into branched hydrocarbons, mainly $C_4–C_8$ paraffins. The main reason for developing this process was production of isobutane for producing high-octane gasoline by alkylation to isooctane (Shah and Perrotta, 1976).

Certain tetravalent nonreducible oxides such as ThO_2, ZrO_2 and CeO_2 exhibit high activity and selectivity in producing isobutene under severe condition in the range of 375–475°C and pressures 300–600 bar. Lower temperatures enhanced alcohol synthesis whereas higher temperatures favored DME and methane formation (Olah et al., 2018).

The catalysts used for isosynthesis are based on practically nonreducible tetravalent oxides (ThO_2, ZrO_2 and CeO_2) and their mixed oxides like ThO_2-ZrO_2, ThO_2-ZnO and ThO_2-Al_2O_3. Despite its higher activity, ThO_2 use is restricted due to its radioactivity (α emmitor) and most research works are devoted to ZrO_2 (Wu et al., 2019). Alumina and other oxides (e.g., chromia, lanthana, titania and manganese oxide) exhibit inferior activity and isobutane selectivity (Shah and Perrotta, 1976). However, Al_2O_3 promotes the activity of ThO_2. The catalysts can be synthesized by precipitation method from nitrate solutions with hot Na_2CO_3.

The test results of micron- and nanoscale zirconia and ceria in isosynthesis showed that the acid–base properties, crystallite size and crystal phase significantly affect the catalytic performance (Tangchupong et al., 2010).

Zirconia is versatile because of its acidic-basic properties, redox properties and different crystalline phases. Zirconia shows high selectivity to i-C_4 paraffins (Postula et al., 1994). ZrO_2 with 7% CeO_2 shows fairly high conversions (25%–40%) and reasonable selectivities (about 20%). Ca salts such as CaF_2 and $CaSO_4$ have been used to increase i-C_4 selectivity (Li et al., 2001).

TABLE 5.3

A Comparison of FTS and Isosynthesis Processes

Parameter/Process	FTS	Isosynthesis
Main product(s)	Normal paraffins	Isoparaffins
Product distribution	ASF	Non-ASF
T and P	Moderate	High
Catalyst reducibility	Reducible	Nonreducible
Catalyst regenabilirty	Non regenerable	Regenerable by oxidation
Sulfur tolerance	Poisonous	Non sensitive

An influence of strong basic active sites has been proposed (Cornils et al., 2013). Basic sites are effective in increasing isobutene selectivity whereas acidic sites increase CO conversion (Su, 2000). A large basic/acidic site ratio has been proposed for achieving high i-C_4 selectivity over ZrO_2-based catalysts (Feng et al., 1995).

The main reaction pathway of CO to isobutene over ZrO_2 was CO insertion and aldol condensation/dehydrogenation (Maruya et al., 2000). Tracer studies with ^{13}C showed that the methoxide species produced by CO hydrogenation gave central carbon of isobutene (Maruya et al., 1996). The methoxide was then reduced to surface methyl species. The insertion of CO to methyl-metal bond formed surface acetyl group. The aldol condensation of C_2 oxygenates with formaldehyde results in formation of i-C_4 products.

Table 5.3 summarizes a comparison between FTS and isosynthesis processes.

A newer approach is incorporating acidic site to FTS catalysts to form bifunctional catalysts such as Ru/alkali promoted zeolites (Oukaci et al., 1987) or hybrid catalysts such as Ru/NaY sulfated zirconia (Song and Sayari, 1994). The selectivity to isoalkanes is attributed to transformation of the primary olefin products of FT on acid sites. However, acidic zeolites are not appropriate supports for FTS catalysts because of low catalytic stability, resulting in low activity and high methane selectivity.

Using supercritical n-butane as reaction medium, light isoparaffin selectivity significantly increased and the formation of the byproduct methane was suppressed with a hybrid Co/SiO_2 and HZSM-5 catalyst. The hydrogenolysis and isomerization reactions were promoted in supercritical-phase, possibly because of the high solubility of heavy hydrocarbons and water in the supercritical medium. Consequently, the byproducts of FTS were removed quickly from acidic sites of zeolite; that promoted the hydrogenolysis and isomerization reactions of hydrocarbons (Yoneyama et al., 2008).

5.4.3 SYNTHESIS OF NITROGEN COMPOUNDS

Still another modification of FTS is the direct synthesis of N-containing products (e.g., amines) by co-feeding of simple nitrogen compounds (e.g., ammonia, HCN):

$$nCO + 2nH_2 + NH_3 \rightarrow C_nH_{2n+1}NH_2 + nH_2O, \Delta H^0_{298} = -164n \text{ kJ/mol} \quad (5.12)$$

Thermodynamic considerations (Gibbs free energy change of reaction) shows that the formation of N-compounds is nearly of the same feasibility as that of the corresponding O-compounds, especially for larger carbon numbers (Sango, 2013).

Though not a novel idea, there are relatively few studies on this topic and many are academic researches (Henkel, 2012; Sango, 2013; de Vries, 2017). Nitrogen-containing compounds such as amines, nitriles and amides are more valuable than hydrocarbons and are also important raw materials for chemical synthesis.

Nitrogen compounds have been considered as poisons of FT catalysts. The reported results are contradictory; while some reported reversible poisoning, others reported irreversible poisoning of Co-based catalysts (Pendyala et al., 2013). For iron, precipitated iron catalyst showed high NH_3 poisoning resistance up to 200 ppmw in a slurry reactor (Ma et al., 2015).

Henkel (2012) investigated the effect of addition of NH_3 (0–0.2 bar) into the feed of FTS over bulk iron and Co catalysts in a fixed-bed reactor ($T = 270°C$, $P = 4$ bar, $H_2/CO = 2$, GHSV = 1,920 mlg⁻¹h⁻¹). Over iron catalyst without the K promoter, the addition of NH_3 to the feed increased catalyst activity, chain growth probability and olefin selectivity, acting the same role as K as promoter. A link between O- and N-containing compounds was observed such that the loss of molar values of the oxygenates was compensated, quantitatively, by the molar values of the N-containing compounds. Consequently, it was concluded that the former are precursors of the latter without loss in paraffin products.

Sango et al. (2015) studied the effect of NH_3 (0–10 vol%) addition to the feed on activity and selectivity of FTS on a precipitated iron catalyst in a low-temperature slurry reactor ($T = 250°C$, $P = 50$ bar, $H_2/C = 2.0$, GHSV = 2,250 mlg⁻¹h⁻¹). At lower ammonia concentrations (2–5 vol%), primary ammines were the main N-containing products. By increasing NH_3 concentration, aliphatic nitriles also formed. An important observation was that with the increasing NH_3 from 0 to 10 vol%, the increase of N-containing products was concomitant with the reduction of oxygenate products such that the total selectivity of N- and O-containing products remained fairly constant. It was thus proposed that oxygenates are the precursor of N-containing products. The total N-compounds for 10 vol. % NH_3 addition was 11.8% (in C wt %) with 7.0% amines, 4.6% nitriles and 2.2% amides.

The formation of nitrogen-containing products can also occur by addition of ammonia under KE synthesis reaction:

$$3nCO + nH_2O + NH_3 \rightarrow C_nH_{2n+1}NH_2 + nCO_2 \quad \Delta H^0_{298} = -218n \text{ kJ/mol} \quad (5.13)$$

Over K- and Cu- promoted iron catalysts in a fixed-bed reactor, primary amines were the main N-containing products (Kölbel and Trapper, 1966).

5.5   SYNTHESIS OF HIGHER ALCOHOLS

Higher alcohols (e.g., ethanol, isopropanol and isobutanol) are valuable as chemical feedstocks and as oxygenate additives in gasoline pool. Higher alcohols are the precursors or reagents for production of olefins, cosmetics, detergents, plasticizers and lubricants (Lu et al., 2017).

Higher alcohols have been traditionally produced by the multistep ALFOL process comprising hydrogenation, ethylation, growth, oxidation and hydrolysis and also by hydroformylation of mixed olefins with H_2/CO using homogeneous catalysts which negatively impact on separation of the product. Alternatively, homogeneous catalysts can be used for hydroformylation and then hydrogenation of FT alpha-olefins. Unfortunately, in addition to α-alcohols, undesired branched alcohols are also produced. Therefore, a direct single-step route with low methanol and hydrocarbon yields using heterogeneous catalyst would be highly valuable and has been subject to intensive research studies (Pei et al., 2015).

Higher alcohols are the byproducts of methanol synthesis and FTS (Pallejà, 2018); however, their selective production form syngas in commercial-scale remains a crucial challenge despite substantial research efforts.

5.5.1 CHEMISTRY AND THERMODYNAMICS

Higher alcohols synthesis (HAS) form syngas is exothermic and proceeds with reduction of volume and, consequently, is thermodynamically favored at lower temperatures and higher pressures:

$$nCO + 2nH_2 \rightarrow C_nH_{2n+1}OH + (n-1)H_2O \tag{5.14}$$

Typical synthesis conditions are 550 K (275°C), 10 MPa and $H_2/CO = 2/1$. The side reactions include WGS and hydrocarbon synthesis by the FT reaction.

Most of the catalysts used in HAS are also active in the WGS reaction, and thus are capable of converting the produced water into CO_2. As a consequence, the overall stoichiometry of HAS may be written as

$$(2n-1)CO + (n+1)H_2 \rightarrow C_nH_{2n+1}OH + (n-1)CO_2 \quad n = 1, 2, 3, \ldots \tag{5.15}$$

It is customary to report the selectivity of oxygenates and hydrocarbons on CO_2-free basis to exclude the conversion of CO into CO_2 via WGS reaction.

Under typical synthesis conditions, the equilibrium conversion of syngas gives hydrocarbons, especially methane, with negligible amount of alcohols such as methanol and ethanol. Therefore, the formation of methane and higher hydrocarbons should be kinetically limited by applying appropriate catalyst.

Thermodynamic equilibrium analysis by Gibbs free energy minimization method with Aspen Plus software using Soave–Redlish–Kwong (SRK) equation of state showed the following order for thermodynamic favorability for CO hydrogenation products (Andersson, 2015):

$$CH_4 > \text{paraffins} > i - BuOH > n - BuOH > n - PrOH > EtOH > MeOH.$$

This illustrates FTS is more thermodynamically favorable than HAS and both are more favorable than methanol synthesis.

5.5.2 CATALYSTS

A higher alcohol synthesis catalyst needs both dissociative and molecular adsorption sites for CO (Subramanian et al., 2009). Thus, an optimum catalyst for HAS is bifunctional containing active sites favoring C–C bond formation (chain growth) and C–O bond formation (termination) and sites for hydrogenation (Ao et al., 2018).

Higher alcohols are produced as by-products of methanol and FT syntheses. Therefore, a natural option would be modification of the catalysts of these processes for improved higher alcohol selectivity. The majority of the catalysts used are modified methanol synthesis catalysts, modified FT synthesis catalysts, supported Rh-based catalysts and MoS$_2$-based catalysts. A remarkable and common feature of most catalysts is promotion by alkali metals to achieve higher selectivities or stimulate HA selectivity.

C$_1$–C$_6$ linear primary alcohols are the major products with the modified Fischer–Tropsch and Mo catalysts, while modified methanol synthesis catalysts produce methanol and isobutyl alcohol (Olah et al., 2018).

5.5.2.1 Modified Methanol Synthesis Catalysts

Higher alcohols, particularly isobutanol, are produced over high-temperature (HT) (ZnO/Cr$_2$O$_3$) and low-temperature (LT) (Cu/ZnO) methanol synthesis catalysts promoted with alkali metal oxides. These catalysts produce linear and branch C$_1$–C$_6$ alcohols primarily by aldol condensation or methanol homologation reactions. However, these catalysts exhibit low ethanol selectivity due to the rapid growth of C–C chain of the surface C$_2$ intermediate (Zaman and Smith, 2012).

These catalysts were found in 1930s when it was observed that alkaline impurities in methanol synthesis catalyst favor formation of higher oxygenates. Currently, two main categories of modified methanol synthesis catalysts exist: HT catalysts based on chromia and LT catalysts based on Cu. Cr-based catalysts have longer lifetimes, but also high selectivity of hydrocarbon and have not been studied as Cu-based catalysts which are found to be one of the most active in formation of branched alcohols. The reaction temperature is limited to below 300°C to avoid sintering of copper. Co-promoted catalysts showed enhanced selectivity to higher alcohols especially ethanol. Doping with alkali metals (particularly Cs) and also employing Zn and Zr-based supports (ZnO and ZrO$_2$, respectively) greatly improved the selectivity to higher alcohols (Pallejà, 2018). Over 1.2 wt% Na-doped Co-promoted Cu/ZnO/Al$_2$O$_3$, a 48.2% selectivity was obtained at 280°C, 6 MPa and 9,600 h^{-1} space velocity (Anton et al., 2016).

Another class of catalysts for higher alcohol production is MgO-based Cu catalysts promoted by K (Hilmen et al., 1998).

5.5.2.2 Modified FT Synthesis Catalysts

This type of catalysts was proposed in late 1970s using catalysts containing Co and Cu. It was found that Co favors dissociation of CO and Cu enables CO adsorption and insertion, as a dual site model predicts (Spivey and Egbebi, 2007). Research on Fe-based catalysts started around the same time but was not expanded until early 2000s when Fe was used as promoter of the Cu system. Similar to Co, Fe is assumed to promote dissociative adsorption of CO and offers chain growth sites which in combination with existing Cu provides the two sites required for HAS. Doping with alkali metals is often used to enhance the HA yield. Many supports were found to improve CO conversion and HA selectivity by alkali doping.

The formation of Cu-Co alloys is important for functioning. Thus, to facilitate alloy formation, the spinel structure $CuCo_2O_4$ has been used as precursor because in which Cu and Co are uniformly distributed at atomic level (Fang et al., 2011).

The formation of carbides is well known for Fe and Co FT catalysts. Hägg carbide (Fe_5C_2) is one of the main active carbides in FT whereas cementite (Fe_3C) exhibits lower activity. In contrast, Co_2C formation in FT synthesis is considered as a main cause of catalyst deactivation. The formed carbide impairs dissociative adsorption of CO and slows down HC chain growth rate. For HAS, Wang et al. (2016) showed that $Co-Co_2C$ sites enhance alcohol formation, however, controlling Co_2C formation for adjusting optimal ratio between metallic Co and Co_2C remains a challenge. Different promoters and pretreatment conditions ranging from a few hours to several weeks under CO environment have been reported (Nebel et al., 2019).

Perovsikte mixed-oxides have been used in preparation of Cu-Co catalysts to increase Cu-Co contact. Perovskites are ABO_3 crystal structures where A and B are broad group of cations. They can produce highly dispersed metal phases upon reduction. Tien-Thao (2006) prepared bimetallic catalysts derived from $LaCo_{1-x}Cu_xO_{3-\delta}$ ($x = 0.1–0.3$) prepared by grinding. The optimum value was $x = 0.3$ with Li as the most effective dopant. 26%–49% $C_1–C_7$ alcohols (5%–10% ethanol) was obtained at 275°C, 1000 psi (70 bar) and VVH = 5,000 h^{-1}. However, a large amount of hydrocarbons ($\approx 50\%$) was also obtained. The catalysts were susceptible to sintering because of low surface areas.

Bimetallic core-shell catalysts (e.g., Co @ Cu) have recently received attention due to their high surface areas and multiple surfaces (Subramanian et al., 2009).

5.5.2.3 Mo-Based Catalysts

These catalysts comprise MoS_2, MoP, MoO_x and Mo_2C. The first patent for MoS_2 was given by Dow in 1980s. They are widely used. Similar to Rh-based catalysts, promoter addition strongly influences product range. Support type is also very important (Pallejà, 2018).

Molybdenum sulfide catalysts (supported on Al_2O_3, promoted with Ni or Co) are widely used for hydrodesulfurization (HDS) in industry, and thus are among the most studied and best described heterogeneous catalysts. However, applying the knowledge of HDS catalysts to HAS is difficult due to the essential role of alkali metal and different reactions occurring (Andersson, 2015). MoS_2 also has been used as sulfur tolerant catalysts in WGS (with low H_2/CO) and methanation of coal-derived CO (Liu et al., 2013).

Doping of MoS_2 with alkali metals is essential for producing HA rather than HC. The alkali tunes the kinetics and energetics of the adsorbed reactants and thereby affecting the fractional coverage in reaction. It has been postulated that the alkali activates CO non-dissociatively and reduce the accessibility of activated H over MoS_2 surface thereby favoring synthesis of HA over HC. There are several reports regarding the most suitable alkali metal, but heavier ones (K, Cs and Rb) are much more preferred than light alkali metals (Li, Na). The optimal alkalinity level is relatively high, alkali/Mo of 0.1–0.7 mol/mol are generally used.

These catalysts are sulfur tolerant (as typical of Mo sulfide catalysts) and less sensitive to CO_2, but need higher operating temperature and pressures. Furthermore, a constant H_2S feed is required which provokes product contamination with organosulfur compounds (Zaman and Smith, 2012).

Mo_2C has been recently investigated as an alternative, but high HC selectivity is still an issue if promotion is not properly done. In addition, deactivation by water and the need for appropriate $CO/CO_2/H_2$ ratios is another limitation (Fang et al., 2011).

5.5.2.4   Rh-Based Catalysts

Rhodium is the sole reported metal exhibiting selective production of $C_{2+}OH$ from synthesis gas. It is attributed to the ability of Rh in catalyzing both dissociation of CO and insertion of CO. However, the dissociation of CO over surfaces such as Rh(1 1 1) is impossible or very slow and the presence of steps/kinks is necessary for enhancing dissociation rate, which is in agreement with Rh hydrogenation behavior since it has been proposed that metals adsorbing CO sufficiently strong but do not dissociate it rapidly are active catalysts for the synthesis of oxygenates (Lu et al., 2017).

These catalysts were developed by the Union Carbide Corporation in 1980s for syngas to HA conversion (Ellgen and Bhasin, 1976). The catalysts were supported on SiO_2 and promoted with alkali metals. However, for enhancing oxygenate production and reducing HC formation, doping with alkali metals or transition metals is necessary (Pallejà, 2018).

The prohibitively high cost and scarecy of Rh makes its use in heterogeneous catalysts for large-scale production unlikely.

5.5.3   Mechanism

The CO insertion mechanism proposed by Xu et al. (1987) is widely accepted for modified FT catalysts. Alcohol formation involves insertion of CO* into a metal-alkyl bond to form an acyl (C_nH_yCO*) intermediate which form alcohol upon hydrogenation. The intermediates predominantly depend on catalyst type, structure and reaction conditions. The study of the reaction is complicated by simultaneous occurrence of chain growth and insertion.

This reaction is an example of prototypical reactions including synergic effect between proximate active sites with diverse functionalities such as CO dissociation (direct or H-assisted), and insertion of CO/HCO into CH_x ($x = 1$–3) species which are required to occur simultaneously to produce CH_xCO/CH_xCHO ($x = 1$–3), thereafter undergoing stepwise hydrogenation until selective production of $C_{2+}OH$ (Lu et al., 2017).

The mechanism of C_2 oxygenate formation has been extensively studied and is believed to comprise (a) dissociative adsorption of CO and H_2 (b) formation of surface hydrocarbon $(CH_x)_{ads}$ and hydroxyl $(OH)_{ads}$ and (c) insertion of CO for C–C bond formation (Subramanian et al., 2009). The formation of ethanol is enhanced by catalysts promoting CO insertion selectively rather than hydrogenation of surface

$(CH_x)_{ads}$ species since hydrogenation of $(CH_x)_{ads}$ leads to hydrocarbons (Subramanian et al., 2009).

Generally, catalysts active in higher oxygenates synthesis should form both adsorbed molecular CO and surface carbon species resulted from dissociative adsorption of CO. Therefore, the catalyst should be able to dissociate only part of CO molecules and balance it with hydrogenation of the intermediates to alcohol. Hydrogenation of $(CH_x)_{ads}$ produces undesired methane.

The active centers of various catalysts of HAS from syngas and the reaction pathways have been recently reviewed (Ao et al., 2018).

5.5.4 PROCESSES

The catalysts used in the processes developed for HAS are typically Cu, Zn, Mo and Cr promoted with alkali metals. A joint venture between Enichem-Snamprogetti-Haldor Topsoe, for example, is producing 15,000 tpy in Italy using a Zn-Cr catalyst at 350–400°C, 12–16 bar, GHSV = 3,000–15,000 h^{-1} for producing mixed alcohols (68%–72% C_1, 2%–3% C_2, 3%–5% C_3s, 10%–15% C_4 and 7%–12% C_{5+} alcohols) (Bartholomew and Farrauto, 2006).

5.6 SYNTHESIS OF ETHYLENE GLYCOL

Ethylene glycol (EG), or simply glycol, $(HOCH_2–CH_2OH)$ is an important intermediate in production of a variety of chemicals including polyester resins, polymers and anti-freezing agents.

It has been produced by two-step epoxidation of ethylene over Ag catalyst followed by reaction with excess water. Even with improved catalysts, the overall yield is limited to about 82% (Wittcoff et al., 2013). Furthermore, the raw material ethylene is mostly petroleum-based. To use other carbon sources, C_1-based methods has received attention.

Ethylene glycol can be produced by hydrogenation of CO under high reaction temperatures and pressures (Eq. 5.16).

$$2CO + 3H_2 \rightarrow HOCH_2CH_2OH \tag{5.16}$$

The reaction is very attractive in terms of atom economy. No carbon and/or hydrogen atom is wasted as CO_2, water or any other product, giving an atom economy of 100%.

Complexes of cobalt, rhodium and ruthenium have been used as the catalyst (Dombek, 1986). The activity depends on many factors including temperature, pressure and promoter and thus a precise screening among the catalysts is not meaningful. However, in general terms, rhodium-based catalysts show higher activity and selectivity for EG synthesis (Dombek, 1983).

Dombek (1985) studied the synergic effect of homogeneous Rh and Ru catalysts at 190–240°C and 400–850 bar. Iodine was identified as an effective promoter. While the overall activity of the bimetallic system did not depend on rhodium concentration, the EG selectivity was affected by the addition of this metal. The soluble complexes observed in the catalytic solutions were $[HRu_3(CO)_{11}]^-$, $[Ru(CO)_3I_3]^-$ and $[Rh(CO)_2I_2]^-$.

Union Carbide (now a part of Dow Chemicals) has been pioneering in this reaction. Many catalysts have been studied and homogenous complexes of Rh and Ru were found to be active. Over Ru carbonyl clusters, at 240°C and 1,000–3,000 bar, EG selectivity of 60%–65% has been achieved. The main by-product is methanol. The catalysts are sensitive and expensive (Rebsdat and Mayer, 2012).

The direct route (Eq. 5.16) is not thermodynamically favorable (Yu and Chien, 2017). Furthermore, due to the slow rate of reaction, severe operating conditions and low selectivity, Union Carbide was led to cooperate with Ube (Japan) that commercialized production oxalate esters from CO that can be used as the precursor of EG (Weissermel and Arpe, 1997). In this oxidative carbonylation route CO, an alcohol (methanol or n-butanol, for example) and an oxidant (e.g., N_2O_3) react over a Pd catalyst to form the corresponding oxalate ester that subsequently, by hydrogenolysis with H_2, produces EG.

Only the Du Pont method that used formaldehyde attained commercial importance in 1940 to 1963 (Rebsdat and Mayer, 2012). Nevertheless, at present, the C_1-based methods are not still competitive with the conventional ethylene-based method.

5.7 HYDROFORMYLATION

5.7.1 CHEMISTRY OF REACTION

Hydroformylation (also called oxo reaction or Roelen reaction) is the reaction of carbon monoxide and dihydrogen with an, preferably, alkene substrate to form (isomeric) aldehydes with one additional carbon in an atom-economic manner. It is commercially important for production of detergent range C_3–C_{15} aldehydes principally used as intermediates to alcohols, acids, polyols, and esters formed by the appropriate reduction, oxidation, or condensation chemistry. (Billig and Bryant, 2000). The most important oxo product is butyraldehyde, the starting material for the manufacture of 2-ethylhexanol through aldol condensation followed by hydrogenation (Olah et al., 2018).

The reaction entails formal addition of a formyl group and hydrogen to a carbon-carbon double bond (Scheme 5.1).

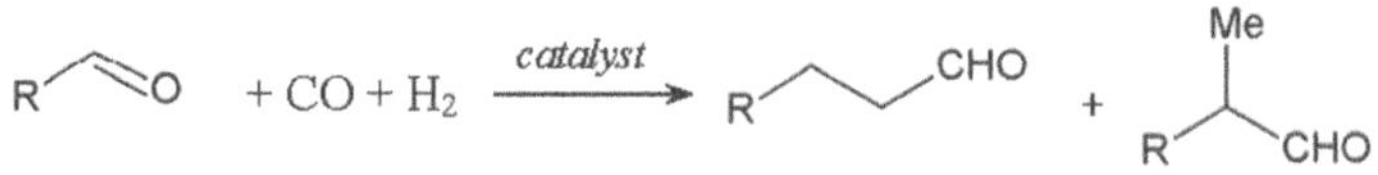

SCHEME 5.1 Hydroformylation of an α-olefin.

The reaction is exothermic and spontaneous ($\Delta H = -129$ kJ/mol, $\Delta G = -56.9$ kJ/mol at 100°C for ethylene hydroformylation). However, the reaction rate is limited due to significant kinetic barrier (i.e., high activation energy) (Navidi et al., 2016).

Side reactions include alkane hydrogenation and isomerization which degrade the product value. The normal to iso (linear to branched) ratio is important criterion for product as the linear product is generally of a higher market value.

5.7.2 CATALYST SYSTEMS

Hydroformylation is one of the most important applications of homogeneous catalysis in chemical industry. The typical catalyst is a transition metal complex with the general formula $HM(CO)_xL_y$ (M = central atom of transition metals; L = ligands, $x + y = 3, 4$) with the M as central atom enables the formation of metal-carbonyl hybrid optionally modified by the ligand L (Cornils, 2018). Thus, replacing some of the CO ligands with other substituent such as phosphine ligands gives the so-called "modified" catalysts.

All group VIII metals, Mn, Cr and Cu show activity in this reaction. The relative activity of the central metals, compared to Co =1 is (Olah et al, 2018)

$$Rh(10^3 - 10^4) \gg Co(1) > Ir > Ru > Os > Pt > Pd \gg Fe > Ni(10^{-4})$$

Currently, only Rh and Co are used with the former has been the preferred central atom. Strong acids, hydrogen cyanide, sulfur, hydrogen sulfide, COS, oxygen, and dienes are catalyst poisons and severely reduce catalyst lifetime (Bahrmann et al., 2013). Therefore, the feed should be purified.

This reaction was discovered in 1938 by Roelen who added ethylene to FTS feed over Co-based catalysts. The original $HCo(CO)_4$ catalyst discovered by Roelen was formed *in situ* from reaction of cobalt metal (or cobalt(II)) with syngas under the high operating temperatures and pressures used.

The activity of rhodium complexes with phosphine or phosphite ligands is about three to four orders of magnitude higher than that of cobalt catalysts. However, Rh catalysts exhibit comparatively poor performance with branched alkenes or otherwise complex ones (Hood et al., 2020).

5.7.3 REACTION MECHANISM

The currently accepted mechanism for hydroformylation was proposed by Heck and Breslow (1960, 1961) for Co-based catalysts. A simplified catalyst cycle for ethylene hydroformylation is shown in Figure 5.7.

The cycle starts with the hydrogenation of cobalt carbonyl ($Co_2(CO)_8$) to $HCo(CO)_4$, reaction (1). The pre-equilibrium gives the effective catalyst, i.e., a 16e complex $HCo(CO)_3$ (cobalt hydrotricarbonyl) in step (2). It reacts with the alkene (3) to form a π-complex that rearranges to a 18e alkyl metal tricarbonyl hydride species (4). Further coordination with carbon monoxide yields the respective tetracarbonyl hydride in step 5. The tetracarbonyl hydride is transformed into the acylmetal tricarbonyl hydride by intramolecular insertion of a CO ligand (6). This species finally (7) undergoes cleavage by H_2 into the aldehyde as the reaction product and recycles hydridometal carbonyl $HCo(CO)_3$.

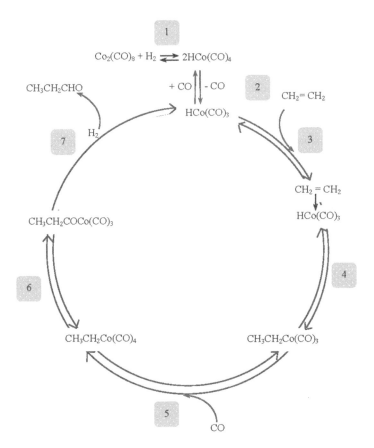

FIGURE 5.7 Mechanism for the unmodified cobalt oxo reaction which produces propional-dehyde from ethylene (Redrawn based on Cornils, 2018 and Billig and Bryant, 2000).

For higher alkenes, isomerization can occur for the products of alkylmetal and acylmetal tricarbonyl hydride, steps (4) and (6), respectively (Cornils et al., 2013). For an Rh-based catalyst, the catalytic cycle is essentially the same.

In the olefinic substrate, the position of the double bond determines the reactivity in hydroformylation reaction (Figure 5.8); with terminal olefins (in comparison to internal or branched alkenes) are hydroformylated faster (Kämper et al., 2016).

Recently, developing efficient and regioselective catalysts for the production of linear aldehydes are receiving interest as research topics (Gehrtz et al., 2016). The tuning of regioselectivity, controlling side reactions (such as hydrogenation and isomerization, i.e., chemoselectivity) and enhancing reaction rate can be achieved by

FIGURE 5.8 Alkene activity in hydroformylation.

a combination of metal center and ligand (Kämper et al., 2016). In the commercial hydroformylation of aliphatic olefins, high *n*/iso ratios can be obtained by modification of the catalysts by phosphine ligands. The ratio of normal/branched isomers also depends on operating condition, especially partial pressure of CO and temperature. High partial pressures of CO and low temperatures favor higher n/iso products ratio from aliphatic olefins (Mika and Ungváry, 2003).

5.7.4 COMMERCIAL PROCESSES

Oxo synthesis has been commercialized by several companies including BASF, Exxon, Shell, Union Carbide and Ruchemie.

The processes utilizing cobalt catalysts are often known as high pressure processes because they need high CO pressures to avoid decomposition of the catalyst to metallic cobalt (Hood et al., 2020).

Currently, Co-based processes of Exxon, BASF, Shell, Cdf Chimie and Nissan are still in operation making use of the advantage of high stability of the Co catalyst toward poisons (Cornils et al., 2013).

5.8 CONCLUSIONS

Carbon monoxide shows a remarkable activity in various hydrogenation and C–C bond formation reactions for the production of valuable chemicals and liquid fuels. Therefore, it is an important platform molecule in C_1 chemistry. Furthermore, syngas as its main source, is the primary product of gasification of low-valued solid and heavy oil feedstocks. Therefore, syngas chemistry provides pathways for valorization of these materials.

As CO is more active than CO_2, many proposed CO_2 conversion processes rely on tandem conversion of CO_2 into CO (via RWGS reaction), followed by conversion of the CO into the desired products. Over hybrid catalysts, these reactions might occur in a single reactor with possible reduction in thermodynamic limitations and heat integration. In this case, the traditional CO conversion catalysts should be improved for possibly more severe operating conditions, most notably formation of large amounts of water formed by RWGS.

NOMENCLATURE

ASF	Anderson-Schulz-Flory
FTS	Fischer–Tropsch synthesis
GTL	gas to liquid
HA	higher alcohol
HAS	higher alcohol synthesis
HDS	hydrodesulfurization
HPS	heavy paraffin synthesis
HT	high-temperature
HTFT	high temperature Fischer–Tropsch
KE	Kölbel–Engelhardt

LT	low-temperature
LTFT	low temperature Fischer–Tropsch
MOF	metal organic framework
MW	molecular weight
RWGS	reverse water-gas shift
SMDS	shell middle distillate synthesis
SMSI	strong metal-support interactions
SPD	slurry phase distillate
SRK	Soave-Redlish-Kwong
TPR	Temperature programmed reduction
TON	turn-over number
WGS	water-gas shift

REFERENCES

Adesina, A. A., Hydrocarbon synthesis via Fischer-Tropsch reaction: travails and triumphs, Appl. Catal. A. G., 138 (1996) 345–367.

Ail, S. S., Dasappa, S., Biomass to liquid transportation fuel via Fischer-Tropsch synthesis – technology review and current scenario, Renew. Sust. Energy Rev., 58 (2016) 267–286.

Anderson, R. B., The Fischer-Tropsch Synthesis. Academic Press, Orlando, FL (1984).

Andersson, R., Catalytic Conversion of Syngas to Higher Alcohols over MoS₂-based Catalysts, KTH Royal Institute of Technology, Stockholm (2015).

Anton, J., Nebel, J., Song, H., Froese, C., Weide, P., Ruland, H., Muhler, M., Kaluza, S., The effect of sodium on the structure-activity relationships of cobalt-modified Cu/ZnO/Al₂O₃ catalysts applied in the hydrogenation of carbon monoxide to higher alcohols, J. Catal., 335 (2016) 175–186.

Ao, M., Pham, G. H., Sunarso, J., Tade, M. O., Liu, S., Active centers of catalysts for higher alcohol synthesis from syngas: a review, ACS Catal. 8 (2018) 7025–7050.

Bahrmann, H., Bach, H., Frey, G. D., Oxo Synthesis, Ullmanns, Wiley-VCH Verlag (2013).

Bartholomew, C. H., Farrauto, R. J., Fundamentals of Industrial Catalytic Processes, 2nd Ed., John Wiley & Sons, Inc., Hoboken, NJ. (2006) 398–486.

Bedel, L., Roger, A. C., Estournes, C., Kiennemann, A., CoO from partial reduction of La(Co,Fe)O₃ perovskites for Fischer-Tropsch synthesis, Catal. Today, 85 (2003) 207–218.

Bedel, L., Roger, A. C., Rehspringer, J. L., Zimmermann, Y., Kiennemann, A., La₍₁₋ᵧ₎Co₀.₄Fe₀.₆O₃₋δ perovskite oxides as catalysts for Fischer-Tropsch synthesis, J. Catal., 235 (2005) 279–294.

Bezemer, G. L., Bitter, J. H., Kuipers, H. P. C. E., Oosterbeek, H., Holewijn, J. E., Xu, X., Kapteijn, F., van Dillen, A. J., de Jong, K. P., Cobalt particle size effects in the Fischer-Tropsch reaction studied with carbon nanofiber supported catalysts, J. Am. Chem. Soc., 128 (2006) 3956–3964.

Billig, E., Bryant, D. R., OXO Process, Kirk-Othmer Encyclopedia of Chemical Technology, Wiley, Germany (2000).

Biloen, P., Helle, J. N., Sachtler, W. M. H., Incorporation of surface carbon into hydrocarbons during Fischer-Tropsch synthesis: mechanistic implications, J. Catal., 58 (1979) 95–107.

Bukur, D. B., Todic, B., Elbashir, N., Role of water-gas-shift reaction in Fischer-Tropsch synthesis on iron catalysts: a review, Catal. Today, 275 (2016) 66–75.

Cheng, K., Kang, J., King, D. L., Subramanian, V., Zhou, C., Zhang, Q., Wang, Y., Advances in Catalysis for Syngas Conversion to Hydrocarbons, In Advances in Catalysis, Volume 60, Elsevier, India (2017).

Chorkendorff, I., Niemantsverdriet, J. W., Concepts of Modern Catalysis and Kinetics, Wiley-VCH, Germany (2017).

Cornils, B., Herrmann, W. A., Wong, C. H., Zanthoff, H. W., Catalysis from A to Z, 4th ed., Wiley-VCH, Weinheim (2013).

Cornils, B., Hydroformylation, In Cornils, B., Herrmann, W. A., Beller, M., Paciello, R. (Eds.), Applied Homogeneous Catalysis with Organometallic Compounds: A Comprehensive Handbook, 3rd Ed., Wiley-VCH, Weinheim (2018).

Dombek, B.D., Homogeneous catalytic hydrogenation of carbon monoxide: ethylene glycol and methanol from synthesis gas, Adv. Catal., 32 (1983) 325–416.

Dombek, B. D., Synergistic behavior of homogeneous ruthenium-rhodium catalysts for hydrogenation of carbon monoxide, Organometallics, 4 (1985) 1707–1712.

Dombek, B. D., Direct routes for synthesis gas to ethylene glycol, J. Chem. Edu., 63(3) (1986) 210–212.

Dry, M. E., Advances in Fischer-Tropsch chemistry, Ind. Eng. Chem. Prod. Res. Dev., 15 (1976) 282–286.

Dry, M. E., The Fischer-Tropsch process: 1950-2000, Catal. Today, 71 (2002) 227–241.

Dry, M. E., Fischer-Tropsch synthesis-Industrial, In Harvath, I. T. (Ed.) Encyclopedia of Catalysis, Wiley, New York (2003).

Dry, M. E., FT catalysts, Stud. Surf. Sci. Catal., 152 (2004) 533–600.

Ellgen, P., Bhasin. M. (1976) US-4096164.

Escalona, N., Fuentealba, S., Pecchi, G., Fischer-Tropsch synthesis over $LaFe_{1-x}Co_xO_3$ perovskites from a simulated biosyngas feed, Appl. Catal. A. G., 381 (2010) 253–260.

Eslava, J. L., Sun, X., Gascon, J., Kapteijnb, F., Rodriguez-Ramos, I., Ruthenium particle size and cesium promoter effects in Fischer-Tropsch synthesis over high-surface-area graphite suppored cataltsts, Catal. Sci. Technol., 7 (2017) 1235–1244.

Fang, Y. Z., Liu, Y., Zhang, L. H., $LaFeO_3$-supported nano Co-Cu catalysts for higher alcohol synthesis from syngas, Appl. Catal. A. G. 397 (2011) 183–191.

Feng, Z., Postula, W. S., Akgerman, A., Anthony, R. G., Characterization of zirconia-based catalysts prepared by precipitation, calcination, and modified sol-gel methods, Ind. Eng. Chem. Res., 34 (1995) 78–82.

Fischer. F., Tropsch, H., Synthesis of petroleum from gasification products of coal at normal pressure, Brennst. Chem., 7 (1926a) 97–116.

Fischer, F., Tropsch, H., The synthesis of petroleum at atmospheric pressures from gasification products of coal, Brennst. Chem., 7 (1926b) 97–104.

Franssen, N. M. G., Walters, A. J. C., Reek, J. N. H., de Bruin, B., Carbene insertion into transition metal-carbon bonds: a new tool for catalytic C-C bond formation, Catal. Sci. Technol., 1 (2011) 153–165.

Friedel, A., Anderson, R. B., Composition of synthetic liquid fuels I. Product distillation and analysis of C_5-C_8 paraffin isomers from cobalt catalyst, J. Am. Chem. Soc., 72 (1950) 1212–1215.

Gaube, J., Klein, H.-F., The promoter effect of alkali in Fischer-Tropsch iron and cobalt catalysts, Appl. Catal. A. G., 350 (2008) 126–132.

Gehrtz, P. H., Hirschbeck, V., Ciszek, B., Fleischer, I., Carbonylations of alkenes in the total synthesis of natural compounds, Synthesis, 48 (2016) 1573–1596.

Gholami, Z., Tišler, Z., Rubáš, V., Recent advances in Fischer-Tropsch synthesis using cobalt-based catalysts: a review on supports, promoters, and reactors, Catal. Rev. Sci. Eng., (2020).

Heck, R. F., Breslow, D. S., Reactions of alkyl-and acyl-cobalt carbonyls with triphenylphosphine, J. Am. Chem. Soc., 82 (1960) 4438–4439.

Heck, R. F., Breslow, D. S., The reaction of cobalt hydrotetracarbonyl with olefins, J. Am. Chem. Soc., 83 (1961) 4023–4027.

Henkel, R., The influence of ammonia on Fischer-Tropsch synthesis and formation of N-containing compounds, PhD thesis, Carl von Ossitzky Universitat (2012).

Hilmen, A. M., Xu, M., Gines, M. J. L., Iglesia, E., Synthesis of higher alcohols on copper catalysts supported on alkali-promoted basic oxides, Appl. Catal. A., 169 (1998) 355–372.

Hindermann, J. P., Hutchings, G. J., Kiennemann, A., Mechanistic aspects of the formation of hydrocarbons and alcohols from CO hydrogenation, Catal. Rev. Sci. Eng., 35 (1993) 1–127.

Hood, D. M., Johnson, R. A., Carpenter, A. E., Younker, J. M., Vinyard, D. J., Stanley, G. G., Highly active cationic cobalt(II) hydroformylation catalysts, Science, 367 (2020) 542–548.

Kämper, A., Kucmierczyk, P., Seidensticker, T., Vorholt, A. J., Franke, R., Behr, A., Ruthenium-catalyzed hydroformylation: from laboratory to continuous miniplant scale, Catal. Sci. Technol., 6 (2016) 8072–8079.

Khodakov, A. Y., Chu, W., Fongarland, P., Advances in the development of novel cobalt Fischer-Tropsch catalysts for synthesis of long-chain hydrocarbons and clean fuels, Chem. Rev., 107 (2007) 1692–1744.

de Klerk, A., Hydroprocessing peculiarities of Fischer-Tropsch syncrude, Catal. Today, 130 (2008) 439–445.

de Klerk A, Fischer-Tropsch Process, Kirk-Othmer Encyclopedia of Chemical Technology, Wiley, New York (2013).

Kölbel, H., Trapper, J., Aliphatic arnines from carbon monoxide, steam, and ammonia, Angew. Chem. Int. Ed., 5 (1966) 843–844.

Kuipers, E. W., Vinkenburg, I. H., Oosterbeek, H., Chain length dependence of α-olefin re-adsorption in Fischer-Tropsch synthesis, J. Catal., 152 (1995) 137–146.

Li, Y., He, D., Cheng, Z., Su, C., Li, J., Zhu, Q., Effect of calcium salts on isosynthesis over ZrO_2 catalysts, J. Mol. Catal. A Chem., 175 (2001) 267–275.

Li, X., Luo, M., Asami, K., Direct synthesis of middle iso-paraffins from synthesis gas on hybrid catalysts, Catal. Today, 89 (2004) 439–446.

Liu, J., Wang, E., Lv, J., Li, Z., Wang, B., Ma, X., Qin, S., Sun, Q., Investigation of sulfur-resistant, highly active unsupported MoS_2 catalysts for synthetic natural gas production from CO methanation, Fuel Proc. Tech., 110 (2013) 249–257.

Lu, Y., Zhang, R., Cao, B., Ge, B., Tao, F. F., Shan, J., Nguyen, L., Bao, Z., Wu, T., Pote, J. W., Wang, B., Yu, F., Elucidating the copper-Hagg iron carbide synergistic interactions for selective CO hydrogenation to higher alcohols, ACS Catal., 7 (2017) 5500–5512.

Ma, W., Jacobs, G., Sparks, D. E., Pendyala, V. R. R., Hopps, S. G., Thomas, G. A., Hamdeh, H. H., MacLennan, A., Hu, Y., Davis, B. H., Fischer-Tropsch synthesis: effect of ammonia in syngas on the Fischer-Tropsch synthesis performance of a precipitated iron catalyst, J. Catal. 326 (2015) 149–160.

Maruya, K. I., Takasawa, A., Haraoka, T., Domen, K., Onishi, T., Role of methoxide species in isobutene formation from CO and H_2 over oxide catalysts methoxide species in isobutene formation, J. Mol. Catal. A. Chem., 112 (1996) 143–151.

Maruya, K. I., Komiya, T., Hayakawa, T., Lu, L., Yashima, M., Active sites on ZrO_2 for the formation of isobutene from CO and H_2, J. Mol. Catal. A. Chem., 159 (2000) 97–102.

Mehariya, S., Iovine, A., Casella, P., Musmarra, D., Figoli, A., Marino, T., Sharma, N., Molino, A., Fischer-Tropsch Synthesis of Syngas to Liquid Hydrocarbons, In Yousuf, A., Pirozzi, D., Sannino, F. (Eds.), Lignocellulosic Biomass to Liquid Biofuels, Elsevier, Chennai (2020), pp. 217–248.

Melaet, G., Lindeman, A. E., Somorjai, G. A., Cobalt particle size effects in the Fischer-Tropsch synthesis and in the hydrogenation of CO_2 studied with nanoparticle model catalysts on silica, Top Catal., 57 (2014) 500–507.

Méndez, C. I., Ancheyta, J., Kinetic models for Fischer-Tropsch synthesis for the production of clean fuels, Catal. Today, 353 (2020) 3–6.

Mika, L. T., Ungváry, F., Hydroformylation-Homogeneous, In Encyclopedia of Catalysis, John Wiley and Sons, Germany (2003).

Navidi, N., Marin, G. B., Thybaut, J. W., A single-event microkinetic model for ethylene hydroformylation to propanal on Rh and Co based catalysts, Appl. Catal. A. G., 524 (2016) 32–44.

Nebel, J., Schmidt, S., Pan, Q., Lotz, K., Kaluza, S., Muhler, M., On the role of cobalt carbidization in higher alcohol synthesis over hydrotalcite-based Co-Cu catalysts, Chin, J. Catal. 40 (2019) 1731–1740.

Niwa, M., Lunsford, J. H., The catalytic reactions of CO and H_2O over supported rhodium, J. Catal., 75 (1982) 302–313.

Novak, S., Madon, R. J., Suhl, H., Secondary effects in the Fischer-Tropsch synthesis. J. Catal., 77 (1982) 141–151.

Olah, G. A., Molnar, A., Surya Prakash, G. K., Hydrocarbon Chemistry, 3rd Ed. Wiley, New York (2018) 125–236.

Oukaci, R., Wu, J. C. S., Goodwin Jr, J. G., Secondary reactions during CO hydrogenation on zeolite-supported metal catalysts: Influence of alkali cations, J. Catal., 107 (1987) 471–481.

Pallejà, J. P., Catalytic conversion of syngas to alcohols and hydrocarbons over transition metal-based micro/mesoporous catalysts, Doctoral Thesis (2018).

Pei, Y. P., Liu, J. X., Zhao, Y. H., Ding, Y. J., Liu, T., Dong, W. D., Zhu, H. J., Su, H. Y., Yan, L., Li, J. L., Li, W. X., High alcohols synthesis via Fischer-Tropsch reaction at cobalt metal/carbide interface, ACS Catal., 5 (2015) 3620–3624.

Pendyala, V. R. R., Gnanamani, M. K., Jacobs, G., Ma, W., Shafer, W. D., Davis, B. H., Fischer-Tropsch synthesis: effect of ammonia impurities in syngas feed over a cobalt/alumina catalyst, Appl. Catal. A. G., 468 (2013) 38–43.

Pendyala, V. R. R., Jacobs, G., Gnanamani, M. K., Hu, Y., MacLennan, A., Davis, B. H., Selectivity control of Cu promoted iron-based Fischer-Tropsch catalyst by tuning the oxidation state of Cu to mimic K. Appl. Catal. A. G., 495 (2015) 45–53.

Pichler, H., Buffleb, H., Behavior of ruthenium catalysts in synthesis of paraffin hydrocarbons of high molecular weight, Bre. Chem., 21 (1940) 273–280.

Pichler, H., Schulz, H., Recent results in the synthesis of hydrocarbons form carbon monoxide and hydrogen, Chem. Eng. Tech., 42 (1970) 1162–1174.

Postula, W. S., Feng, Z., Philip, C., Akgerman, A., Anthony, R. G., Conversion of synthesis gas to isobutylene over zirconium dioxide based catalysts, J. Catal., 145 (1994) 126–131.

Rebsdat, S., Mayer, D. Ethylene Glycol, In Ullmann's Encyclopedia of Industrial Chemistry, Wiley-VCH, Germany (2012).

Reuel, R. C., Bartholomew, C. H., Effects of support and dispersion on the CO hydrogenation activity/selectivity properties of cobalt. J. Catal., 85 (1984) 78–88.

Sango, T., Nitrogen-containing compounds from ammonia co-feed to the Fischer-Tropsch synthesis, Master's thesis, University of Cape Town (2013).

Sango, T., Fischer, N., Henkel, R., Roessner, F., van Steen, E., Claeys, M., Formation of nitrogen containing compounds from ammonia co-fed to the Fischer-Tropsch synthesis, Appl. Catal. A. G., 502 (2015) 150–156.

Schulz, G. V. Z., Phys. Chem. Abt. B., 43 (1939) 25–46.

Schulz, H., Beck, K., Erich, E., Mechanism of the Fischer-Tropsch process, Stud. Surf. Sci. Catal., 36 (1988) 457–471.

Schulz, H., Short history and present trends of Fischer-Tropsch synthesis, Appl. Catal. A. G., 186 (1999) 3–12.

Schulz, H., Major and minor reactions in Fischer-Tropsch synthesis on cobalt catalysts, Top. Catal., 26 (2003) 73–85.

Schulz, H., Principles of Fischer-Tropsch synthesis – constraints on essential reactions ruling FT-selectivity, Catal. Today, 214 (2013) 140–151.

Shah, Y. T., Perrotta, A. J., Catalysts for Fischer-Tropsch and Isosynthesis Ind. Eng. Chem. Prod. Res. Dev., 15 (1976) 123–131.

Song, X., Sayari, A., Direct synthesis of isoalkanes through Fischer-Tropsch reaction on hybrid catalysts, Appl. Catal. A. G., 110 (1994) 121–136.

Spivey, J. J., Egbebi, A., Heterogeneous catalytic synthesis of ethanol from biomass-derived syngas, Chem. Soc. Rev., 36 (2007) 1514–1528.

Su, C., Li, J., He D., Cheng, Z., Zhu, Q., Synthesis of isobutene from synthesis gas over nano-size zirconia catalysts Appl. Catal. A: Gen., 202 (2000) 81–89.

Subramanian, N. D., Balaji, G., Kumar, C. S. S. R., Spivey, J. J., Development of cobalt-copper nanoparticles as catalysts for higher alcohol synthesis from syngas, Catal. Today, 147 (2009) 100–106.

Tangchupong, N., Khaodee, W., Jongsomjit, B., Laosiripojana, N., Praserthdam, P., Assabumrungrat, S., Effect of calcination temperature on characteristics of sulfated zirconia and its application as catalyst for isosynthesis, Fuel Proc. Tech., 91 (2010) 121–126.

Tau, L. M., Dabbagh, H. A., Davis, B. H., Fischer-Tropsch synthesis: comparison of carbon-14 distributions when labeled alcohol is added to synthesis gas, Energy Fuels 5 (1991) 174–179.

Tien-Thao, N., Alamdari, H., Zahedi-Niaki, M. H., Kaliaguine, S, LaCo1-xCuxO3-δ Perovskite catalysts for higher alcohol synthesis, Appl. Catal. A., 311 (2006) 204–212.

Tsakoumis, N. E., Ronning, M., Borg, O., Rytter, E., Holmen, A., Deactivation of cobalt based Fischer-Tropsch catalysts: a review, Catal. Today, 154 (2010) 162–182.

Vannice, M. A., The catalytic synthesis of hydrocarbons from H_2/CO mixtures over the group VIII metals: I. The specific activities and product distributions of supported metals, J. Catal., 37 (1975) 449–461.

Van Berge, P. J., Everson, R. C., Cobalt as an alternative Fischer-Tropsch catalyst to iron for the production of middle distillates. Stud. Surf. Sci. Catal., 107 (1997) 207–212.

van der Laan, G. P., Beenackers, A. A. C. M., Kinetics and selectivity of the Fischer-Tropsch synthesis: a literature review, Catal. Rev. Sci. Eng., 41 (1999) 255–318.

van der Laan, G. P., Kinetics, selectivity and scale up of the Fischer-Tropsch synthesis, PhD Thesis, University of Groningen (1999).

van Dijk, H. A. J., Hoebink, J. H. B. J., Schouten, J. C., Steady state isotopic transient kinetic analysis of the Fischer-Tropsch synthesis reaction over cobalt-based catalysts, Chem. Eng. Sci., 56 (2001) 1211–1219.

de Vries, C., Adding Ammonia during Fischer-Tropsch Synthesis: Pathways to the Formation of N-containing compounds. PhD Thesis, University of Cape Town, Cape Town, South Africa (2017).

Wang, J., Chernavskii, P. A., Wang, Y., Khodakov, A. Y., Influence of the support and promotion on the structure and catalytic performance of copper–cobalt catalysts for carbon monoxide hydrogenation, Fuel, 103 (2013) 1111–1122.

Wang, Z., Kumar, N., Spivey, J. J., Preparation and characterization of lanthanum-promoted cobalt-copper catalysts for the conversion of syngas to higher oxygenates: Formation of cobalt carbide, J. Catal., 339 (2016) 1–8.

Weissermel, K., Arpe, H.-J, Industrial Organic Chemistry, 3rd Ed., Wiley-VCH, Germany (1997).

Wezendonk, T. A., Sun, X., Dugulan, A. I., van Hoof, A. J. F., Hensen, E. J. M., Kapteijn, F., Gascon, J., Controlled formation of iron carbides and their performance in Fischer-Tropsch synthesis, J. Catal., 362 (2018) 106–117.

Wittcoff, H. A., Reuben, B. G., Plotkin, J. S. Industrial Organic Chemicals, 3rd Ed., Wiley, Germany (2013).

Wu, X., Tan, M., Tian, S., Song, F., Ma, Q., He, Y., Yang, G., Tsubaki, N., Tan, Y., Designing ZrO_2-based catalysts for the direct synthesis of isobutene from syngas: the studies on Zn promoter role, Fuel, 243 (2019) 34–40.

Xu, X., Doesburg, E. B. M., Scholten, J. J. F., Synthesis of higher alcohols from syngas-recent patented catalysts and tentative on the mechanism. Catal. Today, 2 (1987) 125–170.

Yates, I. C., Satterfield, C. N., Intrinsic kinetics of the Fischer-Tropsch synthesis on a cobalt catalyst, Energy Fuels, 5 (1991) 168–173.

Yoneyama, Y., San, X., Iwai, T., Tsubaki, N., One-step synthesis of isoparaffin from synthesis gas using hybrid catalyst with supercritical butane, Energy Fuels, 22 (2008) 2873–2876.

Yu, B. Y., Chien, I. L., 2017. Design and optimization of dimethyl oxalate (DMO) hydrogenation process to produce ethylene glycol (EG), Chem. Eng. Res. Des., 121, 173–190.

Zaman, S., Smith, K. J., A review of molybdenum catalysts for synthesis gas conversion to alcohols: catalysts, mechanisms and kinetics, Catal. Rev. Sci. Eng., 54 (2012) 41–132.

Zhang, Q., Kang, J., Wang, Y., Development of novel catalysts for Fischer-Tropsch synthesis: tuning the product selectivity, ChemCatChem, 2 (2010) 1030–1058.

6 Carbon Dioxide Conversions

6.1 INTRODUCTION

The increased level of the anthropogenic carbon dioxide in the atmosphere and the resulting greenhouse affect promoted worldwide research for reducing emissions by capturing and utilization of carbon dioxide.

One route of carbon dioxide capture and utilization (CCU) is CO_2 conversion into hydrocarbon fuels that are compatible with the present storage and distribution network (Sahebdelfar and Takht Ravanchi, 2020). This is a CO_2-neutral path as by combustion, fuels turn to CO_2 again. If fossil fuels were replaced by carbon-neutral fuels, negative CO_2 emissions or net removal of CO_2 from the atmosphere could be achieved (Choi et al., 2017). Polymer production is another sector for mass utilization of carbon dioxide. Carbon dioxide can involve as a co-monomer in several polymers.

The effective catalytic conversion of carbon dioxide as a building block into valuable chemicals and fuels is an important issue in C_1 chemistry. This is due to environmental problems resulted from increased emission of CO_2 to the atmosphere and the potential of CO_2 as a sustainable source of carbon.

Carbon dioxide has many advantages as a chemical feedstock. It is nontoxic, cheap and abundant. Some chemical plants (e.g., ammonia and ethylene epoxidation plants) produce large amounts of high purity carbon dioxide. Furthermore, the biomass- and coal-derived syngas constitutes significant amounts of carbon dioxide. In coal gasification plants, some gasifiers (like conventional Lurgi process) produce CO_2-rich syngas (containing 32% CO_2, 17% CO, 38% H_2 and the rest for other components) (Rao et al., 1992).

In Figure 6.1, direct CO_2 conversion into liquid fuels is depicted. In this route, a renewable hydrogen source is considered that is based on solar water splitting. As it is obvious in this figure, CO_2 is emitted from industrial sources (such as steel mills, coal power plants or chemical plants) and reacted with hydrogen produced from solar hydrogen plant and liquid fuels are obtained in a single step.

Light alkanes production from CO_2 is a modification of the Fischer–Tropsch (FT) process, in which the catalyst is tailored in a way to maximize light alkanes production that will be used as synthetic LPG (liquefied petroleum gas) or petroleum feedstocks. In the FT process, there is a competition between CO_2 and CO that must be taken into account (Saeidi et al., 2014).

For effective utilization of CO_2 relevant to reduced global emissions, only conversion into products with large or even unlimited demands such as polymers and fuels should be targeted. In this chapter, hydrogenation and polymerization of carbon dioxide will be discussed as potentially large-scale processes for using CO_2 as a chemical building block. It includes chemistry, catalyst and processing aspects.

DOI: 10.1201/9781003279280-6

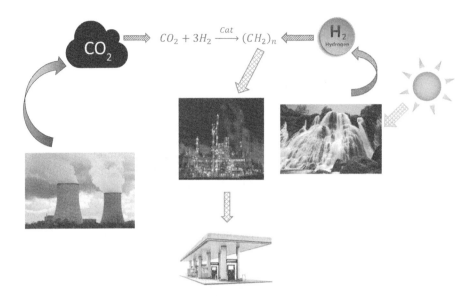

$$CO_2 + 3H_2 \xrightarrow{Cat} (CH_2)_n$$

FIGURE 6.1   Carbon capture and catalytic conversion of CO_2 by renewable energy.

6.2   CHEMICAL UTILIZATION OPTIONS

Despite the advantages of using of carbon dioxide as a chemical feedstock, only a small portion of the produced carbon dioxide is utilized in chemical industry, mostly in manufacture of urea (about 1.5×10^8 tCO_2/y), but also in the production of methanol, salicylic acid and cyclic carbonates.

The challenge of utilization of carbon dioxide in chemical synthesis, especially with heterogeneous catalysts, is the high stability of carbon dioxide which results in high thermodynamic barrier in activation of carbon dioxide molecule. Thus, typically external energy sources (heat, light, etc.) or energetic co-reactants (hydrogen, epoxides, etc.) are necessary to achieve acceptable conversions (Takht Ravanchi and Sahebdelfar, 2021).

Carbon dioxide is a highly stable compound with linear and apolar molecules (Table 6.1), although individual carbon-oxygen bonds are polar due to the higher electronegativity of oxygen, giving partial positive charge to carbon and negative charge to oxygen atoms ($^\delta O{=}C^{2\delta+}{=}O^\delta$). Consequently, the CO_2 molecule is susceptible to both nucleophilic attack to the electron-deficient carbon (e.g., by electron donating reagents or surface sites of organometallic complexes) and electrophilic

TABLE 6.1
Thermochemical Properties of CO_2

Normal b.p. (°C)	Dipole Moment (D)	ΔH^0_f (kJ/mol)	ΔG^0_f (kJ/mol)
−74.5	0	−393.5	−394.4

TABLE 6.2
Possible Producible Substance Groups with CO_2 as a Feedstock, Depending on the Nature of the Reactants (Partly Based on Otto et al., 2015)

Reactant	Product
"Active hydrogen" bearing organic compounds	Carboxylic acids and coumarins
Organometallic compounds (Grignard reagent)	Carboxylic acids
Alkenes	Unsaturated carboxylic acids
Conjugated dienes	Lactones and unsaturated esters
Alkynes	Lactone and pyrones
Hydrogen	Hydrocarbons, alcohols, ethers, aldehydes, carboxylic acids
Epoxides or diols	Cyclic carbonates
Alcohols	Linear carbonates
Ammonia	Urea
Amines	Linear urea derivatives and isocynates
Diamines	Cyclic urea derivatives
Aziridines	Cyclic carbamates
Amines or ammonia + alcohols	Linear carbamates

attack to the oxygens, which can lead to activation and reaction of CO_2. This is why often a bifunctional catalyst is required for CO_2 conversion. However, the electrophilicity of carbon atom is higher than the nucleophilicity of each oxygen atom and thus CO_2 behaves prevalently as an electrophile (Aresta, 2006).

Therefore, despite its high stability, carbon dioxide is a reactive molecule and several complexes and reaction of CO_2 are known (Keim 1986). Table 6.2 summarizes typical reactions of CO_2 with different reagents.

Carbon dioxide reacts with organic compounds containing the so-called "active hydrogen" acting as a carboxylating agent to form valuable products such as ethylene and propylene carbonates under relatively mild conditions (Alper and Orhan, 2017).

The reactants possessing strong electron rich atom in their molecules such as ammonia, alkylureas, phenol, olefins, dienes, epoxides and methanol could be used to activate CO_2 (Scheme 6.1). The hydrogen located at vicinity of the electron-rich atom of the molecules, the "active hydrogen" such as the hydrogen in α site, is capable of attacking the oxygen atom of CO_2. Consequently, these molecules can act as nucleophilic reagents and react with the carbon atom of CO_2 under mild conditions (Sun, 2003).

With phenol salts (alkali phenolates), CO_2 brings about substitution of hydrogen with a carboxyl (–COOH) group (Kolbe or Kolbe-Schmitt reaction) which is important in commercial production of salicylic acid (Scheme 6.2). The ionization of the salt activates the ring for electrophile attack such that feeble electrophiles such as CO_2 can react.

Although urea and salicylic acid syntheses from CO_2 are purely thermal processes, catalysis plays a critical role in most industrially relevant CO_2 transformations.

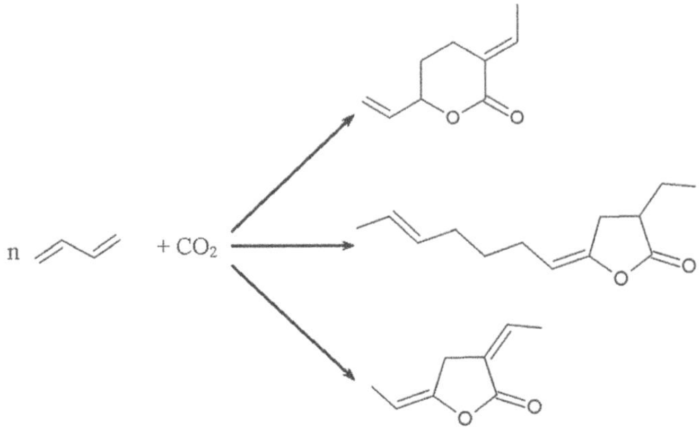

SCHEME 6.1 Reaction of carbon dioxide with 1,3-butadiene (Redrawn from Keim, 1986).

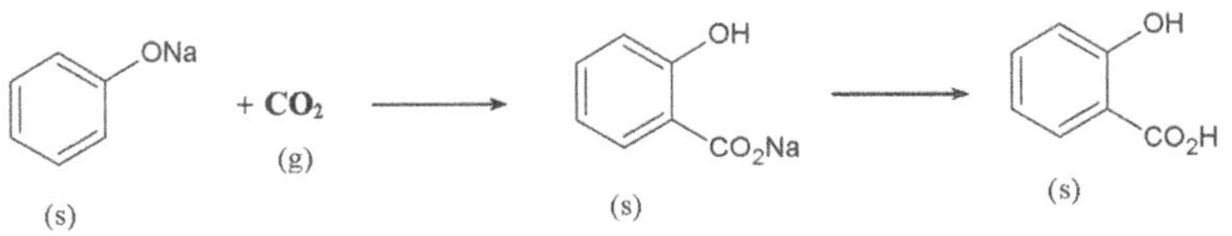

SCHEME 6.2 Synthesis route of salicylic acid.

The synthesis of higher hydrocarbons, oxygenates and polymers are potentially important CO_2 conversion reactions for the production of fuels and bulk chemicals which can realize large-scale chemical fixation of carbon dioxide (Takht Ravanchi and Sahebdelfar, 2021).

6.3 HYDROGENATION OF CARBON DIOXIDE

The catalytic hydrogenation of CO_2 to hydrocarbons and oxygenates is attracting interest as a sustainable route for the production of fuels and basic chemicals (Sahebdelfar and Takht Ravanchi, 2020). Despite the limited product portfolio for heterogeneous catalytic systems, the products are important due to their large-scale markets. Therefore, they could provide a route for recycling carbon dioxide into chemical synthesis.

For CO_2 conversion into hydrocarbons, two challenges must be overcome; that are renewable cheap H_2 source and CO_2 chemical stability. The products of direct catalytic CO_2 hydrogenation are limited to low molecular weight hydrocarbons (C_1–C_4) or oxygenates (HCOOH, CO, CH_3OCH_3, CH_3OH, etc.) (Preti et al., 2011; Razali et al., 2012; Dorner et al., 2010a; Landau et al., 2014; Perathoner and Centi, 2014).

6.3.1 CHEMISTRY

The hydrogenation of carbon dioxide is related to that of carbon monoxide because the former can be converted into the latter under reaction conditions via the reverse water gas shift (RWGS) reaction over most hydrogenation catalysts. The methanol-mediated route over composite catalysts (combinations of methanol synthesis and methanol to hydrocarbon (MTH) catalysts) produces chiefly lower alkanes as the methanol synthesis catalyst can hydrogenate the intermediate alkenes (Sai Prasad et al., 2008). Consequently, direct FT synthesis from CO_2 is the preferable route:

$$CO_2 + 3H_2 \rightarrow 1/n\left(-CH_2-\right)_n + 2H_2O \quad \Delta H_{298}^0 = -111\,kJ/mol \qquad (6.1)$$

CO_2 hydrogenation to hydrocarbons consists of the RWGS reaction (Eq. 6.2) and FT synthesis (Eq. 6.3); that are series reactions (Choi et al., 2017):

$$CO_2 + H_2 \rightarrow CO + H_2O \quad \Delta H_{298}^0 = +41\,kJ/mol \qquad (6.2)$$

$$CO + 2H_2 \rightarrow 1/n\left(-CH_2-\right)_n + H_2O \quad \Delta H_{298}^0 = -152\,kJ/mol \qquad (6.3)$$

The co-presence of direct hydrogenation of CO_2 to hydrocarbons has been also reported in the literature (Fiato et al., 1998).

The chain growth probability (α) in CO_2-FT is smaller than that for traditional FT synthesis because CO is the chain growing agent. Thus, the main products are lower hydrocarbons in the former case, although C_{5+} products are more desirable for transportation fuel applications.

In comparison to the conventional CO-FT, the CO_2-FT reaction utilizes three moles of H_2 per mole of CO_2 and a larger amount of H_2O is produced as by-product that deactivates Fe-based catalysts. Under the same operating conditions, the CO_2-FT is slower than the CO-FT reaction (Sakakura et al., 2007).

6.3.2 CATALYSTS

As Co-based catalysts have high performance to cost ratios, they are widely used in FT process. The selectivity of the Co-based catalysts used for production of high-molecular weight hydrocarbons in the traditional FT reaction shifts to methane by changing the syngas feed to CO_2/H_2 mixture (Gnanamani et al., 2015). On the other hand, the RWGS activity of iron-based catalysts is an advantage for applying them in the CO_2-FT reaction making them capable of operating with a wide range of syngas compositions (Wang et al., 2011). Consequently, iron-based catalysts are more frequently used in the CO_2-FT reaction.

FT synthesis is "sensitive", meaning that product selectivity is influenced by catalyst composition and its synthesis method (Srinivas et al., 2007).

Zhang et al. (2002) evaluated the conversion of CO_2/H_2, CO/H_2 and $(CO + CO_2)/H_2$ on supported Co catalysts for the FT reaction. All these mixtures had similar activities but different selectivities. CO and CO_2 hydrogenation followed different reaction pathways and consequently different catalyst stability. In the case of CO hydrogenation, the catalyst deactivated faster (Centi and Perathoner, 2009).

By Fe-based catalysts, highly olefinic products are obtained (Jin and Datye, 2000). Fe is recommended as the active component of the catalyst, γ-Al_2O_3 is proposed as the best support and K as the favorable promoter. CO_2 hydrogenation occurs via a two-step process; initial CO_2 reduction to CO via the RWGS reaction and CO conversion into hydrocarbons via the FT reaction (Cubeiro et al., 2000; Riedel et al., 2001; Jun et al., 2004).

Riedel et al. (1999) studied CO and CO_2 hydrogenation on Co and Fe catalysts. In the absence of WGS reaction promoters (such as Mn), CO_2 acted as a diluting gas as it neither adsorbed nor hydrogenated on Co catalyst. In the presence of Mn, the RWGS reaction was possible. As CO was strongly adsorbed on the surface, FT chain growth occurred on Co metal. In the low partial pressures of CO, the reaction regime transferred from FT to methanation and more CH_4 was produced. Hence, CO_2 hydrogenation did not occur even in the presence of Mn (Srinivas et al., 2007).

Riedel et al. (1999) reported that H_2/CO and H_2/CO_2 syngas have the same hydrocarbon distributions but CO_2 syngas had a 43% lower reaction rate. If silica was used as support, CO_2 oxidized the catalyst and CO_2 hydrogenation activity was low (Jun et al., 2004). Schulz et al. (1999) reported no deactivation for CO_2-rich syngas; that was due to the low carbon formation tendency.

It was reported that over the Fe-Al_2O_3-Cu-K catalyst, CO and CO_2 hydrogenation yielded the same product distributions. For Fe- and Co-based FT catalysts, there was a difference in methane formation inhibition and desorption of products that is a prerequisite for chain growth. For Fe-based catalysts an irreversible carbiding and alkali surface coverage occurred but for Co-based catalysts, a strong reversible CO adsorption occurred. For Fe-based catalysts, γ-Al_2O_3 was the best support, K was the best promoter and the active phase formed *in situ* (as iron carbide-Fe_5C_2). A higher amount of methane was formed on Co-based catalysts in comparison to Fe-based catalysts (Schulz et al., 2005).

However, when a CO_2-H_2 mixture is used as feed gas, Co acts as a methanation catalyst. A mixture of Co-Fe catalysts also had a low selectivity to the favorable hydrocarbons (Tihay et al., 2002; Dorner et al., 2009). Akin et al. (2002) reported that for Co/Al_2O_3 catalysts, 70 mol.% methane was produced by CO_2 hydrogenation. CO and CO_2 conversions occurred by two different reaction pathways; for the first reaction, C–H and O–H species were produced from hydrogenation and for the second reaction, H–C–O and O–H surface-bound intermediates were formed.

SASOL usually uses iron catalyst (Dry, 2002). A precipitated iron catalyst is used in low temperature reactors operating at 220–250°C and 25 bar in MTFBR (multitubular fixed bed reactor) or SR (slurry reactor) mode. A fused iron catalyst is used for high temperature fluidized bed reactors operating at 300–320°C and 25 bar in fixed or circulating modes.

The main challenges in catalyst development are activity and the wide range of hydrogenation products. Generally, lower olefins and long-chain hydrocarbons are the most desirable products (Yang et al., 2017).

6.3.2.1 Role of Active Site

As iron has higher WGS activity, it is a less attractive choice for traditional FT synthesis. For WGS activity, iron magnetite phase or amorphous oxidic iron phase is responsible; while for chain growth, carbide phase (χ-Fe_5C_2) is essential. However, infra-red (IR) spectroscopic studies revealed that both reactions proceed via the same intermediates (Chakrabarti et al., 2015).

Choi et al. (2017) reported the production of heavy hydrocarbons on a novel Cu-Fe catalyst prepared by reduction of delafossite (a copper iron oxide mineral with the formula $CuFeO_2$). In their catalyst, $\chi\text{-}Fe_5C_2$ (Hägg iron carbide), as active catalytic phase for heavier hydrocarbon production, was formed by reduction and *in-situ* carburization of $CuFeO_2$. Delafossite is a precursor to Cu-Fe catalysts for the production of higher liquid hydrocarbons with 65% selectivity, methane selectivity of 2%–3% and olefin/paraffin ratio of 7.3.

6.3.2.2   Role of Support

The support affects the catalytic performance of the Fe-based catalysts. During catalytic testing, the support acts as a stabilizer that prevents Ostwald ripening or sintering of the active phase. $\gamma\text{-}Al_2O_3$, silica and titania are good supports. Due to SMSI (strong metal-support interactions), by which well dispersion is obtained, $\gamma\text{-}Al_2O_3$ is the best support as it hinders catalyst sintering. Moreover, it was reported that $\gamma\text{-}Al_2O_3$ could also act as a catalytic active phase (Yan et al., 2000; Sai Prasad et al., 2008; Zhao et al., 2008; Dorner et al., 2010b).

An Fe-based CO_2 hydrogenation catalyst with potassium promoter and $\gamma\text{-}Al_2O_3$ support showed improved performance. In the presence of K and $\gamma\text{-}Al_2O_3$ in the catalyst, potassium alanate ($KAlH_4$) phase is formed that has reversible hydrogen sorption at 250–300°C. The K-alanate phase is an H_2 reservoir and causes a reduction in hydrogenation of carbonaceous species that bound to the surface and consequently a higher alkene and lower methane selectivity is observed. On the other hand, this phase has influence on molecular hydrogenation activation. It has been also reported that potassium, in comparison to sodium, is a better promoter for the Fe catalysts (Herranz et al., 2006).

Sodium alanate adsorbs, activates and desorbs H_2 at lower temperatures (150°C) and, consequently, for CO_2-FT hydrogenation, reactor is less useful (Ares et al., 2009).

6.3.2.3   Role of the Promoter

In order to improve the selectivity of Fe-based catalysts, promoters must be utilized. Their first role is enhancing reducibility of the iron oxide species. The promoter must improve the distribution of Fe species and weaken acid-base support properties by metal-support interactions (Li et al., 2001).

In comparison to Co-based catalysts, Fe-based catalysts can convert CO_2 with higher affinity for alkenes and long chain hydrocarbons, but for undoped catalysts, there is a high selectivity to unfavorable products; hence, promoters are needed to optimize the synthesis yield (Dorner et al., 2010a).

Promoters are of electronic and/or structural types. Electronic promoters donate or withdraw electron density near catalyst valence band due to which local electron density around the surface and consequently active site are modified. Electronic promoters increase conversion levels and modify product selectivity. Structural promoters have influence on the formation and stabilization of active phase in catalyst, by which better active phase dispersion and higher conversion rates are obtained.

For Fe-based catalysts, if a suitable promoter is added, its selectivity for C_{5+} hydrocarbons could be improved. Alkaline metals (Rb, Na or K), second metals

(Mo, Mn, Zr, Cu, Ta, Cr or La) and metal oxides (TiO$_2$ and α-Al$_2$O$_3$) are widely studied promoters for tailoring product distribution of Fe catalysts (Li et al., 2013; Rodemerck et al., 2013; Chew et al., 2014; Ding et al., 2014; Al-Dossary et al., 2015).

Potassium (K) influences the catalytic performance. It plays a role in FT synthesis and CO$_2$ conversion and reduces methane selectivity. Potassium is an electronic promoter for the FT reaction. According to the mechanism proposed by Dry et al. (1969), potassium donate electron density to the vacant d orbital of Fe, due to which metal work function is decreased, dissociative CO adsorption is enhanced and hydrogen adsorption ability is decreased. Because of the decrease in H$_2$ adsorption, alkenes hydrogenation is reduced and, consequently, the alkene content in the product is increased. It was reported that larger amounts of K favored the CO$_2$ hydrogenation reaction, as indicated by the increased CO$_2$ conversion and decreased methane yield (Ning et al., 2009; Khobragade et al., 2012).

Manganese (Mn) as a catalyst promoter has both electronic and structural effects on Fe. In the presence of Mn, methane formation was suppressed and alkene/alkane ratio was increased. On the other hand, the reduction, dispersion and carburization of Fe$_2$O$_3$ phase were promoted by Mn and the catalyst surface basicity was increased (Li et al., 2007).

Copper (Cu) can be used (instead of Mn) as a promoter; as it can enhance the ability of CO$_2$ hydrogenation by the Fe catalyst. Copper is an RWGS catalyst and, during carburization, can augment hematite particles reduction and consequently catalytic particle dispersion is increased. For dissociative adsorption of hydrogen, active sites are provided after reduction of copper to its metallic form (Ando et al., 1998).

Ceria (CeO$_2$) is a WGS reaction catalyst. In the beginning of the reaction and before occurring chain growth, CO is produced from CO$_2$ by the RWGS reaction. Ceria is a good promoter for CO$_2$ hydrogenation by Fe-based catalysts. Moreover, it enhances the RWGS reaction at low temperatures. Pérez-Alonso et al. (2008) evaluated unsupported and Ce-promoted Fe catalysts for CO$_2$ hydrogenation and reported that the unpromoted Fe catalyst had lower activity for chain growth but produced more CO. They concluded that in the presence of ceria, Fe active sites were blocked and the catalyst was deactivated. They also reported that when ceria was deposited on the support, better performance was obtained. In the catalyst synthesis, ceria must be added to the support, before the addition of any other metals, to avoid blocking of metal active sites (Dorner et al., 2010c).

Zn, La, Zr, Mg and Ru are other promoters that were evaluated for CO$_2$ hydrogenation on Fe catalysts (Niemelä and Nokkosmäki, 2005).

6.3.2.4   Role of Binder

An extrudate Fe-K/γ-Al$_2$O$_3$ catalyst with alumina binder had acceptable CO$_2$ hydrogenation capability while with silica binder it had an inferior performance. For PVA (polyvinyl alcohol) binder, its presence had no effect on catalyst activity and selectivity. An Fe-K/γ-Al$_2$O$_3$ catalyst with silica binder in comparison to an Fe-K/γ-Al$_2$O$_3$ catalyst had a higher methane (16 mol.%) and C$_2$–C$_4$ (43.5 mol.%) selectivity. For the Fe-K/γ-Al$_2$O$_3$ catalyst with alumina binder, the main products were C$_{5+}$ hydrocarbons, which was related to the increase in strong acid sites of this catalyst. Chemically, alumina is not inert to the catalyst (especially at high temperatures) and its interaction with the catalyst strongly influences the selectivity and activity in CO_2 hydrogenation reaction, by changing the catalyst acidity (Lee et al., 2003).

6.3.3 Mechanisms

For CO_2 hydrogenation to hydrocarbons, researchers proposed different mechanisms. In one proposed by Lee et al. (2004a), CO_2 is first reduced by Fe(II) and in the next step, H radical abstraction occurs on the catalyst surface. By attacking residual H to carbonyl C, OH formic acid and CO are formed. Fe-CH_2 radical (that is, C–C propagation species) is also formed in the same manner. As higher hydrocarbons are the main products, chain propagation is the major route. By less hydrogen uptake and in the absence of excess hydrogen in the reaction media, a higher α-olefin selectivity to paraffin is obtained; hence, reactor configuration must be in a way that hydrogen dosing is adjusted (Saeidi et al., 2014).

6.3.4 Effect of Reaction Conditions

In addition to catalyst formulation, the reaction conditions also influence the catalytic performance and products yield (Kim et al., 2006).

6.3.4.1 Effect of Space Velocity

At low space velocity (that is equivalent to long residence time), higher CO_2 conversion with improved hydrocarbon yield was obtained which is equivalent to the production of more water. Space velocity (SV) is the main tool for increasing conversion (Lee et al., 2004b). The analysis of a large number of experimental reactors showed that a recycle reactor with a space velocity of 2,000 h^{-1} is the best one followed by reactors in series and single reactor, each with a space velocity of 1,000 h^{-1} (Saeidi et al., 2014).

6.3.4.2 Effect of Temperature

There is a thermal conflict between the reactions in CO_2-FT. Therefore, the effect of temperature on hydrocarbon product yield is not so obvious. Higher temperatures favor the endothermic RWGS reaction and increase the catalyst activity but also enhance catalyst deactivation by inert coke deposition. Simultaneously, enhanced C-C bound breakage by cracking (high CH_4 yields) and carburization (and thus, higher olefin/paraffin ratio) are observed (Saeidi et al., 2014).

6.3.4.3 Effect of Pressure

At a high pressure, hydrocarbon distribution shifted toward long-chain products with improved hydrocarbon yield while the olefin/paraffin ratio was decreased. Due to the reversible nature of the RWGS reaction, increasing pressure has no effect on CO_2 conversion unless the produced CO by the RWGS reaction is consumed by the FT reaction. Basically, FT synthesis is an irreversible reaction with product molecules less than those of reactants with water as a byproduct. At high reaction pressures, Fe_3O_4 is the dominant phase of the catalyst. Both FT and RWGS reactions produce water, therefore, the high CO_2 conversion with improved hydrocarbon yield at high pressures lead to high H_2O partial pressures on catalyst bed and consequently the water must be removed in-situ.

6.3.4.4 Role of the Reactor Type

In a single reactor, the endothermic and exothermic steps of a reaction occur. As fixed-bed reactors have some drawbacks, such as heat management, fluidized-bed and slurry reactors are used for increasing CO_2 conversion and obtaining desirable products as they are beneficial for removing the heat of reaction. On an Fe-Cu-Al-K catalyst, for example, the productivity in terms of space-time-yield (STY, $mmol.g_{cat}^{-1} h^{-1}$) was in the order: fluidized-bed (41.8) > slurry (35.2) > fixed-bed (31.8) reactor under 300°C temperature, 1 MPa pressure and 2,000 $ml.g_{cat}^{-1}.h^{-1}$ space velocity (Kim et al., 2006). Moreover, the light olefins and heavy saturated hydrocarbons were selectively formed in fluidized-bed and slurry reactors, respectively.

6.3.5 CATALYST DEACTIVATION

Crystallite size growth and consequent component separation are the main causes of catalyst deactivation and catalyst coking has a minor role.

The carbidization of Fe_5C_3 to the stable and inactive FeC_3 phase causes deactivation of iron-based catalysts (Li et al., 2018). The high amount of water produced during CO_2-FT also decreases the catalyst activity through oxidation of the active phase and promoting sintering. In the case of cobalt catalyst, oxidation of the metal by CO_2 results in deactivation.

According to Riedel et al.'s (2003) research, on Fe-based hydrocarbon synthesis catalysts, five distinct kinetic regime zones can be identified with TOS (Figure 6.2). In zone I, the reactants are adsorbed on catalyst surface and catalyst carbonization occurs. In zones II and III, during deposition of carbon, the products of RWGS reaction are dominant. In zone IV, FT activity develops to the steady-state condition and keeps it to zone V. Before the reaction, for the reduced catalysts, iron phases which mainly consist of α-Fe and Fe_3O_4. Fe_2O_3 and Fe_3O_4 phases are transformed with time and a new amorphous oxidic iron phase is formed that is active in the RWGS reaction. By the reaction between Fe and C (from CO dissociation), FT activity begins and the active Fe_5C_2 (iron carbide) is formed. During the RWGS reaction, a significant deactivation occurs to Fe-based catalysts due to water formation. Hence, water must be removed *in situ*. By Fe_5C_2 carburization, a stable inactive carbide (Fe_3C) is formed that is responsible for catalyst deactivation (Sai Prasad et al., 2008).

A study on deactivation pathway of the Fe-K/γ-Al$_2$O$_3$ catalyst in a packed-bed reactor (300°C, H_2/CO_2=3) showed that Fe-based catalysts were significantly deactivated during CO_2 hydrogenation. X-ray photoelectron spectroscopy (XPS), high-resolution transmission electron microscopy (HR-TEM), temperature programmed oxidation (TPO) and Mössbauer spectroscopy showed that carbon deposits and catalyst poisoning were the main causes of catalyst deactivation. During the time, after hydrogen reductions, Fe_3O_4 (hematite) was formed and gradually carburized to χ-Fe_5C_3; which in turn is converted into Fe_3C (that is an

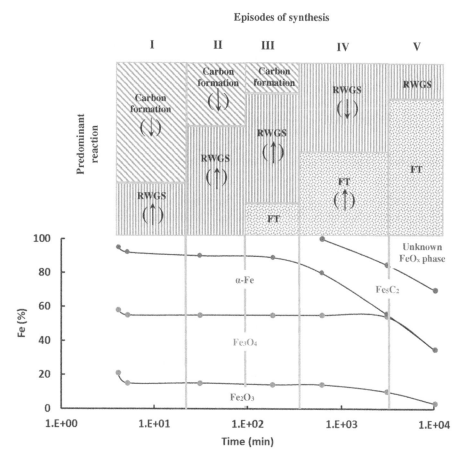

FIGURE 6.2 Iron-phase and catalysts selectivity with time (250°C, 1 MPa, 1,800 ml(NTP) h^{-1} g_{Fe}^{-1}, $H_2/CO_2 = 3$) (Redrawn from Schulz et al., 2005).

inactive species for CO_2 hydrogenation). However, the long-run (500 h) activity was higher than 35% (Lee et al., 2009).

6.3.6 PROCESSES

Gouse et al. (1993) reported a novel process that is the integration of PV (photovoltaic) cells and IGCC (integrated coal gasification combined cycle) technologies for the reduction of CO_2 emissions. Liquid transportation fuels are produced from H_2 obtained by PV water electrolysis and CO_2 emissions of an IGCC plant in a FT process. This process is a closed carbon cycle in which carbon of CO_2 captured from transportation fuels is fixed back in them by H_2 obtained from solar power. This principle can be used for bio-syngas as well (Srinivas et al., 2007).

For a CO_2-rich syngas, a process containing multiple FT reactors with intermediate water removal have better performance in comparison to a single reactor.

6.4 COUPLING WITH OLEFINS

The direct carboxylation of ethylene by carbon dioxide to acrylic acid is a novel route for the synthesis of acrylic acid or acrylates (Scheme 6.3).

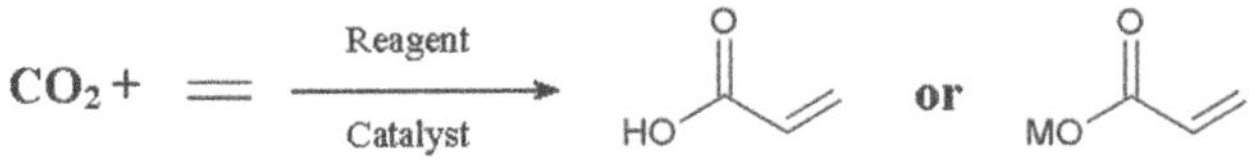

$$CO_2 + \quad = \quad \xrightarrow[\text{Catalyst}]{\text{Reagent}} \quad \underset{HO}{\bigvee} \quad \text{or} \quad \underset{MO}{\bigvee}$$

SCHEME 6.3 Synthesis of acrylic acid and acrylate from CO_2 and ethylene.

Acrylic acid is an important precursor of a large number of polymers and chemicals such as plastics, elastomers, textiles, adhesives, coatings and paints.

The state-of-the art production route is based on oxidation of propylene (via acroleine, $CH_2 = CHCHO$), although less expensive propane has been also considered as a feed.

Since the initial works of Hoberg and coworkers in 1980s (Hoberg et al., 1987), several nickel complexes were found to be active in this reaction. Zero-valent Ni(0) and Pd(0) based complexes have been commonly used as the catalyst for acrylate synthesis although recently low valent-group 10 metal complexes have also been found to be effective (Uttley et al., 2020). Figure 6.3 shows the reaction mechanism.

The formation of sodium acrylate instead of acrylic acid shifts the overall endergonic reaction to an exergonic favorable reaction (−59 kJ/mol; Vavasori et al., 2018).

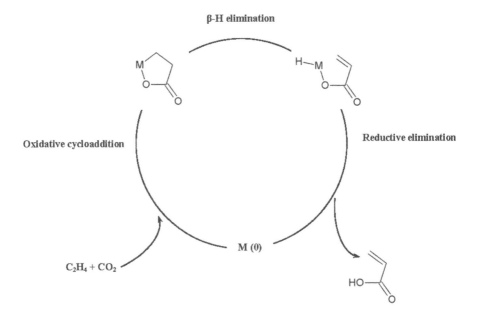

FIGURE 6.3 Simplified catalytic cycle for ethylene carboxylation to acrylic acid via five-membered metallactone intermediate (modified from Liu et al., 2013).

A breakthrough was achieved by Limbach and coworkers who obtained a turn-over number (TON) of 10 or larger with diphosphine nickel complex in the presence of sodium tert-butoxide (NaO^{t}Bu) as the base (Lejkowski et al., 2012). They showed that NaO^{t}Bu accelerates the transformation of Ni-lactones to acrylate complex. The selection of the ligand and base are, therefore, important parameters affecting the catalyst performance.

Despite great research efforts, the chemistry of the reaction is in its early stages and this transformation remains an important challenge in synthesis chemistry. Recent reviews on catalysts and reaction mechanism have been published (Hollering et al., 2016).

6.5 CO$_2$ TO POLYMERS

6.5.1 INTRODUCTION

In everyday life, polymers are present in different applications. As their properties can be tailored, their applications ranged from simple to pure high-tech ones. Recently, researchers are focused on the synthesis and utilization of polymers based on carbon dioxide (Langanke et al., 2015).

In polymer synthesis, carbon dioxide can be used directly as polymer building block or indirectly by using the intermediates or polymer building blocks derived from chemical transformation of carbon dioxide. Researchers proved that both routes are feasible.

CO$_2$ conversion into useful polymers can proceed via different reaction pathways that are mostly endothermic. Some methods have been developed for overcoming high-energy barriers based on reduction, increasing electrophobicity of carbonyl carbon and oxidative coupling with unsaturated compounds on low valent metal complexes (Song, 2006; Sakakura et al., 2007).

6.5.2 POLYMERS BASED ON CO$_2$-DIRECT APPROACH

In the direct approach, CO$_2$ (in combination with suitable reagents and in the presence of appropriate catalysts) is used as a monomer and polymers such as polyols, polyesters, polycarbonates, polyurethanes and polyureas with high CO$_2$ content are produced (Grignard et al., 2019).

An example of direct route for synthesis of CO$_2$-based polymers is catalytic copolymerization of epoxides and CO$_2$, by which the epoxide subjects to ring opening and linear carbonate is produced. Based upon the specific catalyst employed, alternating polycarbonates or polyether carbonates containing an ether motif and carbonate units (CUs) could be produced (Scheme 6.4).

There are different routes for the synthesis of polyesters and polycarbonates, namely step-growth and chain-growth polymerizations. The latter includes ring-opening co-polymerization of two monomers and ring-opening polymerization of a cyclic monomer (Zhang et al., 2018).

For step-growth synthesis, highly toxic and hazardous phosgene (COCl$_2$) must be replaced by eco-friendly renewable carbonyl reagents. DMC (dimethyl carbonate)

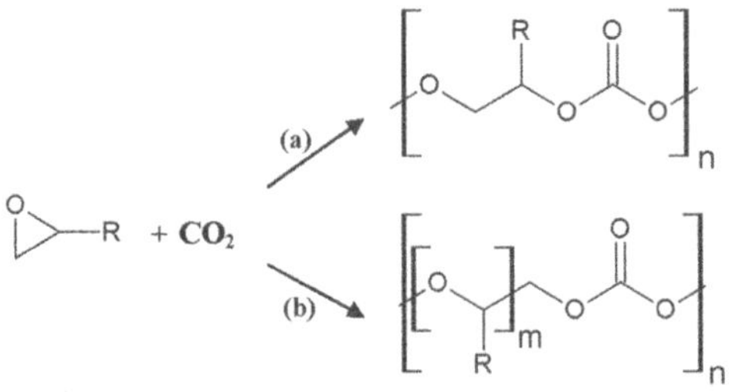

SCHEME 6.4 Catalytic copolymerization of a terminal epoxide and CO_2 to yield (a) alternating polycarbonate or (b) polyether carbonate.

as an organic carbonate is a green safe substitute for phosgene that can be produced from CO or CO_2. CO_2 can be used in polycarbonate synthesis. The direct diol-CO_2 coupling is limited by reaction equilibrium and needs an effective dehydration system (Sakakura and Kohno, 2009; Zhang et al., 2018).

Different epoxides were evaluated for copolymerization with CO_2; such as PO (propylene oxide), ECH (epichlorohydrin) and CHO (cyclohexene oxide) (Qin et al., 2015).

It was reported that for the synthesis of polyether carbonates, double metal cyanide complexes are the best catalytic systems. In the 1960s, these catalysts were developed for epoxide polymerization to polyether polyols (Langanke et al., 2015).

There are other examples of direct CO_2-copolymerization; such as inter alia CO_2 copolymerization with oxetanes (oxygen-containing four-membered rings) or aziridines (nitrogen containing three-membered rings) for the production of aliphatic polycarbonates and polycarbamates, respectively. Moreover, the application of alkenes and phosphorous compounds as co-monomers for direct CO_2-polymerization is also reported (Darensbourg et al., 2011).

6.5.2.1 Polycarbonates

Polymers with -(OCO_2R)- repeating units inside the main backbone are called polycarbonates (PCs). They are traditionally produced by polycondensation of diols with phosgene, which give poor control of the molecular weight and also risks associated with handling highly poisonous phosgene (Zhang et al., 2018). Replacing phosgene with CO_2 is therefore highly desirable.

Poly (ethylene carbonate) (PEC), poly (pentene carbonate), poly (cyclohexene carbonate) (PCHC), poly (cyclohexadiene carbonate), poly (propylene carbonate) (PPC), poly (butylenes carbonate) (PBC), poly (hexane carbonate), poly (cyclopentene carbonate) and poly (styrene carbonate) are examples of aliphatic polycarbonates obtained by alternating copolymerization of CO_2 and epoxides. Among these, PEC, PPC, PBC and PCHC have major industrial applications (Darensbourg et al., 2013; Honda et al., 2014).

Being flame retardant, having good thermal stability, high thermal resistance, toughness, transparence and stiffness, aromatic polycarbonates are used as engineering plastics in construction, automotive, and electronic and electrical devices. Due to poor mechanical properties of aliphatic polycarbonates, their application in packaging is limited. As aliphatic polycarbonates are biocompatible and biodegradable, they can be used in the biomedical sector (Tempelaar et al., 2013; Xu et al., 2014).

These materials are produced from diols (aliphatic PCs), bisphenol A (aromatic PCs), and phosgene or DPC (diphenyl carbonate) by melt polycondensation. Polycarbonates with high molecular weights have applications in rigid plastics, matrices for polymer (nano)composites and blends with petrochemical-based and bio-based polymers (Sun et al., 2016; Zhu et al., 2016).

Synthesis of polycarbonates directly from CO_2 (as monomer) or from CO_2-source building blocks are possible form different pathways. These pathways include polycondensation through transesterification of acyclic carbonates and diols, ROP (ring-opening polymerization) of cyclic carbonates (CCs), ROCOP (ring-opening copolymerization) of oxiranes and CO_2, polyaddition of diols and activated bis-α-alkylidene 5CC (five-membered cyclic carbonates) and direct copolymerization of CO_2 with diols or mixtures of dihalides and diols (Grignard et al., 2019).

The overall polymeric properties of polycarbonates determine their applications. Because of their thermal characteristics, aliphatic polycarbonates cannot be used in the classical thermoplastic applications. They are normally used as binders in adhesives, ceramics and pyrotechnics (Wang et al., 2002).

Recently, it was reported that in the synthesis of cyclic carbonate, CO_2 can be used as a cheap, safe and renewable carbonyl source. Similar to dialkyl carbonate synthesis from CO_2, this dehydrative cyclization reaction has unfavorable thermodynamics as well (Sakakura and Kohno, 2009). Hydroxyl group activation for CO_2 addition with *in situ* formation of a leaving group is a successful strategy for ring-closure facilitation.

6.5.2.1.1 Poly(Propylene Carbonate)

For poly(propylene carbonate) (PPC), its birth was in 1969 and a successful running of its pilot plant was in 2004 (Qin et al., 2015).

Poly(propylene carbonate) that is produced by copolymerization of CO_2 and PO, is of academic and industrial interest. According to the final application, the products of PO-CO_2 copolymerization fall into two groups. One is PPC that has high MW (molecular weight) and high carbonate contents with acceptable biodegradability with different applications in packaging materials and gas barrier films. The other is polycarbonate polyols that have low MW and are used as the raw material of polyurethane industry (Qin and Wang, 2010; Qin et al., 2015).

PPC, as an amorphous polymer with low T_g (~35°C), is brittle at temperatures lower than 20°C. Above 30°C, it has a poor dimensional stability. Hence, for modifying PPC into a viable biodegradable plastic, plasticizing and toughening of PPC are of high importance.

One solution for improving the mechanical and thermal properties of PPC is terpolymerization of PO and CO_2 with a third monomer. Of course, this solution is not an economically viable one. Another alternative is physical blending of PPC with other biodegradable polymers that is a simple process for modifying PPC. It is worth mentioning that blends containing PPC and a biodegradable polymer have low miscibility that is because of relatively weak intermolecular interactions.

Hyperbranched poly(ester-amide) (HBP) that can have many peripheral hydroxyl groups is one alternative for solving miscibility issue. HBP can form intermolecular hydrogen bonding with PPC (Chen et al., 2011).

PPC has versatile applications; such as being used in foams, films and fibers. There is an increasing environmental concern for the packaging wastes; hence, the preparation of biodegradable packaging materials is of high interest (Nagiah et al., 2012).

PPC can be processed into film that can be used for preparation of overwrap, mulch film, shopping and composting bags. Since 2010, Nantong Huasheng Co. (China) has developed biodegradable plastics based on PPC (with PCO2 brand mark) (Qin et al., 2015).

Normally, plastics are used for packaging. Oxygen permeability of PPC is lower than 20 $cm^3/m^2/day/atm$; which is better than PLA (polylactide), PBS (poly(butylenes succinate)) and Ecoflex (as biodegradable polymers) (Qin et al., 2015).

High specific strength, light density, good thermal and sound insulation and strong energy absorption capability are advantages of polymeric foams. These foams have versatile applications in daily life, such as in thermal insulation, automotive and airplane parts, acoustic dampening, sporting equipment, microelectronic applications, packaging materials and optical devices. Waste disposal, flammability, blending agents influence on environment and recyclability are the issues that need to be investigated in polymer foam industry. Biodegradable foam materials improve the waste disposal and recyclability issues (Qin et al., 2015).

6.5.2.1.2   Block Polymers PLA-PCHC

Polylactide (PLA) is a bio-derived plastic that is commercially produced and can be biodegraded or recycled at the end of life. For household goods, packaging, biomedicine and automotive, PLA can be replaced for petroleum-derived plastics. One drawback of PLA is its poor temperature stability that is due to its moderate T_g (~60°C) (Hillmyer and Tolman, 2014; Rabnawaz et al., 2017).

Poly(cyclohexene carbonate) is an amorphous polymer that is produced by copolymerization of CO_2-cyclohexene and has a higher T_g value (110–120°C) that is due to its rigid carbonate units and ring structures (Paul et al., 2015).

Recently, researchers studied the performance of block polymers of PLA-PCHC (PLA-b-PCHC) delivered by combined polymerization using a single catalyst. Its formation pathway comprises two catalytic cycles being bridged by di-zinc catalyst. PLA is synthesized in Stage I by controlled ROP of LLA-OCA (O-carboxyanhydride) where one molecule of CO_2 is released per polymer repeating unit. This CO_2 molecule is used in Stage II in which PCHC is produced via controlled ROCOP of CO_2-epoxide. The di-zinc catalyst used in this process was highly selective by which block polymer was produced with a maximum atom economy of 91% (Buchard et al., 2014).

6.5.2.2   Polyureas

Polymers that have urea linkage (that is –NH–C(=O)–NH– groups) in their polymer backbone are called polyureas (PUAs). In automotive industry, household products, construction and marine-related technologies, PUAs have different applications. As

PUAs exhibit good abrasion, corrosion and chemical resistance, they can be used as coating materials for the protection of steel and concrete substrates. Polyaddition by diamine and diisocyanate reagents are industrial procedure of PAUs synthesis (Scheme 6.5) (Buckwalter et al., 2013; Wazarkar et al., 2017; Iqbal et al., 2018).

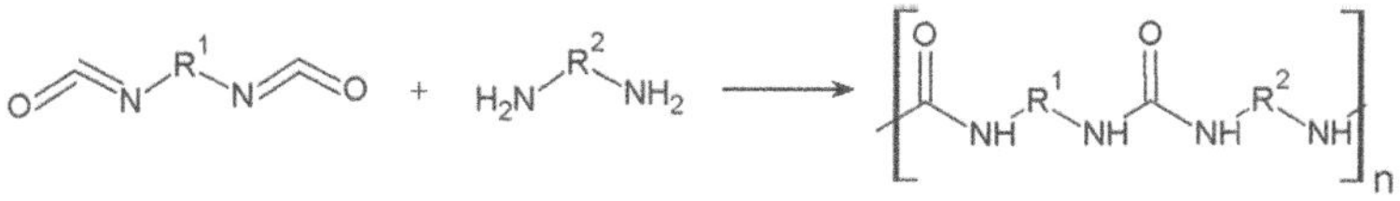

SCHEME 6.5   Industrial synthesis of polyurethanes by addition of isocynates and diamines.

Non-isocyanates route from CO_2-derivatives are urea or (a)cyclic carbonates, or direct CO_2 copolymerization with diamines (Grignard et al., 2019).

6.5.2.3   Polyurethanes

Polyurethanes (PUs) have the –NHCOO– group, but other groups can also be present in their repeating chain. Polyurethane as an important polymer with various applications in everyday life was discovered in 1937 by Otto Bayer. Sealants, heart valves, foams, coatings, cardiovascular catheters, adhesives and elastomers are different applications of PU (Joseph et al., 2018; Groenewolt, 2019). As depicted in Scheme 6.6, PUs are industrially produced by polyaddition of diisocyanates and di- or polyols.

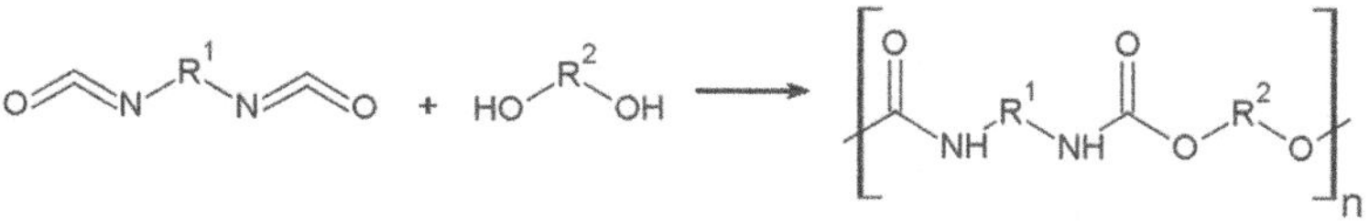

SCHEME 6.6   Industrial synthesis of PUs from isocynates.

Recently, researchers focused on isocyanate-free routes for PU synthesis; carbon dioxide plays an important role in this regard (Delebecq et al., 2013). Several CO_2-based routes with different reagent have been proposed (Grignard et al., 2019). A direct route is ROCOP and step-growth polymerization (SGP), for example, with aziridines (Scheme 6.7).

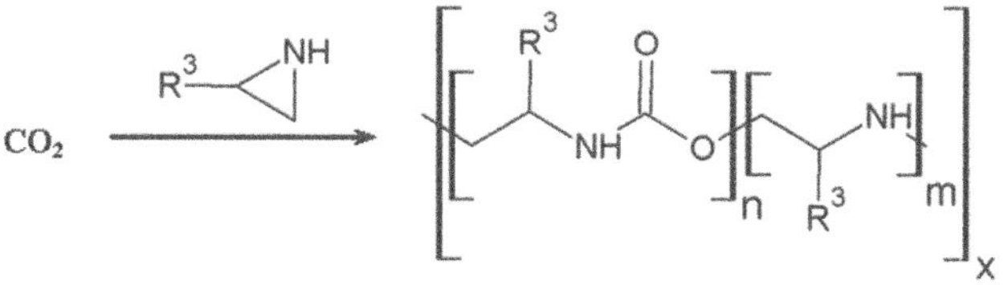

SCHEME 6.7   Synthesis of polyurethanes from CO_2 and aziridies.

6.5.2.4 Polyesters

Polyesters have -C(=O)-O- linkages inside their backbone. Because of excellent bio-compatibility and (bio)degradability, they are used in surgery applications, packaging and fabrics. ROP lactides/lactones, α-hydroxyacids self-polycondensation and dehydrative polycondensation of diols and dicarboxylic acids are industrial routes for polyesters production (Memic et al., 2019; Pappalardo et al., 2019). CO_2/olefin copolymerization is a potential CO_2-based route.

6.5.2.5 Polyols

CO_2-PO copolymerization that produces high MW polycarbonates has received much attention these days, but the controlled synthesis of CO_2-PO based polyols with low MW needs further research. The main application of polyol polymers is in polyurethane industry. As most of the polyol polymers are petroleum based, the production of CO_2-based polyols is a new choice in polyurethane industry (Qin et al., 2015).

Polyols based on aliphatic polycarbonates have low glass transition temperatures and viscosities and can be good alternatives to common petrochemical-based polyols that are used for making adhesives, furniture foams, coatings and clothing (Darensbourg, 2007).

It was reported (Cyriac et al., 2011) that ether linkage incorporation into the polymer chain can tune the mechanical and thermal performance of the polyurethane. Hence, for the industrial applications, poly(carbonate-ether) polyols are more meaningful.

For polyurethanes, the state-of-art synthesis route is the reaction between polyols and isocyanates. In order to decrease the final price of PU, a low-cost feedstock (such as CO_2) is incorporated into the backbone of polyether polyols and polycarbonate polyols or CO_2-polyols are produced. Based upon life cycle analysis, oligoethercarbonates containing 20 wt.% CO_2 reduce GHG emissions by 11%–19% and saves fossil resources by 13%–16%. The PU obtained from CO_2-polyols has better hydrolysis/oxidation resistance (Langanke et al., 2014; Wang et al., 2016).

By immortal copolymerization of CO_2-PO with the Zn-Co-DMC catalyst, in the presence of a proton-containing starter, CO_2-polyols are produced (Scheme 6.8). This catalyst is active for random copolymerization of CO_2-PO. Besides the catalyst, for determining the feature of CO_2-polyols, the selection of starter is important. In the presence of oligomeric alcohol starters, long copolymerization time is necessary and it is difficult to produce CO_2-polyols with high CU content and low M_n. To overcome these problems, organic dicarboxylic acids were used as the starter (Zhang et al., 2011).

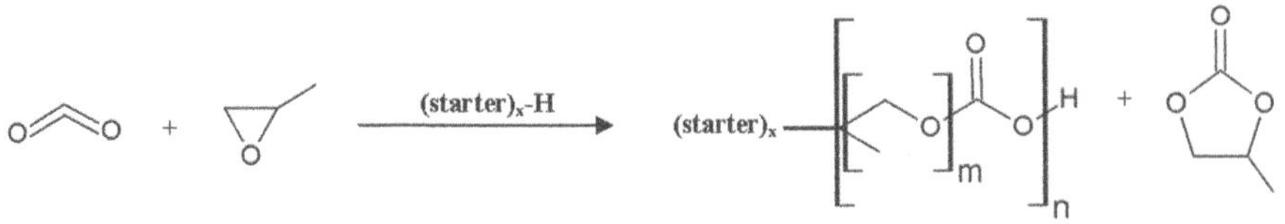

SCHEME 6.8   Synthesis of polyols from copolymerization of CO_2 and PO.

When sebacic acid was used as the starter (with catalytic activity > 1 kg$_{polymer}$/g$_{cat}$), CO_2-diol was synthesized in a controllable manner with M_n lower than 2,000 g/mol and CU content in the range of 40%–75%. For the synthesis of CO_2-triol in a controllable manner, TMA (1,3,5-benzenetricarboxylic acid) was used as starter and the produced triol had M_n in the range of 1,400–3,800 g/mol and CU content of 20%–54%. TMA acted as a chain initiation-transfer agent; in the first stage, PO homo-polymerization was initiated by TMA and oligo-ether triol was produced, which in-turn acted as a new chain transfer agent (CTA) and participated in the copolymerization for the formation of oligo(carbonate-ether) triol (Gao et al., 2012).

Normally 10 wt.% of the starter is incorporated in CO_2-polyols and thus its cost must be taken into account. As oxalic acid is the cheapest organic dicarboxylic acid, it is commonly chosen as the starter (Liu and Wang, 2017).

6.5.3 POLYMERS BASED ON CO$_2$-INDIRECT APPROACH

In this approach, carbon dioxide (as a C_1 feedstock) is reacted with energy-rich substrates and the produced building blocks (such as carbamates, (a)cyclic carbonates and ureas) are transformed to polymeric materials (with controlled specification) by homo- or co-polymerization. It is worth mentioning that urea (and its derivatives) synthesis from carbon dioxide and its subsequent poly-condensation is in operation on an Mt/y scale (Langanke et al., 2015).

Urea is commercially produced by reaction of carbon dioxide with ammonia being mostly used as a fertilizer. Urea can be used for the synthesis of urea-formaldehyde (UF) resins which accounts for about 80% of the so-called amino resins (also called aminoplasts) (Peters et al., 2011). Amino resins are a class of thermosetting polymers being formed by combining an aldehyde with $-NH_2$ group. They are used as binding agent (for woody material), making glues and varnishes.

The synthesis of amino resins from formaldehyde and amino compound involves two successive steps (Scheme 6.9). The first step is the addition of formaldehyde to introduce the hydroxymethyl group, known as methylolation or hydroxymethylation. Then, a condensation reaction occurs which involves the linking of monomer units with the liberation of water to form a dimer, a polymer chain or a vast network. This is usually referred to as methylene bridge formation, polymerization, resinification or simply cure (Williams, 2002).

CO_2-based organic carbonates (e.g., cyclic carbonates) can be used as green phosgene replacements or as (co)monomers in ring-opening (co)polymerization for producing polycarbonates and polyurethanes (Langanke et al., 2015). This route enables good control of the molecular weight and polydispersity (a measure of nonuniformity of molecular sizes) of the product but is limited for the most part to 6,7 membered rings (Zhang et al., 2018). Furthermore, the synthesis of the monomers from renewable sources is not trivial.

In the step-growth synthesis of polycarbonates, dialkyl carbonate (which in turn can be produced from CO_2) can be used as a substitute for phosgene. A green straightforward route for the synthesis of dialkyl carbonates is dehydrative condensation

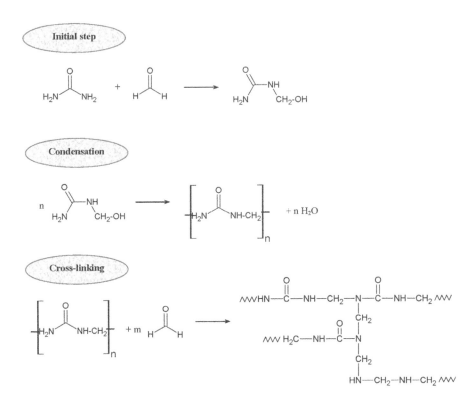

SCHEME 6.9 Synthesis of amino resins from formaldehyde and urea.

of alcohols and CO_2. Unfortunately, this reaction is limited by unfavorable equilibrium for which a catalyst with high performance (such as a metal alkoxide), an acidic co-catalyst and an effective dehydrating agent (such as acetal) is needed to shift the equilibrium toward carbonate formation. By this method and based on acetal, DMC (dimethyl carbonate) can be produced with 40%–60% yield (Choi et al., 2008; Bhatia et al., 2016). The resulting dialkyl carbonate can undergo polycondensation with a diol through transesterification to form a polycarbonate as shown in Scheme 6.10 (Grignard et al., 2019).

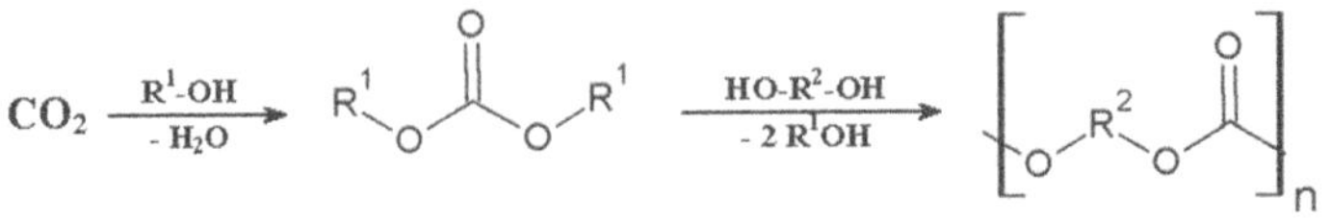

$$CO_2 \xrightarrow[-\,H_2O]{R^1\text{-OH}} \quad R^1{\small\diagdown}O{\overset{O}{\underset{}{\diagup}}}O{\small\diagup}R^1 \xrightarrow[-\,2\,R^1OH]{HO\text{-}R^2\text{-OH}} \left[O{\small\diagdown}R^2{\small\diagup}O{\overset{O}{\underset{}{\diagup}}} \right]_n$$

SCHEME 6.10 Polycarbonate synthesis pathway from CO_2-sourced dialkyl carbonate.

Polycondesation of a linear (or cyclic) polycarbonate with diamines produces polyurea, e.g., as in Scheme 6.11.

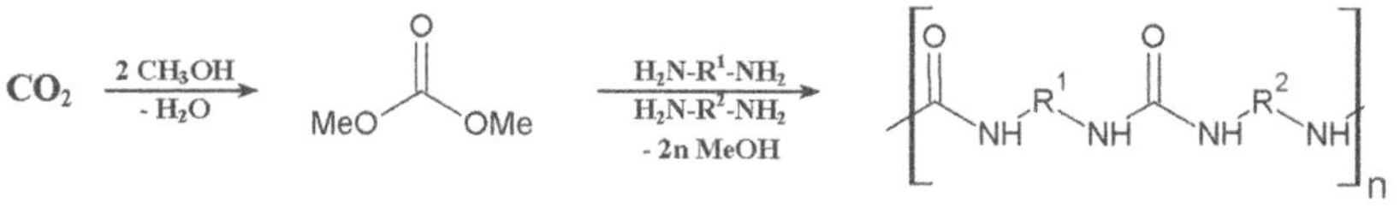

SCHEME 6.11   Polyurea synthesis pathway from CO_2-sourced dimethyl carbonate.

Similarly, step-wise polycondensation with a diol and a diamine produces polyurethane via dicarbamate (Scheme 6.12) (Grignard et al, 2019).

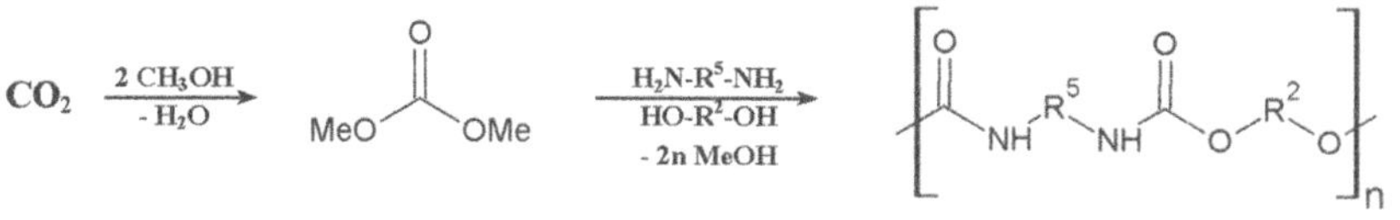

SCHEME 6.12   Polyurethane synthesis from CO_2-sourced dimethyl carbonate.

The (co)polymerization of the CO_2-derived light olefins (from tandem CO_2 hydrogenation/methanol to olefins reactions) and acrylic acid (from CO_2 coupling with ethylene) are still other indirect routes for fixation of carbon dioxide into polyolefins and polyacrylate (PA) polymers, respectively, using the traditional polymerization technologies.

6.5.4 CATALYSTS

Different homogeneous and heterogeneous metal complexes were developed as catalysts for the synthesis of CO_2-based copolymers. ZnEt2-active hydrogen containing compounds, double metal cyanide complexes, zinc dicarboxylates and rare-earth metal coordination ternary catalysts are examples of the heterogeneous catalysts. For industrial purposes, synthesis and handling of heterogeneous catalysts are more convenient; although their chemical and crystal structures are not clear (Tao et al., 2008; Klaus et al., 2011; Dong et al., 2012).

Homogeneous catalysts are attractive for academic research. Metal tetraphenylporphyrin complex as the first single-site homogeneous catalyst was developed and used for the copolymeriztion of CO_2-epoxide (Wu et al., 2013). Metal-porphyrin, single or binary bi-functional catalysts based on (metal)-Salen complexes (such as Co-di(ketoiminato) and (III)-Salen, Cr(III)-Salen and -Salen and Al(III)-Salen complexes), zinc phenolate and discrete β-diiminate zinc are other examples of the homogeneous catalysts (Koning et al., 2001; Wu et al., 2011).

In the co-polymerization of PO and CO_2, zinc gluterate is an effective chemical for commercial use; as it has acceptable catalytic activity and selectivity, low cost and easy synthesis route. It was reported that Zn(II)-gluterate and -diiminateethylsulfinate as specific Lewis acidic catalysts and some rare-earth compounds presented very high carbonate selectivities (> 99%) (Ree et al., 2006; Zhang et al., 2008).

For the copolymerization of CO_2-PO, the bi-functional Salen-Co catalytic systems were introduced that produced copolymers with high MW (M_n>300,000 g/mol) (Qin et al., 2015).

The homogeneous catalysts have well-defined structure, by which a high catalytic activity can be obtained for the copolymerization of CO_2-epoxide even under dilute conditions. The main drawback of these homogeneous catalysts is the presence of a toxic metal (such as Co, Ni or Cr). Hence, developing environmental-friendly catalysts with no toxic metal center is of high importance. Zn, Al, Fe, Mg and Ti are examples of non-toxic metals that can be used in homogeneous catalysts. Surface area, morphology, crystallinity and size play important role in catalyst activity (Meng et al., 2002).

Researchers reported that catalyst plays an important role in ring opening copolymerization of epoxides and CO_2 (Darensbourg and Yeung, 2014). Heterogeneous and homogeneous organo-metallic catalysts containing Co (III), Zn (II) and Cr (III) are mainly used. The initiation step is when a metal bound initiator ligand and an epoxide molecule are displaced with each other. In the subsequent step, due to nucleophilic attack of a ligand or carbonate group ring opening occurs and a metal alkoxide intermediate is formed into which CO_2 is inserted and a metal carbonate species is formed that is a nucleophile for succeeding chain propagation. Then, the ring opening continues and reaction propagates cycling between the carbonate intermediates and metal alkoxide. Hydrolysis of the growing chain occurred when the reaction conditions were changed or reagents were used. When a polymer chain end-capped with a hydroxyl group was formed, polymerization was terminated (Trott et al., 2016).

6.5.5 MECHANISMS

6.5.5.1 Ring-Opening Copolymerization

Generally, ring-opening copolymerization of an epoxide with a cyclic anhydride or CO_2 follows the mechanism of coordination-insertion (Coates and Moore, 2004). In the beginning, by nucleophilic attack of an exogenous nucleophile or a ligand (X) onto a coordinated epoxide, polymerization starts and metal bound to alkoxide and cyclic anhydride or CO_2 inserts to it and, respectively, a metal carboxylate or carbonate species form. In the next step, by alternative insertion of epoxide and cyclic anhydride or CO_2 to oxygen-metal chain end, chain propagation proceeds and desired polyester or polycarbonate backbones forms. For compromising the yield and selectivity of these polymerizations, different competitive side reactions may occur. Consecutive epoxide molecule insertion to metal alkoxide intermediate is a common side reaction, due to which ether linkages in the resulting polymer backbone occur. As many ROCOP catalysts can mediate epoxide homo-polymerization, this side reaction is their problem. If special metal catalysts being highly selective for copolymerization reaction are used, this polyether formation can be mitigated. In the presence of alcohol and water as protic impurities, chain transfer of the propagating chains occurs as another side reaction, due to which molecular weight of the resulting polymer decreases (Nakano et al., 2009; Ian Childer et al., 2014).

For one-pot poly(ester-carbonate) synthesis, the terpolymerization of CO_2, epoxide and cyclic anhydride is a facile strategy. Due to the superior properties of poly(ester-carbonate), in comparison to polyesters and polycarbonates, this class of polymers as biodegradable ones gained attention. At first, cyclic anhydride is incorporated and then CO_2; consequently, diblock or tapered block poly(ester-*co*-carbonates) are formed. As the rate of anhydride insertion to metal alkoxide intermediate is faster than that of CO_2 insertion, a good selectivity is observed. CO_2 addition to an epoxide/cyclic anhydride copolymerization is a method for improving polyester thermal properties (Paul et al., 2015; Longo et al., 2016).

6.5.6 INDUSTRIAL EXAMPLES

For CO_2 utilization in the polymer field, there are successful cases. Direct epoxide-CO_2 copolymerization for the production of polyether carbonate polyols is an industrial case developed at Bayer. The energy required for CO_2 utilization is provided by exothermic epoxide ring opening. For the commercialization of this technology, the main technological barrier is increase in CO_2 content of polymer. As this reaction is highly exothermic, the temperature control of the reactor is of high importance. One solution to this problem is the application of a semi-batch reactor with controlled addition of epoxide, in order to avoid thermal runaways. Better heat removal, lower epoxide concentration and faster overall reaction rates can be obtained by continuous reactors; hence, they are good options as well (Langanke et al., 2015).

In the manufacturing plant of Covestro (near Cologne, Germany), a polyurethane foam product is produced based on polyether polycarbonate polyol precursor that is partially derived from carbon dioxide (Muthuraj and Mekonnen, 2018).

In the PPC market, Empower Materials, Nontong Huasheng, Jiangsu Jinlong-CAS Chemical Co., Ltd, SK Energy, BASF, Tianguan, Novomer, Cardia Bio-Plastics™ and Bangfeng are the key industrial players (Muthuraj and Mekonnen, 2018).

In Nantong city, Jiangsu province, China, a 10,000 t/y production of CO_2-polyols was under construction (Scott, 2015).

QPAC and Converge are commercial examples of sustainable polycarbonate production from CO_2 (Zhang et al., 2018).

6.6 CHALLENGES

In addition to technical problems for developing effective catalysts for CO_2 conversion, there are challenges for scale-up of CO_2-based processes. These include providing high purity CO_2 from typically diluted process streams and effluents and providing energy required for the chemical transformation. The former necessitates developments of novel and cost-effective methods for CO_2 capture and storage.

The energy required for reaction might be provided and/or removed as heat of reaction or energy required for the production of energetic co-reactants (hydrogen, epoxides, etc.). To be effective in CO_2 reduction and sustainable, renewable energy sources (solar, wind, etc.) should be sought.

These problems have delayed large-scale utilization of CO_2 as a feedstock in chemical industries.

6.7   CONCLUDING REMARKS

The use of carbon dioxide as a building block in synthesis of valuable commodity and bulk products is an active research field. Different reaction schemes and processes have different CO_2 footprints. Life cycle assessment (LCA) is the most reliable metrics for assessing the effectiveness of the net effect of a particular CO_2 fixation process in reducing its emissions.

The cost of the FT process can be reduced by CO_2 hydrogenation. In the FT process (based upon coal or biomass), a clean-up process is needed in advance in which CO_2, NH_3 and H_2S are omitted from the syngas feed. In the CO_2-FT process, CO_2 removal is not needed as it is hydrogenated by means of a suitable catalyst.

Among the mentioned catalyst parameters, "promoters" improve CO_2 conversion while "support", "active site" and "binder" improve product selectivity. The reactor type has influence on CO_2 conversion while the catalyst type and operating conditions have influence on product selectivity.

An emerging approach in CO_2 utilization is its incorporation into polymers. This approach (as a non-petroleum route) combines substantial carbon integration, high added value and long duration CO_2 fixation. For sustainable value creation, direct CO_2 copolymerization is of high importance.

NOMENCLATURE

5CC	five-membered cyclic carbonates
CC	cyclic carbonate
CCU	carbon capture and utilization
CHO	cyclohexene oxide
CTA	chain transfer agent
CU	carbonate unit
DMC	dimethyl carbonate
DPC	diphenyl carbonate
ECH	epichlorohydrin
FT	Fischer–Tropsch
HBP	hyper-branched poly(ester-amide)
HC	hydrocarbon
HR-TEM	high-resolution transmission electron microscopy
IGCC	integrated coal gasification combined cycle
LCA	life cycle assessment
LPG	liquefied petroleum gas
MTFBR	multi-tubular fixed bed reactor
MTH	methanol to hydrocarbon
MW	molecular weight
OCA	*O*-carboxyanhydride
PBC	poly (butylenes carbonate)
PBS	poly(butylenes succinate)
PC	polycarbonate
PCHC	poly (cyclohexene carbonate)

PEC	poly (ethylene carbonate)
PLA	Polylactide
PO	propylene oxide
PPC	poly (propylene carbonate)
PU	polyurethane
PUA	polyurea
PV	photovoltaic
PVA	polyvinyl alcohol
ROCOP	ring-opening copolymerization
ROP	ring-opening polymerization
RWGS	reverse water gas shift
SMSI	strong metal-support interactions
SR	slurry reactor
SV	space velocity
TMA	1,3,5-benzenetricarboxylic acid
TPO	temperature programmed oxidation
UF	urea-formaldehyde
WGS	water gas shift
XPS	X-ray photoelectron spectroscopy

REFERENCES

Akin, A. N., Ataman, M., Aksoylu, A. E., O¨nsan, Z. I., CO_2 fixation by hydrogenation over coprecipitated Co/Al_2O_3, React. Kinet. Catal. Lett., 76 (2002) 265–270.

Al-Dossary, M., Ismail, A. A., Fierro, J. L. G., Bouzid, H., Effect of Mn loading onto Mn-FeO nanocomposites for the CO_2 hydrogenation reaction, Appl. Catal. B., 165 (2015) 651–660.

Alper, E., Orhan, O. Y., CO_2 utilization: developments in conversion processes, Petroleum, 3 (2017) 109–126.

Ando, H., Xu, Q., Fujiwara, M., Matsumura, Y., Tanaka, M., Souma, Y., Hydrocarbon synthesis from CO_2 over Fe-Cu catalysts, Catal. Today, 45 (1998) 229–234.

Ares, J. R., Aguey-Zinsou, K. -F., Leardini, F., Ferrer I. J. M, Fernandez, J. -F., Guo, Z. X., Sánchez, C., Hydrogen absorption/desorption mechanism in potassium alanate ($KAlH_4$) and enhancement by $TiCl_3$ doping, J. Phys. Chem. C., 113 (2009) 6845–6851.

Aresta, M., Carbon Dioxide Reduction and Uses as a Chemical Feedstock, In W. B. Tolman (Ed.), Activation of Small Molecules, Wiley-VCH, Weinheim (2006) 1–41.

Bhatia, S. K., Bhatia, R. K., Yang, Y. H. Biosynthesis of polyesters and polyamide building blocks using microbial fermentation and biotransformation. Rev. Environ. Sci. Bio/ Technol., 15 (2016) 639–663.

Buchard, A., Carbery, D. R., Davidson, M. G., Ivanova, P. K., Jeffery, B. J., Kociok-Ko¨hn, G. I., Lowe, J. P., Preparation of stereoregular isotactic poly(mandelic acid) through organocatalytic ring-opening polymerization of a cyclic O-carboxyanhydride, Angew. Chem., Int. Ed., 53 (2014) 13858–13861.

Buckwalter, D. J., Zhang, M., Inglefield, D. L., Moore, R. B., Long, T. E., Synthesis and characterization of siloxane-containing poly(urea oxamide) segmented copolymers, Polymer, 54 (2013) 4849–4857.

Centi, G., Perathoner, S., Opportunities and prospects in the chemical recycling of carbon dioxide to fuels, Catal. Today, 148 (2009) 191–205.

Chakrabarti D., de Klerk, A., Prasad, V., Gnanamani, M. K., Shafer, W. D., Jacobs, G., Sparks, D. E., Davis, B. H., Conversion of CO_2 over a Co-based Fischer-Tropsch catalyst, Ind. Eng. Chem. Res., 54 (2015) 1189–1196.

Chen, L. J., Qin, Y. S., Wang, X. H., Li, Y. S., Zhao, X. J., Wang, F. S., Toughening of poly(propylene carbonate) by hyperbranched poly(ester-amide) via hydrogen bonding interaction, Polym. Int., 60 (2011) 1697–1704.

Chew, L. M., Kangvansura, P., Ruland, H., Schulte, H. J., Somsen, C., Xia, W., Eggeler, G., Worayingyong, A., Muhler, M., Effect of nitrogen doping on the reducibility, activity and selectivity of carbon nanotube-supported iron catalysts applied in CO_2 hydrogenation, Appl. Catal. A., 482 (2014) 163–170.

Ian Childer, M., Longo, J. M., Van Zee, N. J., LaPointe, A. M., Coates, G. W., Stereoselective epoxide polymerization and copolymerization. Chem. Rev., 114 (2014) 8129–8152.

Choi, J. C., Kohno, K., Ohshima, Y., Yasuda, H., Sakakura, T., Tin- or titanium-catalyzed dimethyl carbonate synthesis from carbon dioxide and methanol: large promotion by a small amount of triflate salts. Catal. Commun., 9 (2008) 1630–1633.

Choi, Y. H., Jang, Y. J., Park, H., Kim, W. Y., Lee, Y. H., Choi, S. H., Lee, J. S., Carbon dioxide Fischer-Tropsch synthesis: a new path to carbon-neutral fuels, Appl. Catal. B: Env., 202 (2017) 605–610.

Coates, G. W., Moore, D. R., Discrete metal-based catalysts for the copolymerization CO_2 and epoxides: Discovery, reactivity, optimization, and mechanism. Angew. Chem., Int. Ed., 43 (2004) 6618–6639.

Cubeiro, M. L., Morales, H., Goldwasser, M. R., Pe´rez-Zurita, M. J., Gonza´lez-Jime´nez, F., Promoter effect of potassium on an iron catalyst in the carbon dioxide hydrogenation reaction, React. Kinet. Catal. Lett., 69 (2000) 259–264.

Cyriac, A., Lee, S. H., Varghese, J. K., Park, J. H., Jeon, J. Y., Kim, S. J., Lee, B. Y., Preparation of flame-retarding poly(propylene carbonate), Green Chem., 13 (2011) 3469–3475.

Darensbourg, D. J., Making plastics from carbon dioxide: salen metal complexes as catalysts for the production of polycarbonates from epoxides and CO_2, Chem. Rev., 107 (2007) 2388–2410.

Darensbourg, D. J., Moncada, A. I., Wei, S. H., Aliphatic polycarbonates produced from the coupling of carbon dioxide and oxetanes and their depolymerization via cyclic carbonate formation, Macromolecules, 44 (2011) 2568–2576.

Darensbourg, D. J., Wei, S. H., Yeung, A. D., Ellis, W. C., An efficient method of depolymerization of poly (cyclopentene carbonate) to its comonomers: cyclopentene oxide and carbon dioxide, Macromolecules 46 (2013) 5850–5855.

Darensbourg, D. J., Yeung, A. D., A concise review of computational studies of the carbon dioxide epoxide copolymerization reactions, Polym. Chem., 5 (2014) 3949–3962.

Delebecq, E., Pascault, J. P., Boutevin, B., Ganachaud, F., On the versatility of urethane/urea bonds: reversibility, blocked isocyanate, and non-isocyanate polyurethane, Chem. Rev., 113 (2013) 80–118.

Ding, F., Zhang, A., Liu, M., Guo, X., Song, C., Effect of SiO_2-coating of Fe-K/Al_2O_3 catalysts on their activity and selectivity for CO_2 hydrogenation to hydrocarbons, RSC Adv., 4 (2014) 8930–8938.

Dong, Y. L., Wang, X. H., Zhao, X. J., Wang, F. S., Facile synthesis of poly(ether carbonate)s via copolymerization of CO_2 and propylene oxide under combinatorial catalyst of rare earth ternary complex and double metal cyanide complex, J. Polym. Sci. A: Polym. Chem., 50 (2012) 362–370.

Dorner, R. W., Hardy, D. R., Williams, F. W., Davis, B. H., Willauer, H. D., Influence of gas feed composition and pressure on the catalytic conversion of CO_2 to hydrocarbons using a traditional cobalt-based Fischer-Tropsch catalyst, Energy Fuels, 23 (2009) 4190–4195.

Dorner, R. W., Hardy, D. R., Williams, F. W., Willauer, H. D., Heterogeneous catalytic CO_2 conversion to value-added hydrocarbons, Energy Environ. Sci., 3 (2010a) 884–890.

Dorner, R. W., Hardy, D. R., Williams, F. W., Willauer, H. D., K and Mn doped iron-based CO_2 hydrogenation catalysts: detection of $KAlH_4$ as part of the catalyst's active phase, Appl. Catal. A: Gen., 373 (2010b) 112–121.

Dorner, R. W., Hardy, D. R., Williams, F. W., Willauer, H. D., Effects of ceria-doping on a CO_2 hydrogenation iron manganese catalyst, Catal. Commun., 11 (2010c) 816–819.

Dry, M. E., Shingles, T., Boshoff, L. J., Oosthuizen, G. J., Heats of chemisorption on promoted iron surfaces and the role of alkali in Fischer-Tropsch synthesis, J. Catal., 15 (1969) 190–199.

Dry, M. E., Fischer-Tropsch process: 1950-2000, Cat. Today, 71 (2002) 227–241.

Empower Materials. http://www.empowermaterials.com/ (accessed October 2020).

Fiato, R. A., Iglesia, E., Rice, G. W., Soled, S. L., Iron catalyzed CO_2 hydrogenation to liquid hydrocarbons, Stud. Surf. Sci. Catal., 114 (1998) 339–344.

Gao, Y. G., Gu, L., Qin, Y. S., Wang, X. H., Wang, F. S.,Dicarboxylic acid promoted immortal copolymerization for controllable synthesis of low-molecular weight oligo(carbonate-ether) diols with tunable carbonate unit content, J. Polym. Sci. Part A Polym. Chem., 50 (2012) 177–5184.

Gnanamani, M. K., Jacobs, G., Keogh, R. A., Shafer, W. D., Sparks, D. E., Hopps, S. D., Thomas, G. A., Davis, B. H., Fischer-Tropsch synthesis: Eefect of pretreatment conditions of cobalt on activity and selectivity for hydrogenation of carbon dioxide, Appl. Catal. A: G., 499 (2015) 39–46.

Gouse, S. W., Gray, D., Tomlinson, G., Integration of fossil and renewable energy technologies to mitigate carbon dioxide, Energy Convers. Mgmt., 34 (1993) 1023–1030.

Grignard, B., Gennen, S., Jerome, C., Kleij, A. W., Detrembleur, C., Advances in the use of CO_2 as a renewable feedstock for the synthesis of polymers, Chem. Soc. Rev., 48 (2019) 4466–4514.

Groenewolt, M., Polyurethane coatings: a perfect product class for the design of modern automotive clearcoats, Polym. Int., 68 (2019) 843–847.

Herranz, T., Rojas, S., Pérez-Alonso, F. J., Ojeda, M., Terreros, P., Fierro, J. L. G., Hydrogenation of carbon oxides over promoted Fe-Mn catalysts prepared by the micro-emulsion methodology, Appl. Catal. A: Gen., 311 (2006) 66–75.

Hillmyer, M. A., Tolman, W. B., Aliphatic polyester block polymers: renewable, degradable, and sustainable, Acc. Chem. Res., 47 (2014) 2390–2396.

Hoberg, H., Peres, Y., Krüger, C., Tsay, Y. H., A 1-Oxa-2-nickela-5-cyclopentanone from ethene and carbon dioxide: preparation, structure, and reactivity, Angew. Chem., 26 (1987) 771–773.

Hollering, M., Dutta, B., Kühn, F. E., Transition metal mediated coupling of carbon dioxide and ethene to acrylic acid/acrylates, Coord. Chem. Rev., 309 (2016) 51–67.

Honda, S., Mori, T., Goto, H., Sugimoto, H., Carbon-dioxide-derived unsaturated alicyclic polycarbonate: synthesis, characterization, and post-polymerization modification, Polymer 55 (2014) 4832–4836.

Iqbal, N., Tripathi, M., Parthasarathy, S., Kumar, D., Roy, P. K., Polyurea spray coatings: tailoring material properties through chemical crosslinking, Prog. Org. Coat., 123 (2018) 201–208.

Jin, Y., Datye, A. K., Phase transformations in iron Fischer-Tropsch catalysts during temperature-programmed reduction, J. Catal., 196 (2000) 8–17.

Joseph, J., Patel, R. M., Wenham, A., Smith, J. R., Biomedical applications of polyurethane materials and coatings, Trans. IMF, 96 (2018) 121–129.

Jun, K. W., Roh, H. S., Kim, K. S., Ryu, J. S., Lee, K. W., Catalytic investigation for Fischer–Tropsch synthesis from bio-mass derived syngas, Appl. Catal. A: G., 259 (2004) 221–226.

Keim, W., C_1 chemistry: potential and developments, Pure Appl. Chem., 58 (1986) 825–832.

Khobragade, M., Majhi, S., Pant, K. K., Effect of K and CeO_2 promoters on the activity of Co/SiO_2 catalyst for liquid fuel production from syngas, Appl. Energy, 94 (2012) 385–394.

Kim, J. S., Lee, S., Lee, S. B., Choi, M. J., Lee, K.W., Performance of catalytic reactors for the hydrogenation of CO_2 to hydrocarbons, Cat. Today, 115 (2006) 228–234.

Klaus, S., Lehenmeier, M. W., Herdtweck, E., Deglmann, P., Ott, A. K., Rieger, B., Mechanistic insights into heterogeneous zinc dicarboxylates and theoretical considerations for CO_2-epoxide copolymerization, J. Am. Chem. Soc., 133 (2011) 13151–13161.

Koning, C., Wildeson, J., Parton, R., Plum, B., Steeman, P., Darensbourg, D. J., Synthesis and physical characterization of poly(cyclohexane carbonate), synthesized from CO_2 and cyclohexene oxide, Polymer 42 (2001) 3995–4004.

Landau, M. V., Vidruk, R., Herskowitz, M., Sustainable production of green feed from carbon dioxide and hydrogen, ChemSusChem, 7 (2014) 785–794.

Langanke, J., Wolf, A., Hofmann, J., Bohm, K., Subhani, M. A., Muller, T. E., Leitner, W., Gurtler, C., Carbon dioxide (CO_2) as sustainable feedstock for polyurethane production, Green Chem., 16 (2014) 1865–1870.

Langanke, J., Wolf, A., Peters, M., Polymers from CO_2 – An Industrial Perspective, In Styring, P., Quadrelli, E. A., Armstrong, K. (Eds.), Carbon Dioxide Utilization, Chapter 5, Elsevier, Amsterdam, The Netherlands (2015), pp. 59–71.

Lee, S. C., Jang, J. H., Lee, B. Y., Kang, M. C., Kang, M., Choung, S. J., The effect of binders on structure and chemical properties of Fe-K/γ-Al_2O_3 catalysts for CO_2 hydrogenation, Appl. Catal. A: G., 253 (2003) 293–304.

Lee, S.-B., Kim, J.-S., Lee, W.-Y., Lee, K.-W., Choi, M.-J., Product distribution analysis for catalytic reduction of CO_2 in a bench scale fixed bed reactor, Stud. Surf. Sci. Catal., 153 (2004a) 73–78.

Lee, S. C., Jang, J. H., Lee, B. Y., Kim, J. S., Kang, M., Lee, S. B., Choi, M. J., Choung, S. J., Promotion of hydrocarbon selectivity in CO_2 hydrogenation by Ru component, J. Mol. Catal. A: Chem., 210 (2004b) 131–141.

Lee, S. C., Kim, J. S., Shin, W. C., Choi, M. J., Choung, S. J., Catalyst deactivation during hydrogenation of carbon dioxide: effect of catalyst position in the packed bed reactor, J. Mol. Catal. A: Chem., 301 (2009) 98–105.

Lejkowski, M. L., Lindner, R., Kageyama, T., Bodizs, G. E., Plessow, P. N., Mueller, I. M., Schaefer, A., Rominger, F., Hofmann, P., Futter, C., Schunck, S. A., Limbach, M., The first catalytic synthesis of an acrylate from CO_2 and an alkene-a rational approach, Chem. Eur. J., 18 (2012) 14017–14025.

Li, S., Li, A., Krishnamoorthy, S., Iglesia, E, Effects of Zn, Cu, and K promoters on the structure and on the reduction, carburization, and catalytic behavior of iron-based Fischer-Tropsch synthesis catalysts, Catal. Lett., 77 (2001) 197–205.

Li, S., Krishnamoorthy, S., Li, A., Meitzner, G. D., Iglesia, E., Promoted iron-based catalysts for the Fischer-Tropsch synthesis: design, synthesis, site densities, and catalytic properties, J. Catal., 206 (2002) 202–217.

Li, T., Yang, Y., Zhang, C., An, X., Wan, H., Tao, Z., Xiang, H., Li, Y., Yi, F., Xu, B., Effect of manganese on an iron-based Fischer-Tropsch synthesis catalyst prepared from ferrous sulfate, Fuel, 86 (2007) 921–928.

Li, S., Guo, H., Luo, C., Zhang, H., Xiong, L., Chen, X., Ma, L., Effect of iron promoter on structure and performance of K/Cu-Zn catalyst for higher alcohols synthesis from CO_2 hydrogenation, Catal. Lett., 143 (2013) 345–355.

Li, W., Wang, H., Jiang, X., Zhu, J., Liu, Z., Guo, X., Song, C., A short review of recent advances in CO_2 hydrogenation to hydrocarbons over heterogeneous catalysts, RSC Adv., 8 (2018) 7651–7669.

Liu, A. H., Gao, J., He, L. N., Catalytic Activation and Conversion of Carbon Dioxide into Fuels/Value-Added Chemicals Through C-C Bond Formation, In Suib, S. L. (Ed.), Future Developments in Catalysis: Activation of Carbon Dioxide, Elsevier, Amsterdam (2013).

Liu, S., Wang, X., Polymers from carbon dioxide: polycarbonates, polyurethanes, Curr. Opinion Green Sust. Chem., 3 (2017) 61–66.

Longo, J. M., Sanford, M. J., Coates, G. W., Ring-opening copolymerization of epoxides and cyclic anhydrides with discrete metal complexes: structure-property relationships. Chem. Rev., 116 (2016) 15167–15197.

Memic, A., Abudula, T., Mohammed, H. S., Joshi Navare, K., Colombani, T., Bencherif, S. A., Latest progress in electrospun nanofibers for wound healing applications, ACS Appl. Bio Mater., 2 (2019) 952–969.

Meng, Y. Z., Du, L. C., Tiong, S. C., Zhu, Q., Hay, A. S., Effects of the structure and morphology of zinc glutarate on the fixation of carbon dioxide into polymer, J. Polym. Sci. Part A Polym. Chem., 40 (2002) 3579–3591.

Muthuraj, R., Mekonnen, T., Recent progress in carbon dioxide (CO_2) as feedstock for sustainable materials development: co-polymers and polymer blends, Polymer 145 (2018) 348–373.

Nagiah, N., Sivagnanam, U. T., Mohan, R., Srinivasan, N. T., Sehgal, P. K., Development and characterization of electropsun poly(propylene carbonate) ultrathin fibers as tissue engineering scaffolds, Adv. Eng. Mater., 14 (2012) B138–B148.

Nakano, K., Nakamura, M., Nozaki, K., Alternating copolymerization of cyclohexene oxide with carbon dioxide catalyzed by (salalen)CrCl complexes. Macromolecules, 42 (2009) 6972–6980.

Niemelä, M., Nokkosmäki, M., Activation of carbon dioxide on Fe-catalysts, Catal. Today, 100 (2005) 269–274.

Ning, W., Koizumi, N., Yamada, M., Researching Fe catalyst suitable for CO_2-containing syngas for Fischer-Tropsch synthesis, Energy Fuels, 23 (2009) 4696–4700.

Otto, A., Grube, T., Schiebahn, S., Stolten, D., Closing the loop: captured CO_2 as a feedstock in the chemical industry, Energy Environ. Sci., 8 (2015) 3283–3297.

Pappalardo, D., Mathisen, T., Finne-Wistrand, A., Biocompatibility of resorbable polymers: a historical perspective and framework for the future, Biomacromolecules, 20 (2019) 1465–1477.

Paul, S., Zhu, Y., Romain, C., Brooks, R., Saini, P. K., Williams, C. K., Ring-opening copolymerization (ROCOP): synthesis and properties of polyesters and polycarbonates, Chem. Commun., 51 (2015) 6459–6479.

Perathoner, S., Centi, G., CO_2 recycling: a key strategy to introduce green energy in the chemical production chain, ChemSusChem, 7 (2014) 1274–1282.

Pérez-Alonso, F. J., Ojeda, M., Herranz, T., Rojas, S., González-Carballo, J. M., Terreros, P., Fierro, J. L. G., Carbon dioxide hydrogenation over Fe-Ce catalysts, Catal. Commun., 9 (2008) 1945–1948.

Peters, M., Kçhler, B., Kuckshinrichs, W., Leitner, W., Markewitz, P., Müller, T. E., Chemical technologies for exploiting and recycling carbon dioxide into the value chain, ChemSusChem, 4 (2011) 1216–1240.

Preti, D., Resta, C., Squarcialupi, S., Fachinetti, G., Carbon dioxide hydrogenation to formic acid by using a heterogeneous gold catalyst, Angew. Chem. Int. Ed., 50 (2011) 12551–12554.

Qin, Y. S., Wang, X. H., Carbon dioxide-based copolymers: environmental benefits of PPC, an industrially viable catalyst, Biotechnol. J., 5 (2010) 1164–1180.

Qin, Y., Sheng, X., Liu, S., Ren, G., Wang, X., Wang, F., Recent advances in carbon dioxide based copolymers, J. CO_2 Util., 11 (2015) 3–9.

Rao, V. U. S., Stiegel, G. J., Cinquegrane, G. J., Srivastava, R. D., Iron-based catalysts for slurry phase FTS: technology review", Fuel Proc. Tech., 30 (1992) 83–107.

Rabnawaz, M., Wyman, I., Auras, R., Cheng, S., A roadmap towards green packaging: the current status and future outlook for polyesters in the packaging industry, Green Chem., 19 (2017) 4737–4753.

Razali, N. A. M., Lee, K. T., Bhatia, S., Mohamed, A. R., Heterogeneous catalysts for production of chemicals using carbon dioxide as raw material: a review, Renew. Sust. Energy Rev., 16 (2012) 4951–4964.

Ree, M., Hwang, Y., Kim, J. S., Kim, H., Kim, G., Kim, H., New findings in the catalytic activity of zinc glutarate and its application in the chemical fixation of CO_2 into polycarbonates and their derivatives, Catal. Today, 115 (2006) 134–145.

Riedel, T., Claeys, M., Schulz, H., Schaub, G., Nam, S. S., Jun, K. W., Choi, M. J., Kishan, G., Lee, K. W., Comparative study of FTS with H_2/CO and H_2/CO_2 syngas using Fe and Co catalysts, App. Catal. A,. 186 (1999) 201–213.

Riedel, T., Schaub, G., Jun, K. W., Lee, K. W., Kinetics of CO_2 Hydrogenation on a K-Promoted Fe Catalyst, Ind. Eng. Chem. Res. 40 (2001) 1355–1363.

Riedel, T., Schulz, H., Schaub, G., Jun, K. W., Hwang, J. S., Lee, K. W., Fischer-Tropsch on iron with H_2/CO and H_2/CO_2 as synthesis gases: the episodes of formation of the Fischer-Tropsch regime and construction of the catalyst, Top. Catal., 26 (2003) 41–54.

Rodemerck, U., Holena, M., Wagner, E., Smejkal, Q., Barkschat, A., Baerns, M., Catalyst development for CO_2 hydrogenation to fuels, ChemCatChem, 5 (2013) 1948–1955.

Saeidi, S., Amin, N. A. S., Rahimpour, M. R., Hydrogenation of CO_2 to value-added products – a review and potential future developments, J. CO_2 Util., 5 (2014) 66–81.

Sahebdelfar, S., Takht Ravanchi, M., Heterogeneous Catalytic Hydrogenation of CO_2 to Basic Chemicals and Fuels, In Kumar, A., Sharma, S. (Eds.) Chemo-Biological Systems for CO_2 Utilization, CRC Press, Boca Raton (2020).

Sai Prasad, P. S., Bae, J. W., Jun, K. W., Lee, K. W., Fischer-Tropsch synthesis by carbon dioxide hydrogenation on Fe-based catalysts, Catal. Surv. Asia, 12 (2008) 170–183.

Sakakura, T., Choi, J. C., Yasuda, H., Transformation of carbon dioxide, Chem. Rev., 107 (2007) 2365–2387.

Sakakura, T., Kohno, K., The synthesis of organic carbonates from carbon dioxide., Chem. Commun., 11 (2009) 1312–1330.

Saudi Aramco. http://www.novomer.com (accessed October 2021).

Schulz, H., Schaub, G., Claeys, M., and Riedel, T., Transient initial kinetic regimes of Fischer-Tropsch synthesis, App. Cat. A., 186 (1999) 215–227.

Schulz, H., Riedel, T., Schaub, G., Fischer-Tropsch principles of co-hydrogenation on iron catalysts, Top. Catal., 32 (2005) 117–124.

Scott, A., Learning to love CO_2, C&E News, 93 (2015) 10–16.

Song, C., Global challenges and strategies for control, conversion and utilization of CO_2 for sustainable development involving energy, catalysis, adsorption and chemical processing, Catal. Today, 115 (2006) 2–32.

Srinivas, S., Malik, R. K., Mahajani, S. M., Fischer-Tropsch synthesis using bio-syngas and CO_2, Energy Sust. Dev., 4 (2007) 66–71.

Sun, Y., Chemicals from CO_2 via heterogeneous catalysis at moderate conditions, Proceedings of the 7th International Conference on Carbon Dioxide Utilization, Seoul, Korea, 12-16 October (2003) 9–16.

Sun, Q., Mekonnen, T., Misra, M., Mohanty, A. K., Novel biodegradable cast film from carbon dioxide based copolymer and poly(lactic acid), J. Polym. Environ., 24 (2016) 23–36.

Takht Ravanchi, M., Sahebdelfar, S., Catalytic conversions of CO_2 to help mitigate climate change: Recent process development, Proc. Safety Env. Prot., 145 (2021) 172–194.

Tao, Y. H., Wang, X. H., Chen, X. S., Zhao, X. J., Wang, F. S., Regio-regular structure high molecular weight poly(propylene carbonate) by rare earth ternary catalyst and Lewis base cocatalyst, J. Polym. Sci. A.: Polym. Chem., 46 (2008) 4451–4458.

Tempelaar, S., Mespouille, L., Coulembier, O., Dubois, P., Dove, A. P., Synthesis and post-polymerization modifications of aliphatic poly(carbonate)s prepared by ring-opening polymerization, Chem. Soc. Rev., 42 (2013) 1312–1336.

Tihay, F., Roger, A. C., Pourroy, G., Kiennemann, A., Role of the alloy and spinel in the catalytic behavior of Fe-Co/cobalt magnetite composites under CO and CO_2 hydrogenation, Energy Fuels, 16 (2002) 1271–1276.

Trott, G., Saini, P. K., Williams, C. K., Catalysts for CO_2/epoxide ring-opening copolymerization, Phil. Trans. R. Soc. A., 374 (2016) 20150085.

Uttley, K. B., Shimmei, K., Bernskoetter, W. H., Ancillary ligand and base influences on nickel-catalyzed coupling of CO_2 and ethylene to acrylate, Organometallics, 39 (2020) 1573–1579.

Vavasori, A., Calgaro, L., Pietrobon, L., Ronchin, L., The coupling of carbon dioxide with ethene to produce acrylic acid sodium salt in one pot by using Ni(II) and Pd(II)-phosphine complexes as precatalysts, Pure Appl. Chem., 90 (2018) 315–326.

Wang, S. J., Du, L. C., Zhao, X. S., Meng, Y.Z., Tjong, S. C., Synthesis and characterization of alternating copolymer from carbon dioxide and propylene oxide, J. Appl. Poly. Sci., 85 (2002) 2327–2334.

Wang, W., Wang, S., Ma, X., Gong, J., Recent advances in catalytic hydrogenation of carbon dioxide, Chem. Soc. Rev., 40 (2011) 3703–3727.

Wang, J., Zhang, H., Miao, Y., Qiao, L., Wang, X., Wang, F., Waterborne polyurethanes from CO_2 based polyols with comprehensive hydrolysis/oxidation resistance, Green Chem., 18 (2016) 524–530.

Wazarkar, K., Kathalewar, M., Sabnis, A., High performance polyurea coatings based on cardanol, Prog. Org. Coat., 106 (2017) 96–110.

Williams, L. L., Amino resins and plastics, In Herman F. Mark (Ed.), Encyclopedia of Polymer Science and Technology, Wiley, New York (2002), pp. 340–371.

Wu, G. P., Wei, S. H., Ren, W. M., Lu, X. B., Xu, T.Q., Darensbourg, D. J., Perfectly alternating copolymerization of CO_2 and epichlorohydrin using cobalt(III)-based catalyst systems, J. Am. Chem. Soc., 133 (2011) 15191–15199.

Wu, W., Qin, Y. S., Wang, X. H., Wang, F. S., New bifunctional catalyst based on cobalt-porphyrin complex for the copolymerization of propylene oxide and CO_2, J. Poly. Sci. A: Poly. Chem., 51 (2013) 493–498.

Xu, J., Feng, E., Song, J., Renaissance of aliphatic polycarbonates: new techniques and biomedical applications, J. Appl. Poly. Sci., 131 (2014) 39822–39837.

Yan, S. R., Jun, K. W., Hong, J. S., Choi, M.J., Lee, K. W., Promotion effect of Fe–Cu catalyst for the hydrogenation of CO_2 and application to slurry reactor, Appl. Catal. A.: G., 194–195 (2000) 63–70.

Yang, H., Zhang, C., Gao, P., Wang, H., Li, X., Zhong, L., Wei, W., Sun, Y., A review of the catalytic hydrogenation of carbon dioxide into value-added hydrocarbons, Catal. Sci. Technol., 7 (2017) 4580–4598.

Zhang, Y., Jacobs, G., Sparks, D. E., Dry, M. E., Davis, B. H., CO and CO_2 hydrogenation study on supported cobalt Fischer-Tropsch synthesis catalysts, Catal. Today, 71 (2002) 411–418.

Zhang, Z., Cui, D., Liu, X., Alternating copolymerization of cyclohexene oxide and carbon dioxide catalyzed by noncyclopentadienyl rare-earth metal bis(alkyl) complexes, J. Poly. Sci. Part A: Poly. Chem., 46 (2008) 6810–6818.

Zhang, X. H., Wei, R. J., Sun, X. K., Zhang, J. F., Du, B. Y., Fan, Z. Q., Qi, G. R., Selective copolymerization of carbon dioxide with propylene oxide catalyzed by a nanolamellar double metal cyanide complex catalyst at low polymerization temperatures, Polymer, 52 (2011) 5494–5502.

Zhang, X., Fevre, M., Jones, G. O., Waymouth, R. M., Catalysis as an Enabling Science for Sustainable Polymers, Chem. Rev., 118 (2018) 839–885.

Zhao, G., Zhang, C., Qin, S., Xiang, H., Li, Y., Effect of interaction between potassium and structural promoters on Fischer-Tropsch performance in iron-based catalysts, J. Mol. Catal. A.: Chem., 286 (2008) 137–142.

Zhu, Y., Romain, C., Williams, C. K., Sustainable polymers from renewable resources, Nature, 540 (2016) 354–362.

7 Methanol Conversions

7.1 INTRODUCTION

Methanol is a versatile multi-source multi-purpose chemical. It is widely used as a solvent and raw material for the manufacturing of formaldehyde, methyl t-butyl ether (MtBE, an octane booster of gasoline), acetic acid and olefins with the latter being an emerging and growing sector. Methanol can be used as a fuel and/or fuel additive in fuel cells and in IC engines. Methanol can be readily reformed to hydrogen and thus is a potential hydrogen carrier in the future.

Methanol is a C_1 building block for the production of olefins and other hydrocarbons. Consequently, as a platform chemical, it can be used as a substitute for petroleum for the production of fuels and various chemicals. The boiling point of methanol is higher than that of other C_1 molecules and thus it is liquid under normal conditions. This is an advantage for its transportation and storage and thus it can be used as a feedstock ex-situ.

Nowadays, methanol is produced in mega plants mostly from natural gas and coal. However, in the future, renewable sources such as biomass and captured CO_2 may also acquire significant share. This secures the supply of methanol in long-term virtually without limitation. Furthermore, methanol can be easily transported to the market.

In fact, these conversions are important part of the methanol economy by which methanol is produced from renewable sources and used for the production of a variety of valuable chemicals.

This chapter focuses on the conversion of methanol to hydrocarbons which is of importance for the supply of non-petroleum petrochemical feedstocks (such as olefins and aromatics) and fuels. The carbonylation of methanol to acetic acid is also discussed.

7.2 CHEMICAL PROPERTIES OF METHANOL

The thermochemical properties of methanol are summarized in Table 7.1. It is a polar molecule due to the presence of an –OH (hydroxyl) group.

The chemical properties of methanol, like other alcohols, are determined by its –OH functional group. Its reactions, therefore, may involve the cleavage C–OH or O–H bond, and elimination or substitution of –OH or –H group. Methanol is less reactive than higher alcohols due to low stability of methyl carbocation.

The carbon atom of methanol is electrophilic; thus, it can react with nucleophile carbon such as the carbon atom of CO.

7.3 METHANOL TO HYDROCARBONS

In C_1 chemistry, MTH (methanol-to-hydrocarbons) is an important reaction by which basic petrochemicals are produced from non-oil resources (such as coal and

DOI: 10.1201/9781003279280-7

TABLE 7.1

Thermochemical Properties of Methanol

Compound	Normal Boiling point (°C)	Dipole Moment (D)	ΔH^0_f (kJ/mol)	ΔG^0_f (kJ/mol)
CH₃OH	64.7	1.69	13.97	−162.51

natural gas) (Tian et al., 2015). By the MTH technology, any organic compound can be produced from coal or natural gas.

In refineries, if too much aromatic was produced, enough ethylene and propylene is not produced. Generally, light olefins are produced as gasoline byproducts, directly or by steam cracking of light alkanes. Due to the increasing demand for polyolefins, refineries are not able to meet the demands.

Primarily, MTH is regarded as a powerful technology for the conversion of coal to high-octane gasoline. By steam reforming of natural gas or by coal gasification, synthesis gas and then methanol are produced. According to the catalyst and operating conditions used, methanol can be converted into gasoline (MTG process) or light olefins (MTO process). There is an increasing demand for high-quality gasoline and light olefins are important building blocks in the petrochemical industry.

MTG and MTO processes are two technological breakthroughs in synfuels researches. MTG was first introduced by Mobil in 1977 and MTO was first introduced by Union Carbide in 1981. These two processes are "synfuels factory" (Gogate, 2019).

The discovery of MTH reaction and specifically MTG process is devoted to Mobil Oil Corporation (now Exxon Mobil). The history behind this discovery is interesting; when Mobil researchers worked on unrelated projects, hydrocarbons were formed from methanol on synthetic ZSM-5 zeolites. One group of researchers at Mobil Chemical (Edison, New Jersey) tried to produce EO (ethylene oxide) from MeOH, while others at Mobil Oil's Central Research Laboratory (Princeton) tried to methylate isobutene with methanol over ZSM-5. Both reactions failed and aromatic hydrocarbons were obtained as the main products (Keil, 1999).

Later phosphorous-containing zeolites developed by Union Carbide, especially eight-member ring SAPO-34 with chabazite (CHA) topology were also found to be very effective in the conversion of methanol into olefins.

7.3.1 CHEMISTRY OF MTH

Equation (7.1) represents the overall reaction stoichiometry for methanol to hydrocarbons (Keil, 1999):

$$CH_3OH \rightarrow \left[-CH_2 - \right] + H_2O \tag{7.1}$$

where [CH₂] is the average oligomeric branch of an olefin hydrocarbon. Below 300°C, the main dehydration product is dimethyl ether (DME). The reaction can be catalyzed by solid acids, but only molecular sieve zeotype catalysts give reasonable performance. The upper limit of the product carbon number is determined by shape selectivity of the zeolite catalyst used. Two commonly used catalysts are SAPO-34 and HZSM-5. The former exhibits very high (>90%) selectivities to C_2–C_4 olefins but is rapidly deactivated by coke deposition. HZSM-5 exhibits a lower light olefin selectivity but is much more coke resistant.

Generally, the main reaction steps are as Scheme 7.1.

$$2\,CH_3OH \underset{+\,H_2O}{\overset{-\,H_2O}{\rightleftharpoons}} CH_3OCH_3 \longrightarrow C_2\text{-}C_5\ \text{olefins} \longrightarrow \text{Higher olefins, paraffins, naphthenes, aromatics}$$

SCHEME 7.1 Reaction scheme for methanol to hydrocarboins over HZSM-5 zeolite.

According to reaction Scheme 7.1, methanol is first dehydrated to an equilibrium mixture of MeOH, DME and H_2O. In the next step, the DME is converted into light olefins in the range of C_2–C_4 as the primary hydrocarbon products. In the final reaction step, a mixture of higher olefins, aromatics, n-/iso-paraffins and naphthenes are produced by further transformation. Moreover, in the presence of excess methanol, primary olefinic products (such as propylene) may homologize to the next higher olefins (such as butenes). At last, on stronger acid sites, olefins can be converted by hybrid transfer into a mixture of branched alkanes and methylbenzenes, naphthenes and higher olefins by hydrogen transfer, alkylation and polycondensation (Haw and Marcus, 2003). The product is, thus, a mixture of olefins, naphthenes, aromatics and paraffins.

Thermodynamic equilibrium considerations showed that the favorable products in HZSM-5-catalyzed system are methane and benzene according to the stoichiometry reported in Eq. (7.2) (Sahebdelfar et al., 2019):

$$9CH_3OH \rightarrow 3CH_4 + C_6H_6 + 9H_2O \tag{7.2}$$

According to the reaction Scheme 7.1, the product composition should strongly depend on contact time (or space velocity). The product composition versus 1/LHSV (Figure 7.1) exhibits maxima for intermediate products (DME and light olefins) typical of series reactions (Levenspiel, 1999). The position and intensity of the maxima depends on operating conditions (temperature and catalyst specifications). Thus, the results for H-ZSM-5 catalyzed system at 371°C, with LHSV in the range of 1,080–1 h^{-1}, show the time-resolved progress of the DME forming reaction on at the shortest contact time (Gogate, 2019). In contact times between 0.001 and 0.01 h (LHSVs between 1,080 and 108 h^{-1}), a rapid rise in methanol consumption and water formation rates is observed which coincides with the increased concentration of primary olefin products, an indication of the autocatalytic feature of MTH reaction. The further conversion sequence of DME/CH_3OH to C_2–C_5 olefins and C_5–C_{10} gasoline-boiling range HCs is also clearly evident from the time-resolved profile.

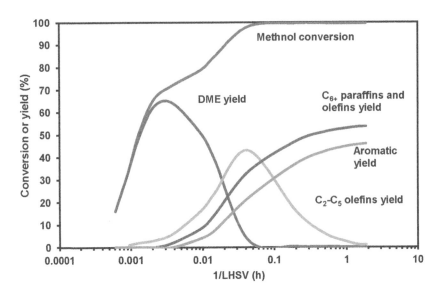

FIGURE 7.1 Effect of space velocity on product composition (carbon-based product yield) in MTG reaction over H-ZSM-5 catalyst at 375°C and 1 atm.

Thus, by carefully controlling the temperature and space velocity, the MTH reaction can be interrupted at maximum light olefin selectivity and the MTO process is obtained, by which light olefins are produced with maximum yield, instead of gasoline.

7.3.2 Catalysts

Primarily, researchers of Mobil Oil Corporation studied MTO process on ZSM-5 catalyst in a fixed-bed reactor at bench-scale. Union Carbide researchers reported for the first time that on SAPO-34 catalyst, shape selectivity and product distribution control on light olefins (C_2–C_4) was obtained. SAPO-34 has unique pore size distribution and milder acidity (in comparison to ZSM-5) (Gogate, 2019).

In many heterogeneous catalytic processes, molecular sieves are used as efficient shape-selective catalysts. Their microporous architecture, strong acidity and large surface area are valuable characteristics for their good performance (Corma, 2003).

Zeolites are crystalline compounds with well-defined cavities or channels (with molecular dimensions). They are microporous aluminosilicates with tetrahedral building units (AlO_4 or SiO_4) which are connected with corner-sharing oxygen atoms and have three-dimensional frameworks. Zeotype materials or zeolitic molecular sieves (ZMSs) are a category of inorganic porous materials with different tetrahedral framework cations, such as Si, Al, B, Ga, Ti, P, Ge, etc. According to IZA (International Zeolite Association), roughly 232 framework type codes were identified, such as CHA, FAU, MFI and MOR. Shape selectivity, as a unique molecular sieving property, separates molecular sieves and zeolites from other porous materials (Di Renzo and Fajula, 2005).

Generally, zeolites have micropores in the range of 0.4–1.2 nm. Micropores are classified to three categories of 8-, 10- and 12-membered ring pore openings for which pore widths are 0.4, 0.55 and 0.7 nm, respectively. Shape selectivity can be classified into three groups, namely reactant shape selectivity (in which bulky molecules larger than pore entrance do not have access to catalytically active sites and consequently cannot convert into products), product shape selectivity (in which molecules produced in the adsorbed phase are too large to be desorbed as products) and transition-state shape selectivity (in which due to geometrical constraints inside the pore channel, transition state formation, which, in turn, causes the production of special chemicals, is prevented) (Zhong et al., 2017).

Zeolites have well-defined micropores (that are suitable for molecular shape selectivity), large specific surface areas, suitable acidity and high physical and chemical stability and can be used in different processes related to detergency, separation, gas adsorption and catalysis. For example, for the removal of Ca^{2+} and Mg^{2+} ions from water (containing high minerals), molecular sieves and zeolites can be used as water softeners. On the other hand, molecular sieves and zeolites can be used as adsorbents for CO_2 separation from N_2 or CH_4. They can also be used for selective NO_x reduction in vehicle exhaust gases. In petrochemical processes and petroleum refining, such as FCC (fluid catalytic cracking) and HC (hydro-cracking), zeolites and molecular sieves are used as efficient heterogeneous catalysts. Isomerization, Friedel-Crafts alkylation, reforming and dewaxing are other examples of catalytic technologies based on zeolites and molecular sieves (Tian et al., 2015).

Framework topology, crystallite size, acid strength and acid site density are the critical zeolite properties affecting their catalytic performance in organic reactions. Crystallite size is an important factor for controlling secondary reactions as rapid mass transfer occurs in small crystallites (Haw and Marcus, 2003).

Based upon the findings obtained in large-scale industrialization of MTO process, R&D researchers did their best to develop MTO catalysts with high performance to improve process efficiency. In order to reach this goal, many scientific and engineering challenges must be overcome:

1. Catalyst design with favorable properties; such as high olefin yield, long-term catalyst stability and minimum side products
2. Cost competitiveness of catalyst
3. The feasibility and reproducibility of catalyst synthesis procedure
4. Environmental considerations for which low waste emissions and high production efficiency is highly needed

7.3.2.1 Aluminosilicates

Zeolites are crystalline aluminosilicates with the general formula of $M_{x/n}[(AlO_2)_x (SiO_2)_y].mH_2O$ in which M is a metal or hydrogen with charge n to neutralize the negative charge of the framework. The replacement of Si^{4+} with Al^{3+} (both called T atoms) produces a negative charge. When neutralization is by H^+, a strong Brønsted acid sites with hydroxyl group located between Si and Al forms. When a metallic cation is the counter ion, a weak Lewis acid site is formed. The Brønsted acid sites can be transformed to Lewis acid sites by dehydration (Figure 7.2).

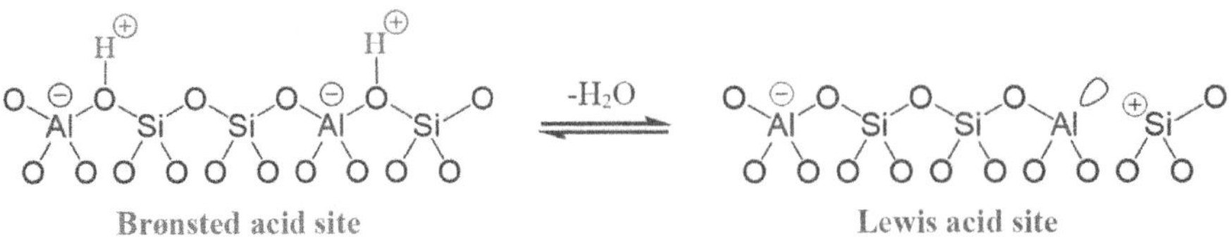

FIGURE 7.2 Structures and interconversion of Brønsted and Lewis acidity in zeolites (Redrawn from Schulman et al., 2020).

The acidity of zeolites, therefore, increases with decreasing framework Si/Al ratio, although the intrinsic acidity is reduced.

Two-dimensional (2D) zeolites have extended sheets with thickness of lower than 2–3 nm. These have large number of active sites on the external surface, therefore, their Brønsted sites can easily be accessed and there is short diffusion distance and fast mass flow for them. Due to these favorable characteristics, 2D-zeolites can be used in catalysis and adsorption (Roth et al., 2016).

HZSM-5 (MFI) and HBEA (as two examples of aluminosilicate zeolites) possess straight channels. HZSM-5 is a zeolite with medium pores and ten-membered rings in both channels, while HBEA is a 12-membered ring one. The commercial catalyst for the MTG process is HZSM-5 and for the production of other hydrocarbon products, this zeolite can be modified. Due to the presence of larger channels in HBEA (that allows larger molecules to be entered and exit), this zeolite is normally used for mechanistic studies in methanol conversions (Haw and Marcus, 2003).

In HZSM-5 and HBEA, the Si/Al ratio (and consequently acid site density) is variable in a wide range. Acid strength of HZSM-5 increases with isolating acid sites. For instance, HZSM-5 with Si/Al = 300 (in comparison to that with Si/Al = 100) is a much stronger acid. Secondary reactions are promoted by higher acid site densities. Zeolites with Si/Al≤20 which have high concentrations of acid site contain "paired" sites that have special activity for some reactions (Haw and Marcus, 2003).

Salvador and Kladnig (1977) evaluated modified Y zeolites for the MTO reaction. At 250°C, light olefins (mainly propylene) were obtained. H-Y zeolite in comparison to Na-Y zeolite showed high activity.

In the MTH reaction, for NaH-Y zeolite samples, catalytic activity was observed at lower temperatures with increase in the degree of proton exchange and at higher temperatures (350°C) with an increase in the degree of dealumination. As researchers reported, this dependence on temperature shows that increasing proton exchange causes more active sites on Na-Y zeolites.

Kubelková et al. (1984) evaluated dealuminated cubic faujasite (Si/Al = 5.7) for MTO reaction at 400°C and reported the production of C_3–C_7 olefins and aromatics up to C_9. Moreover, NaH-Y zeolite with Si/Al = 2.5–8.9 was used as the methanol conversion catalyst and aromatization products were obtained at 300°C.

ZSM-5 with 3D intersected 10-MR channels is another catalyst that can be used in the MTG and MTO processes (Yang et al., 2019).

7.3.2.2 Silicoaluminophosphates (SAPO)

By substituting silicon for phosphorus in aluminophosphate lattice, a framework acid site is generated. At low Si contents, synthesizing HSAPO-34 with one template molecule in each cage and one acid site per cage is fairly easy. Whilst, at high Si concentrations, by forming siliceous islands, the number of acid sites is reduced below one per silicon.

Due to their mild acidity and 3D channels, silicoaluminophosphate (SAPO) molecular sieves are good candidates to obtain high selectivities for light olefins. Synthesis conditions, dealumination and cation exchange, and the ratio of acid sites on the external surface to that on porous structure are the manipulating parameters for making a catalyst with acceptable performance in methanol conversion into light olefins.

HSAPO-34 (CHA) and HSAPO-18 (AEI) as small-pore materials are examples of silicoaluminophosphates with similar topologies. These two configurations have nanometer-sized cages connected internally by eight rings. Aromatics, and branched alkanes or olefins cannot enter or leave these cages. The acid strength of SAPO catalysts differ from HZSM-5 catalysts. For instance, at 35°C, ethylene is very reactive on HZSM-5 catalyst while it can easily flow on fresh HSAPO-34 without production of any other hydrocarbons (Haw and Marcus, 2003).

The microporous zeotype SAPO-34 has good performance in MTO reaction because of the CHA topological structure with small eight-ring pore (0.38 nm × 0.38 nm), large cavity (9.4 Å in diameter), high hydrothermal stability and moderate acidity. The diffusion of branched and heavy hydrocarbons is limited by reduced pore size of SAPO-34, due to which light olefins selectivities are increased. As SAPO-34 has milder acidity, hydrogen transfer reactions are lowered, paraffinic product yield is minimized and olefinic product yield is increased. The main disadvantage of SAPO-34 catalyst is its susceptibility to deactivation due to the retention of coke precursors inside the large chabazite cavities (Gogate, 2019).

Generally, MTO catalysts are deactivated by external coke formation that poisons the active sites or blocks pores. SAPO-34 catalysts have big supercage and narrow pore opening and after methanol conversion, quick coke formation occurs. Haw et al. (2003) reported that during methanol conversion, the produced methylbenzenes are converted into methylnaphthalenes and polyaromatics, phenanthrene derivatives and pyrenes in the cages of SAPO-34. As these are bulky coke species, their formation inhibits the mass transport of reactants and products and consequently, the catalyst deactivates. In order to maintain catalyst activity, in the industrial MTO process, fluidized bed reactors with reaction-regeneration cycles are considered (Haw et al., 2003; Guisnet et al., 2009).

Researchers of Union Carbide (now UOP) discovered H-SAPO-34 by which selectivity towards light olefins was considerably improved and olefin yield was more than 90% with selectivity to ethylene-propylene exceeding 80%. By tuning operating conditions, propylene could be produced as the favored product. The main disadvantage of H-SAPO-34 is its rapid coking which requires frequent regenerations. UOP and Norsk Hydro (now INEOS) jointly developed an MTO process based on H-SAPO-34 with low-pressure fluidized bed reactor (Olsbye et al., 2012).

7.3.3 MECHANISMS

In the first step of Scheme 7.1, the intermediate is a protonated surface methoxyl that is exposed to a nucleophilic attack by methanol. The third step of the scheme proceeded via classical carbenium ion mechanism with concurrent hydrogen transfer (Spivey et al., 1992). Different researchers studied the second step of Scheme 7.1, that involves the first C–C bond formation from C_1 reactants, for which more than 20 mechanisms were proposed (Yarulina et al., 2018). Examples are carbene mechanism, free radical mechanism, oxonium ylide mechanism and carboncation mechanism which are discussed below.

On a fresh catalyst, the initial conversion of methanol is kinetically slow which implies that the formation of C–C bond occurs through the reaction route with relatively high energy barriers (Tian et al., 2015).

7.3.3.1 Oxonium Mechanism

For the formation of the "first" C–C bond by trimethyloxonium ion, two different routes were proposed in Scheme 7.2; in which oxonium cation is deprotonated to an oxonium-ylide. Their difference is that whether the subsequent methylation step is intra-molecular (Stevens re-arrangement) or inter-molecular. Identification of a basic site to be strong enough for the deprotonation of the oxonium cation is still a problem. In both routes, ethylene elimination occurs rapidly (Haw and Marcus, 2003).

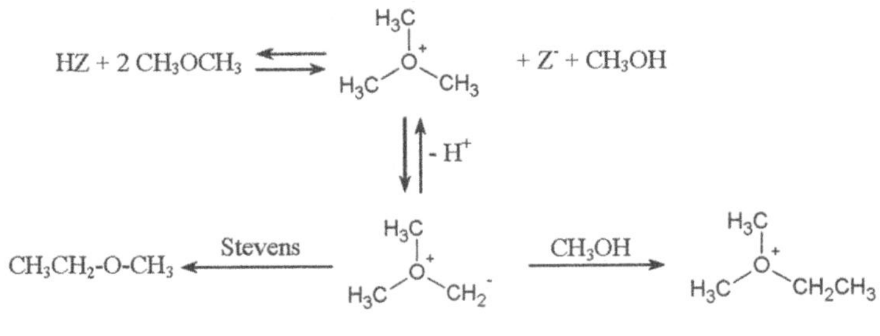

SCHEME 7.2 Two oxonium ylide mechanisms.

Over a zeolite framework, an oxonium ylide mechanism is presented in Scheme 7.3 (Haw and Marcus, 2003).

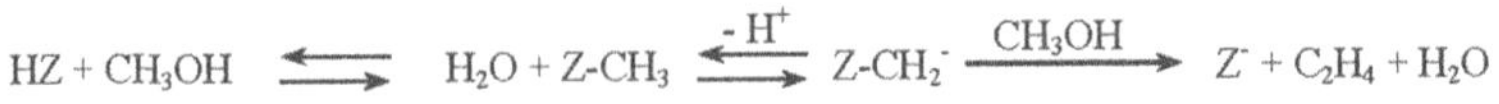

SCHEME 7.3 Framework oxonium ylide mechanism.

7.3.3.2 Carbocation Mechanism

In carbocation mechanism (Scheme 7.4), the equivalent of a carbenium ion (CH_3^+) is transferred to a DME molecule by which a five-coordinate carbonium ion is formed and by deprotonation, ethylmethyl ether is produced. In this mechanism, it is required that the zeolite has strong (superacidic) acid sites (Haw and Marcus, 2003).

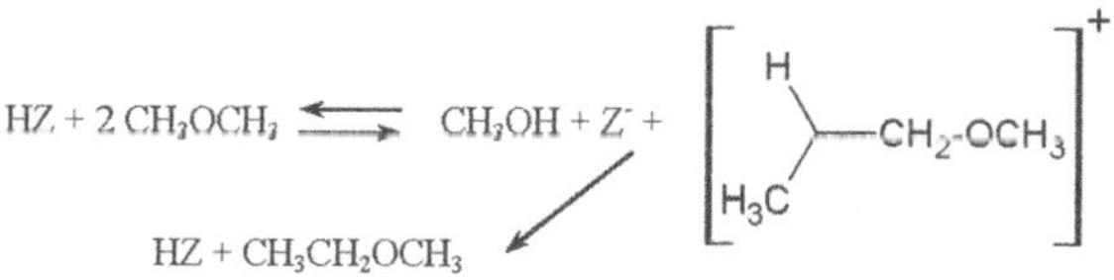

SCHEME 7.4   One of the several carbocation mechanisms.

7.3.3.3   Carbene Mechanism

Under methanol conversion conditions, two mechanisms were proposed for the production of highly reactive carbine (CH_2) intermediates (Schemes 7.5 and 7.6). In Scheme 7.5, it is described how two proximal acid sites could dehydrate methanol to CH_2. Scheme 7.6 shows the formation of a framework methoxonium species (imagined as a CH_3^+, Z^- ion pair) that eliminates carbine. This intermediate is so reactive that C-C bonds could be produced from any of different routes (Haw and Marcus, 2003).

$$Z^- + H\text{-}CH_2\text{-}OH + H\text{-}Z$$
$$\downarrow$$
$$Z\text{-}H + :CH_2 + H_2O + Z^-$$

SCHEME 7.5   A carbine mechanism based on two zeolite acid sites.

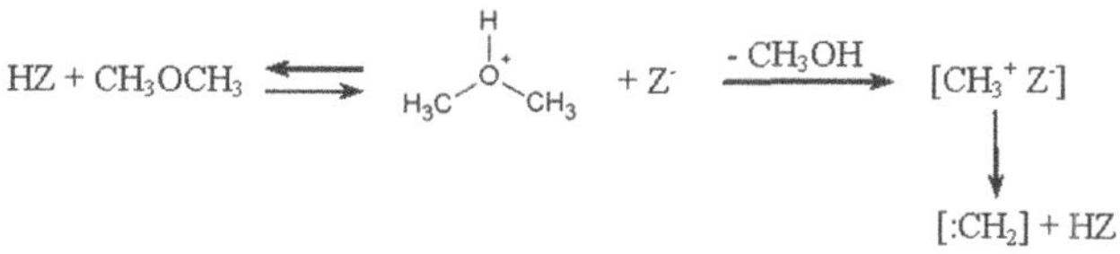

SCHEME 7.6   A mechanism in which a framework methoxonium forms a carbene.

7.3.3.4   Radical Mechanism

A sample of radical chain reaction mechanism is presented in Scheme 7.7 (Haw and Marcus, 2003).

$$S^{\cdot} + CH_3OCH_3 \longrightarrow SH + {}^{\cdot}CH_2OCH_3$$
$$2\,{}^{\cdot}CH_2OCH_3 \longrightarrow CH_3OCH_2CH_2OCH_3$$
$$SH \longrightarrow S^{\cdot} + H^{\cdot}$$

SCHEME 7.7   One of several free radical mechanisms S$^{\cdot}$ is an unspecified surface radical site.

7.3.3.5 Hydrocarbon Pool Mechanism

In the early 1980s, an indirect pathway was proposed for olefin production. In 1993–1996, Dahl and Kolboe (1996) proposed HCP (hydrocarbon pool) mechanism. Different researchers studied the nature of HCP species and proved that in the MTO reaction on SAPO-34, methylbenzenes are HCP species that produce olefin products. According to the HCP mechanism, two reaction routes were proposed for the explanation of MTO reaction pathway, namely side-chain methylation and paring mechanism (Sassi et al., 2002; Wang et al., 2006; Wang and Hunger, 2008; Olsbye et al., 2012).

In 1990s, in another mechanistic evaluation, olefin precursors (such as ethanol) were co-feed with isotopically labeled methanol by which the isotopic composition of products could be labeled with TOS (time-on-stream). As there was inconsistency between isotopic labels distribution and direct reaction mechanism, researchers concluded that in methanol conversion, first olefins are formed on "hydrocarbon pool" the structure of which is loosely specified. According to Scheme 7.8, some sort of alkane or even carbenium ion is also formed. On the other hand, parallel synthesis of olefins is the key feature of this mechanism. For instance, ethylene, propylene or C_4 olefins could be eliminated by this pool. Moreover, each of these products is formed as the first volatile products with C-C bonds. According to other mechanisms, a single olefin (mostly ethylene) is initially formed that can be methylated to form higher olefins (Haw and Marcus, 2003).

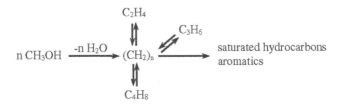

SCHEME 7.8 Kolboe's schematic hydrocarbons pool mechanism (Redrawn from Olah et al., 2018).

An actual methanol reaction network is depicted in Figure 7.3 for acid zeolites or SAPO catalysts. In this network, olefin formation reaction through aromatics- or olefin-based routes was considered. Two reaction modes are related to coke formation; that are further HCP intermediates transformation and further olefin conversion without involving HCP intermediates. If a molecular sieve catalyst with proper cavity size is selected, the first coking reaction can be limited (Tian et al., 2015).

7.3.3.6 Dual-Cycle Mechanism

The generally accepted hydrocarbon pool mechanism cannot explain the high propylene/ethylene ratio observed over HZSM-5 zeolites (e.g., Sahebdelfar et al., 2019) with cavities being too small to activate bulky polymethylbenzenes. To solve this drawback, the HP mechanism has been extended by proposing alkene intermediates in addition to aromatic intermediates of HP, the so called dual-cycle mechanism.

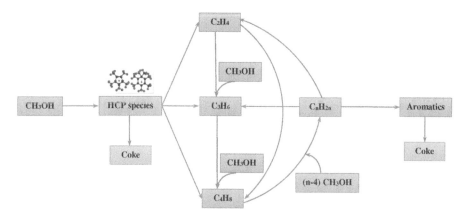

FIGURE 7.3 Reaction networks in MTO in relation to HCP (Redrawn from Tian et al., 2015).

Recently, transient tests with switching $^{12}C/^{13}C$ labeled methanol over HZSM-5 catalysts showed that ethylene and lower methylbenzenes (toluene and xylenes) exhibits comparable rates of ^{13}C incorporation while C_{3+} olefins show a different rate, much higher than ethylene (Svelle et al., 2006; Svelle et al., 2007; Bjorgen et al. 2007). To account for this, a dual-cycle mechanism was proposed by Bjorgen et al. (2007) according to which the reaction follows two simultaneous cycles; an olefin-based cycle and an aromatic-based cycle both being interdependent. The aromatic cycle involves polymethylbenzenes producing predominantly ethylene and aromatics by methylation of aromatic ring and side chain elimination while the olefin-based cycle is an autocatalytic cycle producing propylene and higher alkenes by alkene methylation-cracking (Figure 7.4).

The dual cycle mechanism is an extension of hydrocarbon pool and has received acceptance (Olah et al., 2018).

It was reported that on HZSM-5 catalyst, methanol conversion was accelerated by co-catalytic effect of toluene that was added to the mixture. This effect is because of side-chain alkylation on aromatic rings that causes olefin elimination. This is known as the Mole mechanism (Haw and Marcus, 2003).

In summary, for the MTO reaction over SAPO-34, having relatively large cavities, aromatic-based HCP has reported as the dominant reaction mechanism. For ZSM-5 catalyst, the dual-cycle reactions mechanism and for ZSM-22, olefin methylation and cracking mechanism proceed (Li et al., 2011).

7.3.4 KINETICS

Kinetic equations are essential valuable tools for designing chemical reactors, by which different reactor configurations and operating modes can be simulated (Keil, 1999).

Generally, kinetic models are classified to two groups:

- Lumped models, which are simple and represent the real process
- Detailed models, which consider individual reaction steps

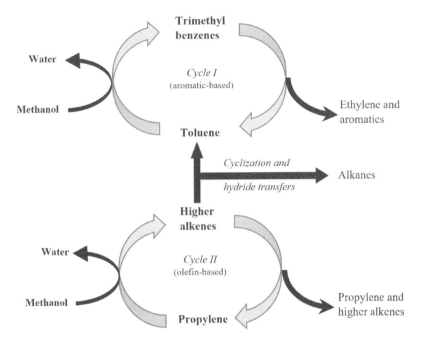

FIGURE 7.4 The scheme of dual-cycle mechanism in MTO reaction (Redrawn from Bjorgen et al., 2007).

The lumped kinetic models are adequate for the design purposes. Developing a detailed kinetic expression is very time-consuming and in some cases impossible.

In developing a kinetic expression, the whole range of pressure, temperature and feed composition of reactor must be considered.

In the 1970s, it was found that the initial ether formation step is faster than the subsequent olefin formation step and is essentially at equilibrium. This finding is valid over wide conversion ranges (Anderson et al., 1980).

In some lumped kinetic models the oxygenates (CH_3OH-DME) disappearance rate are lumped. Chang and Silvestri (1977) developed a mechanism for the formation of hydrocarbons from CH_3OH-DME. They considered a concerted bimolecular process containing carbonoid intermediates, which in turn undergo the insertion of sp^3 into C-H bonds due to which higher alcohols or ethers are formed and subsequently dehydrated for olefin formation.

Chen and Reagan (1979) announced that over ZSM-5 catalyst, the disappearance of oxygenates is autocatalytic. The below reaction scheme was proposed by them:

$$A \xrightarrow{\ k_1\ } B \tag{7.3}$$

$$A + B \xrightarrow{\ k_2\ } B \tag{7.4}$$

$$B \xrightarrow{\ k_3\ } C \tag{7.5}$$

in which A is oxygenates, B olefins and C is aromatics + paraffins. Oxygenates disappearance rate is of first order in oxygenates. Equation (7.6) was used for fitting experimental data:

$$-dA/dt = k_1 A + k_2 AB \tag{7.6}$$

Chang (1980) used the below assumptions for developing a kinetic model:

- Methanol and DME can be considered as a single kinetic species, as they are always at equilibrium
- Production and consumption of the reactive intermediates is of first order in oxygenates
- Olefins disappearance is of first order in olefins

Based upon these assumptions, the below reactions were considered:

$$A \xrightarrow{\ k_1\ } B \tag{7.7}$$

$$A + B \xrightarrow{\ k_2\ } C \tag{7.8}$$

$$B + C \xrightarrow{\ k_3\ } C \tag{7.9}$$

$$C \xrightarrow{\ k_4\ } D \tag{7.10}$$

in which A is oxygenates, B is CH_2 groups, C is olefins and D is paraffins + aromatics. The formal kinetic rates are as below:

$$-dA/dt = k_1 A + k_2 AB \tag{7.11}$$

$$-dB/dt = k_1 A - k_2 AB - k_3 BC \tag{7.12}$$

$$-dC/dt = k_2 AB + k_3 BC - k_4 C \tag{7.13}$$

By considering the steady-state condition for component B, and eliminating time, Eq. (7.14) is obtained:

$$-du/dA = 1/A\left[\left(1 + K_1 u\right)/\left(2 + K_1 u\right)\left(1 - K_2 u\right) + u\right] \tag{7.14}$$

in which u is C/A, $K_1 = k_3/k_2$ and $K_2 = k_4/k_1$.

Doelle et al. (1981) reported the sorption and reaction kinetics of DME and CH_3OH conversion over ZSM-5 catalyst. In the temperature range of 115–200°C, they reported Eq. (7.15) for the kinetics of methanol conversion:

$$r = k_1 P_{CH3O}/\left(1 + k_2 P_{H2O}\right) \tag{7.15}$$

Sedran et al. (1990) reported two lumped models for methanol conversion to hydrocarbons on a ZSM-5 catalyst:

Model I:

$$A \xrightarrow{k_1} B \tag{7.16}$$

$$A + B \xrightarrow{k_2} B \tag{7.17}$$

$$A + C \xrightarrow{k_3} D \tag{7.18}$$

and

$$B + E \xrightarrow{k_4} F \tag{7.19}$$

$$C + E \xrightarrow{k_4} F \tag{7.20}$$

$$D + E \xrightarrow{k_4} F \tag{7.21}$$

Model II:

$$A \xrightarrow{3k_1} B + C + D \tag{7.22}$$

$$B + E \xrightarrow{k_2} F \tag{7.23}$$

$$C + E \xrightarrow{k_3} F \tag{7.24}$$

$$D + E \xrightarrow{k_4} F \tag{7.25}$$

in which A is methanol + DME, B is ethylene, C is propylene, D is butenes and F is paraffins. The models followed the light olefins distributions fairly well.

Benito et al. (1996) proposed a model in which the influence of composition on deactivation was considered. In their kinetic evaluations at 300–375°C, they used ZSM-5 catalyst with an Si/Al ratio of 24. Their fitting models and kinetic equations with their constants are as follows:

$$MEOH/DME(A) \xrightarrow{k_1} light\,olefins\,(C) \tag{7.26}$$

$$2C \xrightarrow{k_2} products\,(D) \tag{7.27}$$

$$A + D \xrightarrow{k_3} D \tag{7.28}$$

$$C + D \xrightarrow{k_4} D \tag{7.29}$$

$$k_1 = 0.773 \times 10^{13} \exp(-33358 / RT) \tag{7.30}$$

$$k_2 = 0.127 \times 10^{8} \exp(-17633/RT) \tag{7.31}$$

$$k_3 = 0.204 \times 10^{12} \exp(-27987/RT) \tag{7.32}$$

$$k_4 = 0.773 \times 10^6 \exp(-15855/RT) \tag{7.33}$$

$$r_{A0} = -k_1 X_A - k_3 X_A X_D = dX_A/d\tau \tag{7.34}$$

$$r_{C0} = k_1 X_A - k_2 X_C^2 - k_4 X_C X_D = dX_C/d\tau \tag{7.35}$$

in which D is products in the gasoline boiling range, τ is the space time and X_i is the weight fraction of lump i.

Bos et al. (1995) proposed a kinetic network comprising ten first-order and two second-order reactions involving 6 component lumps and coke (Scheme 7.9). In this network, reactions (8) and (12) are of second-order.

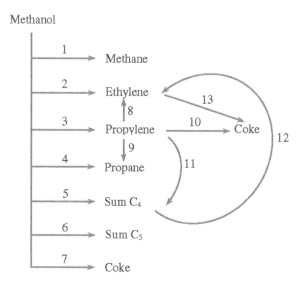

SCHEME 7.9 Reaction network according to Bos et al. (1995) model.

Ethylene formation from propylene is of first-order in methanol and propylene. The rate of ethylene formation from butenes depends on butenes and methanol.

7.3.5 CATALYST DEACTIVATION

A major problem in MTH conversions is rapid catalyst deactivation. Two distinct types of deactivation occur. Rapid catalyst deactivation by coke formation which is reversible and the activity can be recovered by coke removal, and a slow irreversible deactivation for ZSM-5 due to dealumination of the framework by the water product (Gayubo et al., 2002). Therefore, hydrothermal stability is an important issue in the latter case.

Due to catalyst deactivation, methanol conversion decreases with time-on-stream, the yield of light olefins (such as ethylene, propylene and butenes) increases and the yield of heavier olefins (such as aromatics and parrafins) decreases (Benito et al., 1996).

Liederman et al. (1978) developed an exponential model for activity as a function of TOS in which the effects of deactivation and methanol dilution with an inert component were considered.

Schipper and Krambeck (1986) defined Eq. (7.36) for β (overall catalyst activity for a given TOS):

$$\beta = \alpha a \qquad (7.36)$$

in which α is the remaining activity due to deterioration of acidic structure (that is an irreversible deactivation) and a is the remaining activity due to coke deposition (that is a reversible deactivation).

For each individual step i, the reaction rate (r_i) is defined as Eq. (7.37) in which r_{i0} is the reaction rate for the fresh catalyst and β is the overall activity:

$$r_i = r_{i0}\beta \qquad (7.37)$$

The empirical equation for the kinetics of permanent deactivation is given by

$$d\alpha/dt = -k_\alpha \left(\frac{-E_\alpha}{RT} \right) \alpha^m \quad m > 1 \qquad (7.38)$$

While the activity loss due to coke deposition is

$$da/dt = -k_c (\alpha a)^n \quad n > 1 \qquad (7.39)$$

In which k is the rate constant of deactivation and E_α is the activation energy for permanent deactivation.

Combining Eqs. (7.36, 7.38 and 7.39), total deactivation is expressed as follows:

$$d\beta/dt = \alpha \left[-k_d (\alpha a)^n - k_\alpha \alpha^{m-2} (\alpha a) \right] \qquad (7.40)$$

In the model of Schipper and Krambeck (1986), oxygenates, light olefins and the rest of the products (as three lumps) are probable coke precursors. Aromatic compounds play an important role in coke formation as precursors or intermediates. Ethylene and propylene (as light olefins) have high capacity for oligomer formation that after trapping in microporous structure produce carbonaceous compounds. In the MTG process, due to competitive diffusion-reaction mechanism, methanol and DME (as oxygenates) are coke formation precursors on the catalyst surface. Under some reaction conditions, butenes are the first hydrocarbon products of the MTG process, which after oligomerization and posterior light olefin cyclation can be coke precursors.

Sedran et al. (1990a) developed different kinetic models in which olefins distribution in the MTG process was considered as well. In their models, an empirical equation was considered for deactivation in which any decrease in kinetic constants is related to cumulative amount of produced hydrocarbons per catalyst mass $\left(\overline{H}_c/W\right)$.

$$k_i = k_{i0} \exp\left(-\beta_i \frac{\overline{H}_c}{W}\right)^d \tag{7.41}$$

In which β_i is the deactivation coefficient for reaction i.

Benito et al. (1996) proposed Eq. (7.42) as a deactivation kinetic model that considered the concentration of the lump components (namely; oxygenates, light olefins and the rest of hydrocarbon products):

$$-da/dt = \left[\sum\left(k_{di}X_i\right)\right]a^d \tag{7.42}$$

In this model, deactivation due to coke deposition was considered. For each lump, its individual contribution to deactivation was considered as well. The decrease in each reaction rates was described by activity, a.

7.3.6 Catalyst Modifications

The physicochemical characteristics of the catalyst play an important role in its performance (activity, selectivity and stability) in MTH reaction. The properties relevant to MTH conversions can be tuned or modified during hydrothermal synthesis and/or by post-synthesis treatments. Examples of the former are isomorphous substitution of metals in the framework and for the latter are ion exchange, surface passivation, desilication and dealumination treatments (Losch et al., 2016). Both of the above approaches have been employed for the synthesis of mesoporous or hierarchical zeolites to reduce their diffusion limitations.

The Si/Al ratio which is the main factor determining acidity of the zeolite can be tuned during synthesis or can be modified by steaming or chemical dealumination of the synthesized sample. Under proper conditions, steaming process can produce mesopores that facilitate diffusion within catalyst crystallites. A combination of treatments might also be applied.

7.3.6.1 Metal Incorporation

Substitution of other atoms (e.g., B, Cr, Ga, Ge, Fe, Co, Ni, P and Ti) for Al and Si atoms is widely used for modifying zeolite properties (Mokrani and Scurrell, 2009).

To improve the performance of ZSM-5 catalysts, in addition to *in situ* hydrothermal synthesis modifications, post-synthesis steam treatment and phosphorous doping have been commonly applied, mostly, for tuning the acidity and thus catalytic performance (Stöcker, 1999).

The incorporation of boron into ZSM-5 catalyst (starting gel with $20SiO_2:0.05Al_2O_3:xB_2O_3:1TPABr$(tetrapropylammoniumbromide)$:1.5Na_2O:200H_2O$, where $x = 0$, 1 for unmodified and boron modified samples, respectively) during hydrothermal synthesis greatly enhanced catalyst lifetime compared to the unmodified catalyst (1,300 vs. 340 h, respectively, at $T = 480°C$, WHSV = 0.9 h^{-1}). The higher stability of the boron-modified sample was attributed to its reduced strong/medium acid sites ratio and smaller crystallite size (Yaripour et al., 2015).

For SAPO-34 catalysts, metal modifications by both *in situ* and ion exchange have been reported to improve catalyst stability (Varzaneh et al., 2016; Salih et al., 2018). The influence of incorporating metals (Me = Fe, Co, Ni) into framework (MeAPSO34) was studied by Kang (2000). Metal addition enhanced catalyst activity compared to the unmodified catalyst. The ethylene selectivity (%) was in the order Ni-(80.7)>Fe-(71.6)>Co-(70.0)>SAPO-34(58.6) at 450°C. The selectivity of all samples reduced with TOS, but it was slower for the Co-containing catalyst.

Dubois et al. (2003) studied the effect of incorporation of Co, Mn and Ni on the SAPO-34 catalyst performance. The catalysts were prepared by direct hydrothermal crystallization of the respective metal containing gels. The selectivity of C$_2$–C$_4$ olefins of these catalysts were comparable with the unmodified sample. However, the catalyst lifetime was significantly affected by metal incorporation and Mn-SAPO34 showed the longest lifetime.

Zhong et al. (2018) studied the effect of incorporation of metal (Zn, Cu, Co, Ni) cations by a number of post synthesis methods, namely conventional ion exchange (CIE), template-assisted ion incorporation (TII) and alcoholic ion exchange (AIE). The metallic species located in cavity of SAPO-34 caused extra diffusion limitation especially for Zn-modified sample which exhibited a core-shell like structure. This structure enhanced ethylene selectivity by extra diffusion limitation to the more bulky products.

7.3.6.2 Introducing Mesoporosity

There is a direct relationship between catalytic performance and accessibility of active sites of catalysts (which is influenced by pore size and pore connectivity). If there are any limitations in internal diffusion within the pores, coke formation, pore blocking and retarded mass transport may occur. Researchers agreed that configurational diffusion (which is intra-crystalline mass transport from/to active sites inside micropores) is considerably slower than molecular diffusion and Knudsen diffusion and consequently there is a fast deactivation rate (Hartmann et al., 2016). Hence, diffusion limitations must be alleviated.

The effectiveness factor (η) is a criterion for describing the degree of catalyst utilization (Levenspiel, 1999). When the Thiele modulus is low ($\varphi \to 0$), full utilization of catalyst particle ($\eta \to 1$) is observed. In contrast, $\varphi = 10$ leads to $\eta = 0.1$ which is equivalent to 10% of catalyst surface utilization. Characteristic length of diffusion (L), efficient diffusivity (D_{eff}) and intrinsic reaction rate constant (k) are parameters used for the calculation of the Thiele modulus (φ). As for a given reaction and molecular sieve, k is fixed, for decreasing Thiele modulus, diffusion length (L) must be shortened and/or effective diffusivity (D_{eff}) in the pores must be increased (Zhong et al., 2017).

Zeolites are crystalline materials with a regular well-defined microporous structure. For decreasing the intra-crystalline diffusion path length, three strategies are proposed: decreasing crystal size, incorporating an auxiliary pore-structure (namely hierarchization, Figure 7.5) or a combination of both these strategies (Zhong et al., 2017).

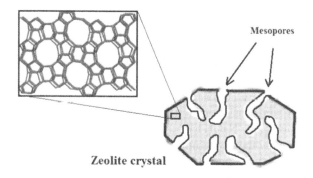

FIGURE 7.5 Schematic representation of a hierarchical zeolite.

The application of nano-sized and hierarchical molecular sieves and zeolites has many advantages; such as decreasing intra-crystalline diffusion path length, overcoming diffusion limitations, accelerating catalytic conversions, prohibiting unfavorable secondary reaction, reducing coke formation and increasing the catalyst lifetime. As these zeolites and molecular sieves have wide pathways, they can process molecules larger than micropores. There are additional mesopores and/or macropores inside hierarchically porous molecular sieves and zeolites. Intra- or inter-crystalline mesopores are two categories of the introduced mesopores. The voids originating from the aggregation of crystals are called inter-crystalline mesopores; they are in the vicinity of the crystalline surface. On the other hand, intra-crystalline mesopores, which are located inside the crystal, are inside zeolites and zeotype materials (Jiang et al., 2010).

For the synthesis of hierarchical zeolites, two strategies namely "top-down" or "bottom-up" methods can be used. In the former approach, which is an *ex situ* post-synthetic demetallation method, an element is extracted from the framework (such as desilication by organic amines or inorganic bases and dealumination by acids). In the latter approach, which is an *in situ* templating method, there are two sub-categories, namely hard- or soft-template methods (Egeblad et al., 2008; Meng et al., 2009; Valtchev et al., 2013).

In the reactions occurring inside the micropores those occurring at the external surface of the catalyst and in shape-selective reactions, nanosized and hierarchical zeolites and molecular sieves have advantages (Yan et al., 2015).

In the MTO reaction, the main reason of deactivation is related to diffusion limitation. The accessible cages of nanosized SAPO-34 are near its external surface, hence, it has slower deactivation. The same is valid for hierarchical SAPO-34, due to the alleviated diffusion limitation (Wang et al., 2011).

7.3.6.2.1 Synthesis of Hierarchical SAPO-34

7.3.6.2.1.1 "Top-Down" Approach   The "top-down" approach is economic, scalable and simple. However, for SAPO-type molecular sieves, the Al or Si cations in the framework might be extracted by acidic or basic post-treatments and consequently Brønsted acid sites destruction occurs. Moreover, any change in the cations of framework causes SAPO-type materials to be less stable (in comparison to aluminosilicates) and consequently after any conventional alkaline or acid treatment, they lose their crystallinity. As a concluding remark, the "top-down" approach that is applied to aluminosilicate zeolites for improving their catalytic performance can be applied to SAPO molecular sieves under special conditions (Verboekend et al., 2014).

Two methods can be used for this approach, namely "basic/acidic etching" and "fluoride etching". In "basic/acidic etching" approach for SAPO-type molecular sieves, when the inorganic NaOH base is used, they may suffer from amorphization. Therefore, for keeping the crystallinity, expensive organic bases must be used, the use of which simplifies the post-treatment steps. Moreover, there is no need to any additional ion-exchange process for obtaining molecular sieves with protonic form. For the case of alkaline post-etching with organic bases, the demetallation process is controllable (Koohsaryan and Anbia, 2016; Qiao et al., 2016).

Over hierarchical SAPO-34 crystals synthesized by a facile tetraethylammonium hydroxide (TEAOH) etching post-treatment, the methanol conversion and light olefin selectivities (C_2–C_4>93%, C_2–C_3>79%) were similar to those of the parent SAPO-34. However, the lifetime significantly prolonged from 320 to 640 min because of diffusion intensification and unchanged active (acid site density) properties (Liu et al., 2016).

The main drawback of the "top-down" approach is high weight loss due to dissolution of crystalline domains. Moreover, for mesopores, their number, shape and size cannot be controlled and crystal defects may be formed due to harsh dissolution conditions. This approach needs a time-consuming washing process that produces a large amount of wastewater (Chen et al., 2016).

7.3.6.2.1.2 "Bottom-Up" Approach Two methods can be used for this approach, namely "hard template" and "soft template". The hard-template approach is costly and needs a post-treatment process after crystallization for solid template removal. This drawback delayed its industrial application. By soft-template approach, hierarchical molecular sieves with mesopore structure can be obtained in one-step manner. Carbon materials are examples of hard templates and polymers and cationic surfactants are examples of soft templates (Perez-Ramirez et al., 2008).

Carbon materials, as hard templates, are hydrophobic and thus exhibit a weak interaction with the synthesis gels.

The polymers that have different functional groups, three-dimensional networks with specific molecular sizes can be used as mesoporous templates. These polymers can form a network inside the molecular sieve matrix and have impact on crystal nucleation and growth. Polyethylene glycol (PEG) is an example of these polymers, as it is a low-toxic linear water-soluble polymer that can be used for the production of secondary mesopores in SAPO-34 (Yao et al., 2010).

For aluminosilicate zeolites, cationic surfactants are of great importance for the incorporation of mesopores. Besides the pure template function of polymers, organosilane surfactants have influence on crystalline growth, texture and morphology and prohibit the separation between microporous and mesoporous structures of SAPO molecular sieves (Choi et al., 2006).

7.3.6.2.1.3 Other Approaches "Hollow SAPO-34", "solvothermal or solventless synthesis", "seed-assisted synthesis" and "mother-liquor-reuse synthesis" are other approaches for the synthesis of hierarchical SAPO-34 (Zhong et al., 2017).

In hollow zeolites and zeotype materials, diffusion limitations are decreased by the small thickness of microporous walls. Generally, these materials synthesized in fluoride medium have fewer defects with better crystallinity (Pagis et al., 2016).

Environmental pollution, low utilization efficiency of autoclaves, high spontaneous pressure and the subsequent requirement for equipment safety factor, high energy consumption and low yield of zeolite products are some disadvantages of solvothermal or hydrothermal synthesis methods. Hence, solvent-free synthesis routes are of high interest. It is worth mentioning that in the beginning of the crystallization process, water is a reactant for hydrolysis of silica source for the formation of active silica species and it is a by-product during the crystallization process because of the condensation of active silica species into zeolite products (Wang and Xiao, 2016).

Crystal seeds are building units that have an important role in crystallization and their introduction into the starting gel is desirable as the nucleation and consequently crystallization rate is accelerated (Sun et al., 2015).

Generally, a large number of reactants is needed for molecular sieve synthesis. After the synthesis procedure, the reaction mixture is filtered and the mother-liquor (as filtrate) that contains toxic and expensive organic species is discarded as a waste solution. It is worth mentioning that this filtration step causes an increase in synthesis cost and causes environmental pollution.

7.3.6.3 Synthesis of Nanosized SAPO-34

Regarding hydrothermal stabilities, nanometer-sized molecular sieves cannot be compared to their micrometer-sized counterparts. There are some difficulties for industrial application of these nanocrystals; such as high synthetic cost, difficult synthesis process and difficult filtration step for nanocrystal separation from their colloidal mother liquors (Schwieger et al., 2016).

Generally, any decrease in the crystal size causes an increase in external surface area and a decrease in diffusion path length by which mass transfer is facilitated and diffusion limitation is reduced. On the other hand, catalytic performance is improved by small particle size, which is because of suppressing side reactions (such as coke deposition). For SAPO-34 catalyst, the number of accessible cages near the external surface of catalyst has an important role in the MTO reaction. When the crystal size of SAPO-34 is decreased its accessible cages increases and catalyst lifetime is improved. For the synthesis of nanosized SAPO-34, reaction parameters must be optimized in a way to increase nuclei number and decrease crystal growth rate (Lee et al., 2009). Different approaches were used for the synthesis of nanosized SAPO-34; namely microwave synthesis, ultrasound-assisted synthesis, mixed-template synthesis, dry gel conversion, fast high-temperature synthesis, post-synthesis treatment and crystal growth inhibitor-assisted synthesis; which are discussed in subsequent sections.

7.3.6.3.1 Microwave Synthesis

Microwave heating technique is an energy-efficient and green method of heating. Its irradiation improves heating rate due to which the desired temperature in the substrate mixture can be obtained rapidly and thus crystallization time of molecular sieves with nanoporous structure is reduced. In contrast to conventional hydrothermal synthesis method, a more homogeneous heating is obtained by microwave irradiation. Moreover, the dissolution rate of precursor gel is improved by microwave energy. Consequently, for the synthesis of defect-free, small size and uniform crystals, microwave heating technique is an effective way (Meng and Xiao, 2014).

Yang et al. (2013) studied the influence of crystal size on the performance of SAPO-34 catalysts synthesized under hydrothermal conditions in conventional or microwave ovens using the same starting gel composition and TEAOH as the structure-directing agent. The catalyst lifetime increased with decreased crystal size, ranging from 150 to 786 min for 8-μm and 20-nm samples, respectively ($T = 400°C$, WHSV = 2 h^{-1}).

7.3.6.3.2 Ultrasound-Assisted Synthesis

In recent years, sono-chemical synthesis gained much attention. In this method, ultrasonic irradiation (20 kHz–10 MHz) is used for the formation, growth and collapse of bubbles in the liquid medium. Hot spots are localized by the collapse of bubbles and after that high pressure (181.8 MPa) and temperature (5,000–25,000°C) are provided which break the chemical bonds and accelerated the reaction. It is worth mentioning that the collapse of bubbles occurs in less than a nanosecond by which temperature decreases very rapidly (with a high rate of 10^{11}°C/s) and further growth and agglomeration of the particles are inhibited (Gedanken, 2004).

7.3.6.3.3 Dry Gel Conversion

The main advantage of dry gel conversion (DGC) method (in comparison to conventional hydrothermal crystallization) is a shorter crystallization time. This method is categorized into two techniques: SAC (steam-assisted conversion) and VPT (vapor-phase transport). In the former technique, a dry gel containing an organic template is crystallized under water-streaming treatment and molecular sieves are obtained. In the latter technique, in the steam flow of SDA and water and under specific temperature, the dry gel is crystallized (Zhong et al., 2017).

7.3.6.3.4 Fast High-Temperature Synthesis

The synthesis of SAPO-34 by hydrothermal method in a batch autoclave in temperature range of 180–200°C needs long crystallization time (approximately 24–72 h) that is an energy-intensive process with high synthetic cost. In dry gel conversion, ultrasound-assisted synthesis and microwave-assisted synthesis crystallization time of SAPO-34 is about several hours. Another solution is using the fast high-temperature synthesis method in which crystallization time is below an hour (Zhong et al., 2017). The high temperature results in decreased particle sizes, presumably, due to the enhanced nucleation and reduced crystal growth rate.

7.3.6.3.5 Post-Synthesis Treatment

Post-synthesis treatment is another route for the production of nano-sized molecular sieves. In this method, the micrometer-sized molecular sieve is milled and re-crystallized (Koohsaryan and Anbia, 2016).

7.3.6.3.6   *Crystal Growth Inhibitor-Assisted Synthesis*

Generally, nano-crystal formation is facilitated if crystal nucleation is preferred to crystal growth. Due to the strong interaction between nuclei and CGI (crystal growth inhibitor), the presence of CGI is beneficial for nanocrystal production. In the synthesis medium, there is an interaction between CGI and reactive sites of inorganic species, by which nucleation is improved and a large number of smaller nuclei are obtained. In other words, the adsorption of CGI on the surface of nuclei causes a separation between nuclei agglomerated by steric hindrance and inhibits the crystal growth (Vuong and Do, 2015).

7.3.7   BYPRODUCT UPGRADING

In MTO processes, in which light (C_2–C_3) olefins are the desired products, some higher olefins are inevitably formed as byproducts. These products can be transformed into light olefins after separation by cracking reaction, thereby further increase in the overall yield of MTO processes is achieved.

The Olefin Cracking Process (OCP), developed by UOP for increasing lower olefin yields by cracking of heavier olefins ($C_4^=$–$C_8^=$), was further developed and demonstrated by Total. It has been reported that for the MTO process in combination with the OCP, the carbon-based lower olefin selectivity can be improved to about 85%–90% (Chen et al., 2005). As OCP increases the propylene yield, propylene to ethylene ratio of 2.1 can be obtained. OCP can also be used to upgrade low-valued higher olefins of refineries and crackers streams.

In OCP, fixed-bed reactors packed with ZSM-5 catalyst, operating at relatively high temperatures (500–600°C), low pressures (0.1–0.5 MPa) and high space velocities are used. The reactors operate in swing mode for catalyst regeneration.

In 2013, the first UOP MTO/OCP plant with 295 kt/y light olefins was commissioned in Nanjing, China (Ye et al, 2015).

7.3.8   PROCESSES

Methanol is a highly reactive chemical and on acidic zeolites, hydrocarbons can be formed from it. The MTH reactions have been recently commercialized by many companies using different catalysts and reactor systems capable of controlling the heat of reaction and operating under rapid catalyst deactivation. Different strategies have been employed for increasing the selectivity to the desired product including reprocessing of the byproducts.

7.3.8.1   MTG Process

In the early 1970s, Mobil researchers synthesized H-ZSM-5 and discovered its unique catalytic properties for methanol conversion into high-octane gasoline (from a non-petroleum source). The produced gasoline is in a narrow boiling range (C_4–C_{10}, no C_{11+}) and its quality and yield is superior to the product of a conventional Fischer–Tropsch process with RONs (research octane number) higher than 90–95 (Gogate, 2019).

In 1973 and 1979, due to the first and second oil crises, R&D programs were started; among which bench- and pilot-scale MTG process are examples. Besides fixed-bed processes, fluidized-bed ones are also reported as advantageous one, because of the excellent temperature control and the possibility of continuous catalyst regeneration. As MTG is an exothermic reaction ($\Delta H^0 = -1.74$ MJ/kg$_{MeOH}$) with approximately 600°C adiabatic temperature rise, the control and dissipation of the generated heat is of high importance. The C$_{5+}$ gasoline yields for fixed-bed and fluidized-bed MTG process are 80 and 60 wt%, respectively, while the RON is larger in the latter (97 versus 93) (Keil, 1999).

Mobile's MTG process has attractions all over the world. During 1981–1984, a plant with capacity 4 bpd was developed in Paulsboro, NJ (Olsbye et al., 2012).

In 1985, the first commercial plant started-up in New Zealand with capacity 14,500 bpd that had a fixed-bed reactor. For controlling reactor temperature, the effluent stream from the reactor was recycled. This plant was based on natural gas that was converted into methanol through syngas.

Due to considerable decrease in oil price in the mid-nineties, the MTG part of that plant was shut down and switched to a methanol production plant (Olsbye et al., 2012).

In fixed-bed MTG process, methanol vapor is fed into a fixed-bed DME reactor, in which methanol is dehydrated to an equilibrium mixture of DME, H$_2$O and MeOH over a special alumina catalyst at 310–320°C and 26 bar pressure. In order to control the temperature rise of the second reactor, the effluent stream of the DME reactor is mixed with recycle gas and then fed to the ZSM-5 reactor at 350–370°C temperature where a mixture of aromatics, olefins and aliphatics (up to C$_{10}$) is produced (Gogate, 2019). The outlet temperature and pressure are 410°C and 20 bar, respectively.

The effluent stream is cooled and separated into gas, liquid water and liquid hydrocarbons. Most of the gas phase is recycled to ZSM-5 reactor. The produced water contains oxygenates that are treated in a subsequent biological wastewater treatment plant. The hydrocarbon is separated in a distillation column to raw gasoline, CO$_2$, dissolved hydrogen and light hydrocarbons. The obtained raw gasoline contains durene (1,2,4,5-tetramethylbenzene), which is removed in HGT (heavy gasoline treating) unit as it causes driveability problems (due to the high freezing point of durene, i.e., 79°C). The treated heavy gasoline and other gasoline components are blended to obtain gasoline with the required specification. In the operating conditions of MTG reaction, ZSM-5 catalyst has two types of ageing: coke formation on catalyst (as a by-product) causes a reversible activity loss and the product stream causes a gradual permanent activity loss. This type of deactivation is enhanced by operating at high temperatures (Keil, 1999).

The H-ZSM-5 catalyst has unique properties due to which coke formation is slow and by controlling coke burn-off, the catalyst activity can be restored. Hence, MTG has parallel reactors for intermittent regenerations.

The unique advantages of MTG process are as below (Gogate, 2019):

- Product stream is hydrocarbons in narrow compositional range, a small amount of methane and hydrocarbons smaller than C$_{11}$ are formed

- High methanol conversion with high selectivities of iso-paraffins and aromatics with higher octane value are achieved
- Low aging rates for ZSM-5 based catalysts were reported

Commercially, the MTG reaction is operated at 400°C and several bars of methanol partial pressures over a ZSM-5 catalyst. These are the optimal conditions for paraffins and aromatics production from the intermediate olefins. In the MTG reaction, product stream normally contains 40% of light olefins. Researchers tried to selectively produce light olefins from methanol over small/medium and large-pore zeolites, and SAPO-type molecular sieves. Olefin yield can be increased by adjusting reaction conditions and catalyst type.

A fluidized-bed MTG demo plant was designed jointly by Mobil R&D Corporation, Uhde GmbH and Union Rheinische Braunkohlen Kraftstoff AG and operated in Wesseling, Germany. In this plant, gasoline was produced with 100 bpd capacity during 1982–1985.

Crude methanol feed is vaporized and enters the reactor. Water removal helps improving the overall process economy. Heat of the reaction is used for the generation of high-pressure steam. Heat exchanger can be used inside/outside of the reactor; the most efficient case is the inside one. There is continuous catalyst regeneration and there are some cyclones for dust removal from the reactor effluent. The produced hydrocarbons are further processed in a fractionation unit by which C_{5+} hydrocarbons, olefin recycle and alkylation feed is obtained. If there is a need to reduce durene content, the heavy gasoline must be treated. By using this plant, hydrocarbons with the desired selectivity were produced. By using fluidized-bed technology, increased product yield, better quality and efficient heat recovery was obtained. Based upon the market situation, liquid fuels can be produced via methanol, by this technology (Keil, 1999).

In 2009, an MTG demo-plant with 100 kt/y capacity was constructed and operated in Shanxi province, China (Olsbye et al., 2012).

7.3.8.2 MTP Process

The MTP process, developed by Lurgi, is basically similar to the fixed bed MTG, although with different operating conditions and a ZSM-5 catalyst with higher Si/Al ratios. In this process, methanol is first pre-heated to 260°C and fed into adiabatic DME reactor over acidic dehydration catalyst and 75% of methanol feed is converted into DME and H_2O. In the next step, the mixture is heated to 470°C and fed to the first MTP reactor. In this reactor, steam is used (0.75–2 kmol per each kmol of reaction mixture) and >99% of methanol-DME mixture is changed to propylene (Figure 7.6). In order to increase propylene yield, the second and third MTP reactors were used in the subsequent steps. The MTP reactor is based on a high siliceous ZSM-5 catalyst and operating at atmospheric pressure. There is a recycling stream of C_2H_4 and C_{4+} in this process which helps maximizing propylene yield. Moreover, by recycling C_2 and C_{4+} olefins, a heat sink for the exothermic reaction is provided. Process condensate water is also recycled that work as diluting agent by which olefin selectivity is increased (Bjorgen et al., 2008). The carbon-based propylene yield is about 70%.

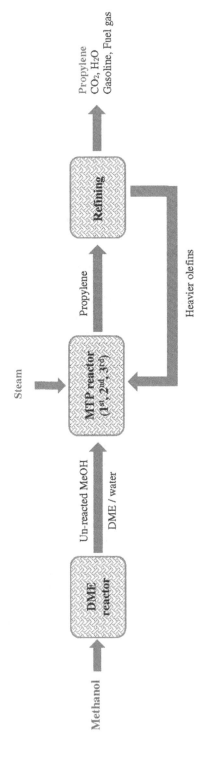

FIGURE 7.6 Flow diagram of Lurgi MTP process.

The catalyst is regenerated by coke burning with the N_2/air mixture after about 400–700 hours-on-stream.

In 2010–2011, in Ningxia, China, the first MTP plant based on Lurgi technology was commissioned with 500 kt/y propylene capacity and 185 kt/y gasoline as its major by-product (Yang et al., 2019).

7.3.8.3 MTO Process

The MTO reaction on SAPO-34 catalyst is exothermic with −819 kJ/kg$_{methanol}$ heat of reaction at 495°C. In order to keep reaction temperature in the desired range, heat of reaction must be removed simultaneously. On the other hand, due to coke deposition, the MTO reaction on SAPO-34 catalyst shows rapid deactivation. In-line coke combustion is needed to keep high catalyst activity. Based upon good heat transfer performance, fluidized-bed reactor (in comparison to fixed-bed one) is superior. For the MTO process, Exxon Mobil and UOP had some patents for fast fluidized bed reactors (Tian et al., 2015).

In 1977, for the first time, Mobil researchers discovered MTO reaction by use of aluminosilicate zeolite as the catalyst. Up to now, different researchers studied reaction principles, catalyst synthesis and processes. The MTO reaction is very complicated as different products can be obtained on various zeolite catalysts (Chen et al., 2012).

Different molecular sieves with various pore networks were evaluated for the production of light olefins among which small pore 8-MR (eight-membered ring) molecular sieves were more shape-selective toward light olefins. Small pore openings hinder the diffusion of higher olefins and aromatics (as large intermediates) from the channels; due to which high ethylene and propylene selectivity is obtained (Dusselier and Davis, 2018).

In 1990, the Dalian Institute of Chemical Physics (DICP) reported the first excellent performance of SAPO-34 for MTO reaction. Moreover, SAPO-34 had good regeneration stability and high-temperature steam resistance. Since then, many researchers have been focusing on explanation of its crystallization mechanism and optimization of its synthesis procedure in order to promote the development of MTO process, technologically.

Despite high catalytic activity and selectivity of SAPO-34 catalyst in the production of light olefins, its rapid deactivation is still a disadvantage. On the other hand, the MTO reaction on SAPO-34 is extremely exothermic. Hence, in order to have efficient heat transfer and continuous catalyst regeneration, fluidized bed reactor is the promising configuration of the industrial MTO process (Yang et al., 2019).

For MTO processes, there are four main technologies, namely DMTO/DMTO-II, S-MTO, MTO by UOP/Norsk Hydro and MTP (methanol-to-propylene) by Lurgi; most of them are used in China's burgeoning coal to olefin industry (Gogate, 2019).

S-MTO technology was developed by Sinopec, as a player in MTO technology. Hydro (based in Oslo, Norway) and UOP (now Honeywell UOP, based in Des Plaines, IL, USA) jointly developed the UOP/Norsk Hydro technology for MTO process (Gogate, 2019). This process is based on SAPO-34 catalyst in a low pressure fluidized bed reactor. In this technology, in the first stage, methanol is pre-heated and entered into the MTO reactor for the synthesis of olefins which is operated at

340–540°C and 0.1–0.3 MPa. The products (ethylene, propylene, heavier olefins and water) are cooled and water is separated from them. In the next step, which is separation, ethylene and propylene are separated and C_4–C_6 heavier olefins are cracked to C_3 (propylene) or C_4 (butadiene). In the next step, propylene is recycled for light olefin recovery and C_4 is used as the fuel for MTO process.

In 2013, in Nanjing, China, a commercial MTO plant with a capacity of 0.3 Mt/y olefins was commissioned which is a combination of UOP/Hydro MTO process and Total/UOP olefin cracking process by which C_2H_4 and C_3H_6 yield was enhanced (Gogate, 2019).

7.3.8.4 MOGD Process

In the MTG process, light olefins are the intermediates, the selectivity to which increases at high temperatures (due to thermodynamic effect) and low pressures (kinetic effect). This concept was extended to produce a combined gasoline/diesel product by subsequent oligomerization of the light olefins. In MOGD (Mobil's olefin-to-gasoline and distillate) process, olefins are converted into different products by a ZSM-5 based process. This is a refinery process coupled with the MTO process; in which a ZSM-5 catalyst oligomerizes light olefins to higher molecular weight olefins in the range of gasoline, distillate and lubricants.

In the MOGD process, the selectivity of distillate and gasoline is > 95% and the ratio of gasoline to distillate is in the range of 0.2–100. Tuning shape-selectivity in a way to produce methyl-branched iso-olefins (C_5–C_{20}), helps obtaining high octane numbers. The C_{10}–C_{20} fraction must be hydrogenated. In 1981, a large-scale MOGD process evaluation was performed in Mobil refinery. A commercial zeolite catalyst was used in this process. This process has four fixed-bed reactors, three of which are on-line, operated in series with inter-stage coolings and recycle of liquids for controlling the heat of reaction and one is for catalyst regeneration. At first, olefin is mixed with gasoline recycle stream and enters the reactors. It is necessary to use fractionation for the production of gasoline-rich stream to be recycled to the reactors (Olsbye et al., 2012).

MOGD and MTO processes can be coupled. Before the MOGD section, MTO gasoline (with high octane number) is split off and blends with MOGD gasoline. In the two plants (with a 15.9 m^3/d capacity) at Wesseling, typical distillate and gasoline yields from olefins yield are 50:50 (wt:wt). In comparison to the gasoline obtained from FCC, this gasoline is olefinic and aromatic and has better quality. The RON and MON (motor octane number) are 85.0 and 93.0, respectively, and durene content is little. As MOGD diesel has a density of 0.8, it can be used as the jet fuel (Keil, 1999).

7.3.8.5 GTO Process

GTO (natural gas to olefins) process was introduced by UOP and Norsk Hydro. In this process, natural gas is first converted into methanol. The subsequent step is the MTO process of UOP-Hydro using SAPO-34 catalyst. Ethylene and propylene are the primary products of this plant with high yields of 48% and 33%, respectively. This MTO process is highly selective to ethylene. Although ethylene oligomerizes to higher compounds (that are thermodynamically favored) but within the cavities of SAPO-34 only small linear olefins and paraffins pass through < 4 Å diameter pores of the catalyst. This is the main reason of obtaining ethylene with high selectivity over SAPO-34. Another advantage of SAPO-34 is its attrition resistance and stability due to which it can pass several reaction-regeneration cycles. In this process, a fluidized-bed reactor accompanying a fluidized-bed regenerator is used. By steam generation, heat of reaction is controlled. Sending the catalyst to the regenerator, burning the coke and steam generation by the heat produced from coke burning is the continuous cycle in this process. The next step is heat removal and cooling the reactor effluent. The effluent stream, after being compressed, passes through a caustic scrubber for CO_2 removal and sent to a dryer for water removal. The reaction section is similar to that of Mobil/Uhde process. The effluent stream of reaction section is sent to the product recovery step that includes demethanizer, deethanizer, depropanizer and C_2 and C_3 splitters. Different products are obtained from these separation columns; namely methane, ethane, propane and C_4.

As the MTO process (for ethylene and propylene production) needs a large amount of methanol feed, methanol production unit and MTO unit must be located in the same site (Keil, 1999).

7.3.8.6 DMTO Process

In DMTO (dimethyl ether or methanol-to-olefins) technology, the reaction is an exothermic one occurring at a high temperature (400–500°C) and medium pressure (0.1–0.3 MPa) over nano-SAPO34 catalyst with 100% methanol conversion. It consists of a fluidized-bed reactor and regenerator, a separation unit for ethylene-propylene and heavier hydrocarbons separation and utility equipment (Gogate, 2019).

In 2006, a DMTO demonstration unit was constructed in Huaxian, Shanxi with 2.5 t/h methanol feed rate. Fluidized-bed reactor and regenerator, quench tower and sour water stripper are parts of this demonstration unit. The C_2–C_3 and C_2–C_4 olefin selectivities were 79% and 89%, respectively (Tian et al., 2015).

In 2008, in Dalian, China, a 2,000 t/y catalyst manufacturing plant was constructed to supply the commercial catalyst of the DMTO unit (Yang et al., 2019).

In 2010, the first commercial DMTO unit was constructed and started up in Baotou, China with a capacity of 600 kt/y ethylene-propylene. In this unit, light olefins are produced from coal. The data obtained showed methanol conversion of 100% and selectivity to C_2H_4 and C_3H_6 of over 80% (Tian et al., 2015).

In China, currently, 64% and 70% in terms of the number of units and capacity of MTO technology is devoted to the DMTO/DMTO-II technology (Gogate, 2019).

DICP developed the second generation of the DMTO process, in which the C_{4+} by-product is separated and converted in a second cracking fluidized-bed reactor into ethylene and propylene. One advantage of this process is that the MTO and C_{4+} conversion reactors use the same catalyst and one regenerator is needed for the regeneration of the catalyst. In 2009, by the addition of a second bubbling fluidized-bed reactor, the DMTO demonstration plant was revamped. According to the results obtained from demo plants, ethylene and propylene selectivity increased from 79.21% (in DMTO plant) to 85.68% (in DMTO-II plant) with 60 wt.% of C_{4+} recycling. For producing 1 ton of ethylene and propylene, methanol consumption was decreased from 2.96 to 2.67 tons; which is an economical improvement for DMTO process. The first DMTO-II plant with 0.67 Mt/y polyethylene and polypropylene capacity was licensed to Pucheng Clean Energy Chemical Co. Ltd in 2010, and in 2014, the first commercial DMTO-II unit was started-up (Tian et al., 2015).

Till the end of 2018, 13 DMTO plants were in operation with a total capacity of 7.16 Mt/y. Moreover, in 2018, DICP presented a new generation of DMTO catalyst and completed the evaluation of the DMTO-III technology at a pilot scale by which demonstrated the yield of DMTO unit was further improved (Yang et al., 2019).

7.3.8.7 TIGAS Process

For the production of either methanol or synthetic gasoline from natural gas, Topsøe presented a process with low investment cost. Due to low price of natural gas, in future, synthetic fuel plants must be constructed in remote areas. TIGAS (Topsøe integrated gasoline synthesis) is an example in this regard. In the 1980s, Haldor Topsøe developed the TIGAS process (which is an alternative to the MTG process) and constructed a plant with capacity 1 t/d. In this process, MeOH, DME and gasoline syntheses were merged into a single loop and no methanol separation occurred. Moreover, by the integration of methanol and DME syntheses, syngas conversion was enhanced and process efficiency was improved (Olsbye et al., 2012).

At Houston, a demo plant of this process was operated for 10,000 h. For the production of syngas, a combination of steam reforming and auto-thermal reforming was selected. A multi-functional catalyst was used for the production of oxygenates mixture. In the conventional methanol synthesis plants, a 50–100 bar syngas compression is needed, but the TIGAS process does not need it and this is also a reduction in capital and operating costs.

7.3.8.8 DTG Process

University of Akron developed a process in which LP-DME synthesis was used for direct production of DME from syngas, followed by dehydration to gasoline, namely DTG (LP-DME to gasoline) process. In this process, a combination of gamma-alumina and $Cu/ZnO/Al_2O_3$ was used as the catalyst by which the reactor productivity and per-pass conversion were improved. Over γ-Al_2O_3 catalyst, due to *in situ* conversion into DME, the liquid phase is lean in MeOH. Because of water production by methanol synthesis and DME synthesis, the WGS reaction is shifted to completion. In comparison to the process of syngas to methanol conversion, the volumetric productivity improved by 100% with this one-step process. The reason is that chemical equilibrium did not limit syngas conversion into DME (Keil, 1999).

7.3.8.9 Others

Researchers (Weitkamp, 1991) reported the application of CMHC (coupled methanol-hydrocarbon cracking) process which is a combination of exothermic MTO process and endothermic hydrocarbon steam cracking. In this process, operating conditions were adjusted in a way that the whole process was thermoneutral. In this process, the appropriate temperature is in the range of 600–700°C and ZSM-5 zeolite is used as the catalyst.

STF (syngas-to-fuels; via methanol) is another version of MTG process that was developed by CAC Chemnitz (Germany) (Joensen et al., 2011).

Total Petrochemicals and UOP jointly developed a process which is the combination of UOP/INEOS MTO process and OCP (olefin cracking process), by which ethylene-propylene selectivity was improved. In 2009, in Feluy, Belgium, a semi-commercial demo plant with a capacity of 10 t/d of methanol feed was strated-up and in 2011, in Nanjing, China, a plant with 295 kt/y capacity was constructed (Olsbye et al., 2012).

Mobil Research and Development Corporation (in response to New Zealand Government request) constructed a plant for GTG (gas-to-gasoline) process. This plant has three main units: two methanol trains with 2,200 t/d production capacity and an MTG conversion unit. In the gas-to-methanol part, the ICI low pressure process was used. The MTG part is based on Mobil ZSM-5 catalyst (Keil, 1999).

7.4 METHANOL CARBONYLATION

Acetic acid is one of the simplest carboxylic acids. It is an important feedstock in chemical industries and is used for manufacture of polyethylene terephthalate (PET or PETE) for bottles, cellulose acetate for photographic films, polyvinyl acetate for the production of wood glues and also synthetic fibers. In household applications, it is used as a condiment. Acetic acid is produced by a variety of methods such as carbonylation of methanol, oxidization of acetaldehyde, oxidation of ethylene and fermentation. The carbonylation of methanol is the most important route for the production of acetic acid (about 80% of total) (Budiman et al., 2016). The process was first described by BASF around 1913.

7.4.1 Reaction Chemistry

The methanol carbonylation for producing acetic acid is based on the following reaction:

$$CH_3OH + CO \rightarrow CH_3COOH \quad \Delta H^0_{298} = -138.6\,\text{kJ/mol} \qquad (7.43)$$

The reaction is exothermic and accompanied by a decrease in number of moles. As both methanol and carbon monoxide are raw materials, carbolylation of methanol has been for a long time an attractive route for the production of acetic acid because of its atom economy.

Henry Drefyus in British Celanse developed a pilot plant for carbonylation of methanol in 1925. But, the extreme corrosiveness of reaction mixture and high operating pressures required (>200 atm) and the absence of materials for maintaining this mixture delayed commercialization of the process. The first commercial plant for the carbonylation of methanol based on cobalt catalysts was developed by the German Company BASF in 1963 (Kalck et al., 2020).

The carbonylation of methanol is catalyzed by the complexes of group VIII, especially, rhodium, iridium, cobalt and nickel. In all carbonylation processes, iodine is an essential co-catalyst. The reaction proceeds via methyl iodide which alkylates the involving transition metal.

The reaction involves methyl iodide as an intermediate and proceeds in three steps. A catalyst, usually a metallic complex, is necessary for the carbonylation step (step 2)

$$(1) \quad CH_3OH + HI \rightarrow CH_3I + H_2O$$

$$(2) \quad CH_3I + CO \rightarrow CH_3COI$$

$$(3) \quad CH_3COI + H_2O \rightarrow CH_3COOH + HI$$

By changing process conditions, acetic anhydride can also be produced. The largest producers of acetic acid are Celanese, BP chemicals, Millennium Chemicals, Sterling Chemicals, Eastman and Svensk Etanolkemi.

7.4.2 Homogeneous Catalytic Systems

Catalytic systems for methanol carbonylation are one of the few commercial systems using homogeneous catalysts, although researches for development of heterogeneous catalysts are underway. A recent review by Kalck et al. (2020) covers the catalyst used for carbonylation of methanol.

7.4.2.1 Catalyst Components

In the liquid phase, the catalyst system is homogeneous and includes active species, co-catalyst and solvent for providing a homogeneous liquid phase. The active component may be an oxide, organo-metallic, coordination compound or salt of a noble metal such as rhodium, palladium, platinum, iridium, osmium or ruthenium. However, the preferred catalyst is a homogeneous combination of one of the above metals and a halide like chlorine, bromine and iodine and to the same fraction of suitable ligands such as organophosphine, organoazerine or organo stibine. Among the above noble metals, rhodium is commercially used. Among rhodium compounds, chlorocarbonyl-bis-triphenyl phosphine rhodium $Rh(CO)[P(C_6H_5)_3]_2Cl$ is thermally stable and is commonly used. The second component of the catalyst system is a halogen containing or halogen compound co-catalyst. The most selected halogen is iodine being mostly used as methyl iodide or aqueous hydrogen iodide. The third catalyst component is a solvent being used for dissolving the metallic catalyst portion. The common solvent is a mixture of acetic acid and water.

Experimental results showed that carbonylation reaction does not occur in the absence of rhodium catalyst. Thus, for example, in an autoclave filled with 420 mmol methyl acetate, 434 mmol water, 27 mmol methyl iodide and 802 mmol acetic acid and heated at a constant pressure of 30 bar to 195°C for 1 h, no CO comes out from CO tank, indicating no reaction occurred (Brad et al., 1991).

Studies showed that the effect of increase in rhodium concentration on reaction rate in the presence of Li is much larger than that in the absence of Li or when present in low concentrations. As might be expected, the rate of reaction increases with rhodium and activator concentrations. It has been shown that in both cases, the reaction rate is linearly correlated to the concentration of these two species.

7.4.2.1.1   *Activators*

Appropriate activators used include Ru, Os, W, Zn, Cd, In, Ca and Hg compounds with ruthenium and osmium compounds are preferably used. Ruthenium is the best activator. Typically, ruthenium is 0.5–1.5 wt% and preferably 2–10 wt%. Also, the appropriate concentration of the activator is about 400–500 ppm.

The molar ratio of each activator to catalysts is about 0.1:1 to 1:15 and preferably 0.5:1 to 1:10 is suitable.

The effect of activator concentration on rate depends on the concentration of the pre-activator as well. Generally, the addition of activator in appropriate (high) concentration of pre-activator results in significant (up to 96%) increase of the reaction rate (Sherman et al., 1994).

7.4.2.1.2   *Pre-activators*

Halogen compounds that can generate halide ion exhibit favorable effect on carbonylation reaction as pre-activators. These compounds include alkali metal iodides, alkali earth metal iodides, complexes of the salts capable of generating Br and I or their mixture. The appropriate alkali metals are lithium iodide and calcium iodide.

Suitable complexes for generating I are complexes of lanthanide series such as La and also Ga, Ni, Fe, Al, Cr, and especially $Al(OAc)_3$ and $Ce(OAc)_3$, quaternary salts of ammonium iodide and phosphonium iodide which can produce iodide ions.

The best pre-activator is lithium iodide. A suitable mixture of two or more of these compounds increases the alkylation reaction rate.

7.4.2.2   Commercial Methanol Carbonylation Processes

A number of technologies have been developed by research. The main aim was improving the yield and moderating operating conditions.

7.4.2.2.1   *High Pressure Carbonylation*

The first methanol carbonylation to acetic acid process was commercialized by BASF. The process employed iodide-promoted cobalt catalyst and operated at high pressure (600 atm) and also high temperature (230°C), but produced acetic acid with about 90% selectivity.

7.4.2.2.1.1   The Chemistry of BASF Process   The BASF catalyst includes cobalt carbonyl with iodide promoter. The dicobalt octacarbonyl ($Co_2(CO)_8$) is formed *in situ* by the reaction of cobalt iodide and carbon monoxide (Eq. 7.44) which then transforms to the active species hydrido cobalt tetracarbonyl $[Co(H)(CO)_4]$ (Eq. 7.45):

$$2CoI_2 + 2H_2O + 10CO \rightarrow [Co_2(CO)_8] + 4HI + 2CO_2 \qquad (7.44)$$

$$[Co_2(CO)_8] + H_2O + CO \rightarrow 2[Co(H)(CO)_4] + CO_2 \qquad (7.45)$$

Figure 7.7 shows the catalytic cycle of the reaction for cobalt-based catalysts.

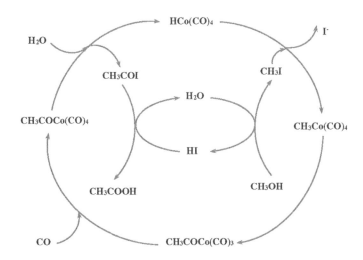

FIGURE 7.7 BASF catalytic cycle for acetic acid synthesis from methanol (based on Cheung et al., 2003).

To achieve a reasonable reaction rate for commercial application, severe operating temperatures and pressers are necessary. The reaction rate depends on partial pressure of carbon monoxide and methanol concentration. The yield of acetic acid based on methanol is 90% and based on carbon monoxide is about 70% (Budiman et al., 2016).

The byproducts of BASF process include methane, ethanol, carbon dioxide, propanoic acid, alkyl acetates and 2-ethyl-1-buthanol (King et al., 1982; Ellwood, 1969). About 3.5% of methanol leaves the system as methane, 4.5% as liquid byproducts and 2% as gaseous products. About 10% of carbon monoxide is converted into CO_2 by WGS reaction.

7.4.2.2.1.2 Process Description   In BASF methanol carbonylation process to acetic acid (Grewer and Schmidt, 1973; Cheung et al., 2003), carbon monoxide, methanol (containing 60% dimethyl ether), catalyst recycle, catalyst make-up and methyl iodide recycle (from the wash column) are sent to the high-pressure reactor (stainless steel coated with Hasetlloy). Part of the relatively low heat of reaction is used for feed preheating. The reaction product is cooled and sent to a high-pressure separator. The off-gas goes to a washing column and the liquid is expanded to 0.5–1 MPa in a medium pressure separator. The released gas is also sent to washing column. The liquid leaving the medium pressure separator goes to expansion chamber and the gas leaves the chamber to the scrubber. The gases leaving the stripper and washing column are vented as off-gas. Both scrubber and wash column use methanol feed for recovery of methyl iodide and other volatile iodine compounds. The methanolic solution is recycled to the reactor. The composition of the off-gas on volume basis is 65%–75% CO, 15%–20% CO_2, 3%–5% CH_4 and the balance is CH_3OH (Cheung et al., 2003).

The crude acid leaving the expansion chamber comprises 45 wt% acetic acid, 35 wt% water and 20 wt% esters (mostly methyl acetate). The first column degases the crude product; the off-gas is sent to a scrubber and then, the catalyst is separated as concentrated acetic acid solution by stripping volatile components in a separation column. The acid is then dried by azeotropic distillation in drying column. The overhead of drying column includes acetic and formic acid, water and byproducts which form azeotrope with water. This overhead is a two-phase system which is separated in a chamber. Part of the organic phase, predominantly esters, is recycled to the column where it acts as an azeotrope agent. The remainder organic phase is sent to the auxiliary column where the heavy residues are separated at the bottom of the column and light esters from overhead are recycled to the reactor. The aqueous phase and catalyst solution are recycled to the reactor. The bottom of drying column is sent to finishing operation where pure acetic acid is obtained from the overhead. The column bottom stream (finishing) is sent to residue column and the overhead of the column is recycled to dewatering column. The bottom of residue column contains 50 wt% propionic acid which may be recovered.

In BASF process, the theoretical carbon monoxide to methanol ratio is 1/1. However, due to side reactions, 5/1.1 ratio is recommended.

7.4.2.2.2 Low Pressure Methanol Carbonylation

Monsanto offered the rhodium catalyzed methanol carbonylation in 1968 (Gaub et al., 1996). Rhodium catalyst is 10^3–10^4 times more active than cobalt catalyst; thus, lower CO pressure and temperature are necessary for the reaction. Rh catalyst also shows higher selectivity to acetic acid. A comparison between the two catalyst systems is given in Table 7.2 (Cheung et al., 2003).

According to Table 7.2, the reaction conditions in reactor (180°C and 3 MPa) are milder than that for BASF process. Methanol-based yield is 99% and based on carbon monoxide is 90% (Paulik and Roth, 1968).

The chemistry of BASF and Monsanto processes is similar but their kinetics differ due to different rate determining steps. In both systems, two important catalytic cycles exist (Figures 7.7 and 7.8). One encompasses metal carbonyl catalyst and the other iodide promoter. Fig. 7.8 shows the catalytic cycle of acetic acid synthesis on rhodium catalyst. This cycle is a classic example of homogeneous catalytic process.

TABLE 7.2
Comparison of Operating Conditions and Performance of Cobalt and Rhodium Catalysts

Parameter	Cobalt	Rhodium
Concentration (M)	~10^{-1}	~10^{-3}
Temperature (°C)	~230	~180
Pressure (bar)	500–700	30–40
Selectivity (%)	90	>99
H_2 effect	Byproducts, CH_4, CH_3CHO, EtOH	none

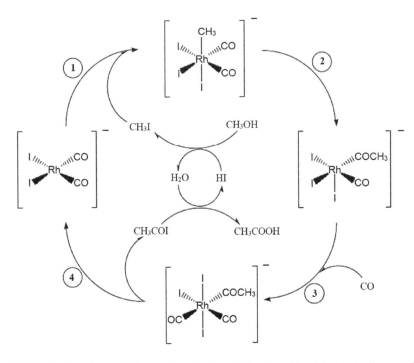

FIGURE 7.8 Reaction cycle for methanol carbonylation to acetic acid with rhodium catalyst (Redrawn from Cheung et al., 2003).

During methanol carbonylaton, methyl iodide is produced from addition of methanol to hydrogen iodide. IR spectroscopic studies showed that the principal species of the rhodium catalyst is [Rh(CO)$_2$I$_2$]$^-$.

The key reaction in Monsanto process is oxidative addition of methyl iodide (step 1) to planar square rhodium metallic center (I) to form hexagonal rhodium (III) species. The insertion of carbon monoxide (step 2) into cis-bond CH$_3$-Rh produces a penta-coordinated acyl intermediate, which upon addition of carbon monoxide (step 3) and reductive elimination of acetyl iodide (step 4) regenerates the original rhodium (I) complex. Hydrolysis of acetyl iodide with water aqueous methanol feed produces acetic acid and HI that reacts with methanol to form methyl iodide and repeat its own cycle.

In the presence of large amounts of water (>8 wt%), the rate determining step is oxidative addition of methyl iodide to rhodium center. Consequently, the reaction is of first order in the catalyst and methyl iodide concentrations and independent of other parameters:

$$rate \propto C_{\text{catalyst}} \times C_{\text{CH}_3\text{I}} \qquad (7.46)$$

If the water level is less than 8 wt%, the reductive elimination of acyl species from the catalyst determines the rate.

The carbonylation process with a rhodium catalyst is highly selective, but side reactions also exist.

7.4.2.2.2.1 Process Description   In Monsanto (now BP) process (Gaub et al., 1996; Cheung et al., 2003), carbon monoxide and methanol are introduced continuously into a liquid-phase mixed reactor operating at 150–200°C and 50–60 bar. Noncondensable byproducts (CO_2, H_2 and CH_4) are purged for controlling CO partial pressure in the reactor. The off-gas from reactor and purification sections of the process are combined and sent to a recovery system from purge exit in which the light terminals such as organic iodides (e.g., methyl iodide) are stripped from vent before non-condensable gases burn. The light terminals from the exit of recovery system purge are recycled to the reactor. The solution from reactor is sent to pressure reduction column where the catalyst is separated as a residue stream from crude acetic acid product and recycled to the reactor. Crude acetic acid which contains methyl iodide, methyl acetate and water is taken from overhead of the pressure reduction column and is sent to light ends column. The light components (methyl iodide, methyl acetate and water) are recycled to the reactor as overhead two-phase stream, while wet acetic acid is separated as a side stream and sent to dewatering column. An overhead aqueous acetic acid from light ends column is recycled to the reactor and a bottom product acetic acid residue is sent to a heavy terminals column. Propionic acid which is the main liquid byproduct is taken as a waste stream in this column along with other carboxylic acids with higher boiling points. The acetic acid product in heavy terminal column is taken as side stream and the overhead is recycled to purification section of the process.

7.4.2.2.2.2 Iridium-Based Cativa Process   The production of acetic acid by iridium catalytic system was commercialized as Cativa process by BP Amaco in 1996 (Sunley and Watson, 2000). Although to achieve similar activity level as Rh, more Ir is required, the catalyst is capable of operating at reduced levels of water (lower than 8 wt% for Cativa compared to 14–15 wt% for traditional Monsanto). Therefore, byproduct formation is reduced, carbon monoxide-based yield improves and steam consumption decreases. One of the main advantages of Ir-based processes is high stability of catalytic species of iridium. Tolerating low water concentrations (0.5 wt%) of the catalyst is especially important and is ideal for optimizing methanol carbonylation process. It has been found that iridium catalyst is active in wide range of conditions in which the rhodium counterparts are decomposed to completely inactive and to large extent nonregenerable salts. In addition to higher stability, iridium catalysts are much more soluble than rhodium catalysts; therefore, higher solution concentrations are achieved which provide much higher reaction rates available.

As the Cativa process is a modification of the original Monsanto process (Sterling plant in Texas), the general process is possibly the same, perhaps with some modifications.

7.4.3 Heterogeneous Carbonylation

The main drawback of homogenous carbonylation of methanol is the high corrosivity of reaction medium ($HI+CH_3COOH$ at $T>150°C$), requiring high-cost corrosion-resistant alloys such as Hastelloy. A green route is direct synthesis of acetic acid by vapor phase carbonylation over a halogen-free and noble metal-free stable hetero-geneous catalyst which also reduces the problems associated with the separation of product and catalyst from reaction medium in homogeneous systems. Therefore, development of a halide-free heterogeneous catalyst in carbonylation is important.

The activity of heterogeneous systems is typically much lower than that of homo-geneous systems. Nevertheless, the advantages of the formers make it interesting for commercial application. The research efforts are directed to approach the activity of heterogeneous systems to homogeneous ones.

3% Rh/C prepared by decomposition of rhodium nitrate over activated carbon is active in gas phase formation of acetic acid and methyl acetate from carbonylation of methanol. Under mild operating conditions (175–250°C, 1–14 atm) and in the presence of CH_3I (or HI), it is highly selective (>99%) (Schultz and Montgomery, 1969).

In supported Ni catalysts, the support plays an important role. Ni/C is highly active and selective in formation of carbonylated products at 300°C and 11 atm. The order of activity is Ni/C > Ni/γ-Al$_2$O$_3$ > Ni/SiO$_2$ (Omata et al., 1988).

It has been found that methanol carbonylation can be catalyzed with strong acids under medium pressure in the absence of a metal without requiring a co-catalyst (Stepanov, 2003). The reaction proceeds via a Koch-type carbonylation comprising adsorption of methanol (or DME) over Brønsted acid sites to form methoxy species which react with CO (i.e., CO-insertion) to form an acyl or acylium species there-after reacting with water to form acetic acid or methanol which can produce methyl acetate as well (Kalck et al., 2020).

Theory and experiments showed that 8 member rings (e.g., in mordenite) and not 12-member ring zeolites are the factor for high activity. The products are acetic acid, methyl acetate and methyl formate, and the main byproduct is DME (Ren et al., 2020).

Over metal-free H-MOR zeolite modified with pyridine, at 50 bar with 0.5% methanol and 99.5% CO in the feed, methanol conversion and acetic acid selectivity of 100% and 95%, respectively, were obtained in vapor-phase (Ni et al., 2017). No significant deactivation was observed after 145 h on stream.

Carbonylation of methanol with CO to methyl acetate over H-MOR and CuHMOR was studied by solid state NMR spectroscopy (Zhou et al., 2016). The latter showed much higher catalytic activity. The presence of Cu^+ enables DME stabilization which suppresses hydrocarbon formation during carbonylation reaction. Furthermore, the CO is absorbed by Cu^+ is not active species for formation of methyl acetate or acetic acid.

The *in-situ* generation of CO via methanol decomposition appears as a promising strategy for the development of a methanol-only halide-free synthesis route. Ormsby et al. (2009) used the combination of methanol decomposition (Pd/CeO$_2$) and metha-nol carbonylation (Cu/MOR) catalysts. At 300°C and 1 bar, acetic acid and methyl acetate yields of 24 and 10 g/kg$_{cat}$/h were obtained, respectively. The main byprod-ucts were DME, methane, ethane, propane and carbon.

Eastman Chemical Co. patented gas phase carbonylation of methanol with CO to acetic acid and methyl acetate over a carbon-supported gold catalyst in the presence of methyl iodide at 240°C and 17 bar (Zoeller et al., 2003a, b).

7.5 METHANOL-BASED CHEMICAL INDUSTRY AND METHANOL ECONOMY

Methanol is a promising C_1 feedstock and a platform molecule for chemical industry. When renewable electric power is available, its molecules can be used for energy storage. Methanol produced from CO_2, water and renewable energy can lead to liquid sustainability, as proposed by methanol economy.

Methanol is near the base of both fossil fuels and biomass value chains. Thus, in can be integrated into the existing value chains.

The main focus points of research are, therefore: to improve methanol synthesis efficiency; to increase the efficiency conversion of methanol to desired products, and to develop novel applications as alternatives to the existing ones.

NOMENCLATURE

AIE	alcoholic ion exchange
BFD	block flow diagram
CGI	crystal growth inhibitor
CIE	conventional ion exchange
CMHC	coupled methanol-hydrocarbon cracking
DEA	diethylamine
DGC	dry gel conversion
DICP	Dalian Institute of Chemical Physics
DME	dimethyl ether
DMTO	dimethyl ether or methanol-to-olefins
DPA	dipropylamine
DTG	LP-DME to gasoline
EO	ethylene oxide
FCC	fluid catalytic cracking
GTG	gas-to-gasoline
GTO	natural gas to olefins
HC	hydro-cracking
HCP	hydrocarbon pool
HGT	heavy gasoline treating
IZA	International Zeolite Association
LHSV	liquid hourly space velocity
MOGD	Mobil's olefin-to-gasoline and distillate
MON	motor octane number
Mor	morpholine
MR	membered ring
MTG	methanol-to-gasoline
MTH	methanol-to-hydrocarbons
MTO	methanol-to-olefins
MTP	methanol-to-propylene
OCP	olefin cracking process
PEG	Polyethylene glycol

PET	polyethylene terephthalate
RON	research octane number
SAC	steam-assisted conversion
SAPO	silicoaluminophosphate
STF	syngas-to-fuels
TEA	triethylmine
TEAOH	tetraethylammonium hydroxide
TIGAS	Topsøe integrated gasoline synthesis
TII	template-assisted ion incorporation
TOS	time-on-stream
TPABr	tetrapropylammonium bromide
VPT	vapor-phase transport
ZMS	zeolitic molecular sieve

SYMBOLS

a	remaining activity due to coke deposition
E	activation energy
k	rate constant
r_i	reaction rate
r_{i0}	reaction rate for fresh catalyst
X	fractional conversion

GREEK SYMBOLS

α	remaining activity due to deterioration of acidic structure
β	catalyst activity for a given TOS
η	effectiveness factor
φ	Thiele modulus

REFERENCES

Anderson, J. R., Mole, T., Christoo, V., Mechanism of some conversions over ZSM-5 catalyst, J. Catal., 61(1980) 477–484.

Benito, P. L., Gayubo, A. G., Aguayo, A. T., Castilla, M., Bilbao, J., Concentration-dependent kinetic model for catalyst deactivation in the MTG process, Ind. Eng. Chem. Res., 35 (1996) 81–89.

Bjorgen, M., Svelle, S., Joensen, F., Nerlov, J., Kolboe, S., Bonino, F., Palumbo, L., Bordiga, S., Olsbye, U., Conversion of methanol to hydrocarbons over zeolite H-ZSM-5: on the origin of the olefinic species, J. Catal., 249 (2007) 195–207.

Bjorgen, M., Joensen, F., Holm, M. S., Olsbye, U., Lillerud, K.P., Svelle, S., Methanol to gasoline over zeolite H-ZSM-5: improved catalyst performance by treatment with NaOH, Appl. Catal. A., 345 (2008) 43–50.

Bos, A. N. R., Tromp, P. J. J., Akse, H. N., Conversion of methanol to lower olefins. kinetic modeling, reactor simulation, and selection, Ind. Eng. Chem. Res., 34 (1995) 3808–3816.

Brad, L., Torrence, G. P., Alder, A. A. J. S., Methanol carbonylation process, U.S. Patent 5144068 (1991).

Budiman, A. W., Nam, J. S., Park, J. H., Mukti, R. I., Chang, T. S., Bae, J. W., Choi, M. J., Review of acetic acid synthesis from various feedstocks through different catalytic processes, Catal. Surv. Asia, 20 (2016) 173–193.

Chang, C. D., Silvestri, A. J., The conversion of methanol and other O-compounds to hydrocarbons over zeolite catalysts, J. Catal., 47 (1977) 249–259.

Chang, C. D., A kinetic model for methanol conversion to hydrocarbons, Chem. Eng. Sci., 35 (1980) 619–622.

Chen, N. Y., Reagan, W. J., Evidence of autocatalysis in methanol to hydrocarbon reactions over zeolite catalysts, J. Catal., 59 (1979) 123–129.

Chen, J. Q., Bozzano, A., Glover, B., Fuglerud, T., Kvisle, S., Recent advancements in ethylene and propylene production using the UOP/Hydro MTO process, Catal. Today, 106 (2005) 103–107.

Chen, D., Moljord, K., Holmen, A., A methanol to olefins review: diffusion, coke formation and deactivation on SAPO type catalysts, Micro. Meso. Mater., 164 (2012) 239–250.

Chen, X., Vicente, A., Qin, Z., Ruaux, V., Gilson, J. P., Valtchev, V., The preparation of hierarchical SAPO-34 crystals via post-synthesis fluoride etching, Chem. Commun., 52 (2016) 3512–3515.

Cheung, H., Tanke, R. S., Torrence, G. P., Acetic Acid, In Ullmann's Encyclopedia of Industrial Chemistry, 6th Ed., Wiley-VCH, Germany (2003) 149–178.

Choi, M., Cho, H. S., Srivastava, R., Venkatesan, C., Choi, D. H., Ryoo, R., Amphiphilic organosilane-directed synthesis of crystalline zeolite with tunable mesoporosity, Nat. Mater., 5 (2006) 718–723.

Corma, A., State of the art and future challenges of zeolites as catalysts, J. Catal., 216(1–2) (2003) 298–312.

Dahl, I. M., Kolboe, S., On the reaction mechanism for hydrocarbon formation from methanol over SAPO-34: 2. Isotopic labeling studies of the co-reaction of propene and methanol, J. Catal., 161 (1996) 304–309.

Di Renzo, F., Fajula, F., Introduction to molecular sieves: trends of evolution of the zeolite community, Stud. Surf. Sci. Catal., 157 (2005) 1–12.

Doelle, H. J., Heering, J., Riekert, L., Sorption and catalytic reaction in Pentasil zeolites: influence of preparation and crystal size on equilibria and kinetics, J. Catal., 71 (1981) 27–40.

Dubois, D. R., Obrzut, D. L., Liu, J., Thundimadathil, J., Adekkanattu, P. M., Guin, J. A., Punnoose, A., Seehra, M. S., Conversion of methanol to olefins over cobalt-, manganese- and nickel-incorporated SAPO-34 molecular sieves. Fuel Process. Technol., 83 (2003) 203–218.

Dusselier, M., Davis, M. E., Small-pore zeolites: synthesis and catalysis, Chem. Rev., 118 (2018) 5265–5329.

Egeblad, K., Christensen, C. H., Kustova, M., Christensen, C. H., Templating mesoporous zeolites, Chem. Mater., 20 (2008) 946–960.

Ellwood, P., Acetic acid via methanol and synthesis gas Chem. Eng., 76(11) (1969) 148–150.

Gaub, M., Seidel, A., Hehymanns, P., Herrmann, W. A., (Eds.), Applied Homogeneous Catalysis with Organomethalic Compounds, Weinheim, Germany (1996).

Gayubo, A. G., Aguayo, A. T., Mor´an, A. L., Olazar, M., Bilbao, J., Role of water in the kinetic modeling of catalyst deactivation in the MTG process, AIChE J., 48 (2002) 1561–1571.

Gedanken, A., Using sonochemistry for the fabrication of nanomaterials, Ultrason. Sonochem., 11 (2004) 47–55.

Gogate, M. R., Methanol-to-olefins process technology: current status and future prospects, Pet. Sci. Tech., 37 (2019) 1–7.

Grewer, T., Schmidt, A., Dampf/Flüssigkeits Gleichgewichte mit Assoziation im Dampf, Chem. Ing. Tech., 45 (1973) 1063–1067.

Guisnet, M., Costa, L., Ribeiro, F. R., Prevention of zeolite deactivation by coking, J. Mol. Catal. A.: Chem., 305 (2009) 69–83.

Hartmann, M., Machoke, A. G., Schwieger, W., Catalytic test reactions for the evaluation of hierarchical zeolites, Chem. Soc. Rev., 45 (2016) 3313–3330.

Haw, J. F., Marcus, D. M., Examples of Organic Reactions on Zeolites: Methanol to Hydrocarbon Catalysis, Marcel Dekker, Boca Raton (2003).

Haw, J. F., Song, W. G., Marcus, D. M., Nicholas, J. B., The mechanism of methanol to hydrocarbon catalysis, Acc. Chem. Res., 36 (2003) 317–326.

Jiang, J., Yu, J., Corma, A., Extra-large-pore zeolites: bridging the gap between micro and mesoporous structures, Angew. Chem., Int. Ed., 49 (2010) 3120–3145.

Joensen, F., Hojlund Nielsen, P. E., Palis Sorensen, M. D., Biomass to green gasoline and power, Biomass Conv. Bioref., 1 (2011) 85–90.

Kalck, P., Le Berre, C., Serp, P., Recent advances in the methanol carbonylation reaction into acetic acid, Coord. Chem. Rev., 402 (2020) 213078.

Kang, M. Methanol conversion on metal-incorporated SAPO-34s (MeAPSO-34s). J. Mol. Catal. A.: Chem., 160 (2000) 437–444.

Keil, F. J., Methanol-to-hydrocarbons: process technology, Micro. Meso. Mater., 29 (1999) 49–66.

King, D. L., Ushiba, K. K., Whyte, T. E., Make oxygenated products from methanol, Hydrocarbon Process, 61 (1982) 131–136.

Koohsaryan, E., Anbia, M., Nanosized and hierarchical zeolites: a short review, Chin. J. Catal., 37 (2016) 447–467.

Kubelková, L., Nováková, J., Jiru, P., In Jacobs, P. A., Jaeger, N. I., Jiru, P., Kazansky, V. B., Schulz-Ekloff, G. (Eds.), Structure and Reactivity of Modified Zeolites, Elsevier, Amsterdam (1984) 217.

Lee, K. Y., Chae, H. J., Jeong, S. Y., Seo, G., Effect of crystallite size of SAPO-34 catalysts on their induction period and deactivation in methanol-to-olefin reactions, Appl. Catal. A., 369 (2009) 60–66.

Levenspiel, O., Chemical Reaction Engineering, 3rd Ed., Wiley, New York (1999).

Li, J., Wei, Y., Liu, G., Qi, Y., Tian, P., Li, B., He, Y., Liu, Z., Comparative study of MTO conversion over SAPO-34, H-ZSM-5 and H-ZSM-22: correlating catalytic performance and reaction mechanism to zeolite topology, Catal. Today, 171 (2011) 221–228.

Liederman, D., Jacob, S. M., Voltz, S. E., Wise, J. J., Process variable effects in the conversion of methanol to gasoline in a fluid-bed reactor, Ind. Eng. Chem. Process Des. Dev., 17 (1978) 340–346.

Liu, X., Ren, S., Zeng, G., Liu, G., Wu, P., Wang, G., Chen, X., Liu, Z., Sun, Y., Coke suppression in MTO over hierarchical SAPO-34 zeolites, RSC Adv., 6 (2016) 28787–28791.

Losch, P., Boltz, M., Bernardon, C., Louis, B., Palčić, A., Valtchev, V., Impact of external surface passivation of nano-ZSM-5 zeolites in the Methanol-To-Olefins reaction, Appl. Catal. A.: G., 509 (2016) 30–37.

Meng, X., Nawaz, F., Xiao, F., Templating route for synthesizing mesoporous zeolites with improved catalytic properties, Nano Today, 4 (2009) 292–301.

Meng, X., Xiao, F., Green routes for synthesis of zeolites, Chem. Rev., 114 (2014) 1521–1543.

Mokrani, T., Scurrell, M., Gas conversion to liquid fuels and chemicals: the methanol route-catalysis and processes development, Catal. Rev., 51 (2009) 1–145.

Ni, Y., Shi, L., Liu, H., Zhang, W., Liu, Y., Zhu, W., Liu, Z., A green route for methanol carbonylation, Catal. Sci. Technol., 7 (2017) 4818–4822.

Olah, G. A., Molnár, A., Prakash, G. K. S., Hydrocarbon Chemistry, 3rd Ed., Wiley, New York (2018) 125–236.

Olsbye, U., Svelle, S., Bjørgen, M., Beato, P., Janssens, T. V. W., Joensen, F., Bordiga, S., Lillerud, K. P., Conversion of methanol to hydrocarbons: how zeolite cavity and pore size controls product selectivity, Angew. Chem. Int. Ed., 51 (2012) 2–24.

Omata, K., Fujimoto, K., Shikada, T., Tominaga, H., Vapor-phase carbonylation of organic compounds over supported transition metal catalyst: on the character of nickel/active carbon as methanol carbonylation catalyst, Ind. Eng. Chem. Res., 27 (1988) 2211–2213.

Ormsby, G., Hargreaves, J. S. J., Ditzel, E. J., A methanol-only route to acetic acid, Catal. Commun., 10 (2009) 1292–1295.

Pagis, C., Morgado Prates, A. R., Farrusseng, D., Bats, N., Tuel, A., Hollow zeolite structures: an overview of synthesis methods, Chem. Mater., 28 (2016) 5205–5223.

Paulik, F. E., Roth, J. F., Novel catalysts for the low-pressure carbonylation of methanol to acetic acid, Chem. Commun., (1968) 1578a–1578a.

Perez-Ramirez, J., Christensen, C. H., Egeblad, K., Christensen, C. H., Groen, J. C., Hierarchical zeolites: enhanced utilisation of microporous crystals in catalysis by advances in materials design, Chem. Soc. Rev., 37 (2008) 2530–2542.

Qiao, Y., Yang, M., Gao, B., Wang, L., Tian, P., Xu, S., Liu, Z., Creation of hollow SAPO-34 single crystals via alkaline or acid etching, Chem. Commun., 52 (2016) 5718–5721.

Ren, Z., Lyu, Y., Song, X., Ding, Y., Review of heterogeneous methanol carbonylation to acetyl species, Appl. Catal. A.: Gen., 595 (2020) 117488.

Roth, W. J., Gil, B., Makowski, W., Marszalek, B., Eliasova, P., Layer like porous materials with hierarchical structure, Chem. Soc. Rev., 45 (2016) 3400–3438.

Sahebdelfar, S., Yaripour, F., Ahmadpour, S., Khorasheh, F., Methanol-to-hydrocarbons product distribution over SAPO-34 and ZSM-5 catalysts: the applicability of thermodynamic equilibrium and Anderson-Schulz-Flory distribution, Iranian J. Chem. Chem. Eng., 38 (2019) 49–59.

Salih, H. A., Muraza, O., Abussaud, B., Al-Shammari, T. K., Yokoi, T., Catalytic enhancement of SAPO-34 for methanol conversion to light olefins using in-situ metal incorporation, Ind. Eng. Chem. Res., 57 (2018) 6639–6646.

Salvador, P., Kladnig, W., Surface reactivity of zeolites type H-Y and Na-Y with methanol, J. Chem. Soc., Faraday Trans. I., 73 (1977) 1153–1168.

Sassi, A., Wildman, M. A., Ahn, H. J., Prasad, P., Nicholas, J. B., Haw, J. F., Methylbenzene chemistry on zeolite HBeta: multiple insights into methanol-to-olefin catalysis, J. Phys. Chem. B., 106 (2002) 2294–2303.

Schipper, P. H., Krambeck, F. J., A reactor design simulation with reversible and irreversible catalyst deactivation. Chem. Eng. Sci., 41 (1986) 1013–1019.

Schulman, E., Wu, W., Liu, D., Two-dimensional zeolite materials: structural and acidity properties, Materials, 13 (2020) 1822.

Schultz, R. G., Montgomery, P. D., Vapor phase carbonylation of methanol to acetic acid, J. Catal., 13 (1969) 105–106.

Schwieger, W., Machoke, A. G., Weissenberger, T., Inayat, A., Selvam, T., Klumpp, M., Inayat, A., Hierarchy concepts: classification and preparation strategies for zeolite containing materials with hierarchical porosity, Chem. Soc. Rev., 45 (2016) 3353–3376.

Sedran, U., Mahay, A., de Lasa, H. I., Modelling methanol conversion to hydrocarbons: Alternative kinetic models. Chem. Eng. J., 45 (1990) 33–42.

Sherman, G. C., Martin, G., Giles, F., Sunley, J. G., Process and catalyst for the production of acetic acid, E.P. Patent 0643034B1 (1994).

Spivey, J. J., Froment, G. F., Dehertog, W. J. H., Marchi, A. J., Zeolite catalysis in the conversion of methanol into olefins, Catalysis, 9 (1992) 1–64.

Stepanov, A. G., Carbonylation-Heterogeneous, In Horvath, I. T. (Ed.), Encyclopedia of Catalysis, John Wiley and Sons, New York (2003).

Stöcker, M., Methanol-to-hydrocarbons: catalytic materials and their behavior, Micro. Meso. Mater., 29 (1999) 3–48.

Sun, Q., Wang, N., Guo, G., Yu, J., Ultrafast synthesis of nano-sized zeolite SAPO-34 with excellent MTO catalytic performance, Chem. Commun., 51 (2015) 16397–16400.

Sunley, G. J., Watson, D. J., High productivity methanol carbonylation catalysis using irid-ium The CativaTM process for the manufacture of acetic acid, Catal. Today, 58 (2000) 293–307.

Svelle, S., Joensen, F., Nerlov, J., Olsbye, U., Lillerud, K. P., Kolboe, S., Bjorgen, M., Conversion of methanol into hydrocarbons over zeolite H-ZSM-5: ethene formation is mechanistically separated from the formation of higher alkenes, J. Am. Chem. Soc., 128 (2006) 14770–14771.

Svelle, S., Olsbye, U., Joensen, F., Bjorgen, M., Conversion of methanol to alkenes over medium- and large-pore acidic zeolites: steric manipulation of the reaction interme-diates governs the ethene/propene product selectivity, J. Phys. Chem. C., 111 (2007) 17981–17984.

Tian, P., Wei, Y., Ye, M., Liu, Z., Methanol to olefins (MTO): from fundamentals to commer-cialization, ACS Catal., 5 (2015) 1922–1938.

Valtchev, V., Majano, G., Mintova, S., Perez-Ramirez, J., Tailored crystalline microporous materials by post-synthesis modification, Chem. Soc. Rev., 42 (2013) 263–290.

Varzaneh, A. Z., Towfighi, J., Sahebdelfar, S., Carbon nanotube templated synthesis of metal containing hierarchical SAPO-34 catalysts: impact of the preparation method and metal avidities in the MTO reaction, Micro. Meso. Mater., 236 (2016) 1–12.

Verboekend, D., Milina, M., Perez-Ramirez, J., Hierarchical silicoaluminophosphates by post-synthetic modification: influence of topology, composition, and silicon distribu-tion, Chem. Mater., 26 (2014) 4552–4562.

Vuong, G. T., Do, T. O., Nanozeolites and nanoporous zeolitic composites: synthesis and applications, In García-Martínez, J., Li, K. (Eds.), Mesoporous Zeolites: Preparation, Characterization and Applications, Wiley-VCH, Weinheim (2015) 79–114.

Wang, W., Jiang, Y. J., Hunger, M., Mechanistic investigations of the methanol-to-olefin (MTO) process on acidic zeolite catalysts by in situ solid-state NMR spectroscopy, Catal. Today, 113 (2006) 102–114.

Wang, W., Hunger, M., Reactivity of surface alkoxy species on acidic zeolite catalysts, Acc. Chem. Res., 41 (2008) 895–904.

Wang, Q., Wang, L., Wang, H., Li, Z., Wu, H., Li, G., Zhang, X., Zhang, S., Synthesis, char-acterization and catalytic performance of SAPO-34 molecular sieves for methanol-to-olefin (MTO) reaction, Asia-Pac. J. Chem. Eng., 6 (2011) 596–605.

Wang, Y., Xiao, F., Understanding mechanism and designing strategies for sustainable syn-thesis of zeolites: a personal story, Chem. Rec., 16 (2016) 1054–1066.

Weitkamp, J., New directions in zeolite catalysis, Stud. Surf. Sci. Catal., 65 (1991) 21–46.

Yan, Y., Guo, X., Zhang, Y., Tang, Y., Future of nano-/hierarchical zeolites in catalysis: gas-eous phase or liquid phase system, Catal. Sci. Technol., 5 (2015) 772–785.

Yang, G., Wei, Y., Xu, S., Chen, J., Li, J., Liu, Z., Yu, J., Xu, R., Nanosize-enhanced lifetime of SAPO-34 catalysts in methanol-to-olefin reactions, J. Phys. Chem. C., 117 (2013) 8214–8222.

Yang, M., Fan, D., Wei, Y., Tian, P., Liu, Z., Recent progress in methanol-to-olefins (MTO) catalysts, Adv. Mater., (2019) 1902181.

Yao, J., Huang, Y., Wang, H., Controlling zeolite structures and morphologies using polymer networks, J. Mater. Chem., 20 (2010) 9827–9831.

Yaripour, F., Shariatinia, Z., Sahebdelfar, S., Irandoukht, A., Effect of boron incorporation on the structure, products selectivities and lifetime of H-ZSM-5 nanocatalyst designed for application in methanol-to-olefins (MTO) reaction, Micro. Meso. Mater., 203 (2015) 41–53.

Yarulina, I., Chowdhury, A. D., Meirer, F., Weckhuysen, B. M., Gascon, J., Recent trends and fundamental insights in the methanol to hydrocarbons process, Nat. Catal., 1 (2018) 398–411.

Ye, M., Li, H., Zhao, Y., Zhang, T., Liu, Z., MTO processes development: The key of meso-scale studies, Adv. Chem. Eng., 47 (2015) 279–335.

Zhong, J., Han, J., Wei, Y., Tian, P., Guo, X., Song, C., Liu, Z., Recent advances of the nano-hierarchical SAPO-34 in the methanol-to-olefin (MTO) reaction and other applications, Catal. Sci. Technol., 7 (2017) 4905–4923.

Zhong, J., Han, J., Wei, Y., Xu, S., Sun, T., Guo, X., Song, C., Liu, Z., Enhancing ethylene selectivity in MTO reaction by incorporating metal species in the cavity of SAPO-34 catalysts, Chin. J. Cataly., 39 (2018) 1821–1831.

Zhou, L., Li, S., Qi, G., Su, Y., Li, J., Zheng, A., Yi, X., Wang, Q., Deng, F., Methanol carbonylation over copper-modified mordenite zeolite: a solid-state NMR study, Solid State Nucl. Mag. Reson., 80 (2016) 1–6.

Zoeller, J. R., Singleton, A. H., Tustin, G. C., Carver, D. L., Vapor phase carbonylation process sing gold catalysts. US Patent 6,506,933 B1 (2003a).

Zoeller, J. R., Singleton, A. H., Tustin, G. C., Carver, D. L., Gold based heterogeneous carbonylation catalysts. US Patent 6,509,293 B1 (2003b).

8 Methane Derivative Routes

8.1 INTRODUCTION

The commercialized routes for converting methane into higher hydrocarbons are mostly via conversion of methane into syngas followed by Fischer–Tropsch synthesis, or the syngas to methanol and methanol to olefins (MTO) reaction. The production of syngas from methane by reforming is, however, highly capital and energy intensive and scale dependent. The syngas facility account for about 50% constructing of a gas to liquid (GTL) plant (Olsbye et al., 2011). Therefore, development of alternative routes operating under milder conditions with lower energy utilization is of great importance from economic and environmental point of view.

In addition to methanol and DME, other heterosubstituted methanes can be condensed to higher hydrocarbons. These compounds may be used as alternative intermediates in indirect conversion of methane without using intermediate methane reforming. The use of halogens (Cl_2 and Br_2) and sulfur as methane activation mediators has been studied recently.

Unlike methane, monosubstituted methanes (e.g., methanol, methyl halide and methanetiol) are asymmetric and polar (Table 8.1). The functional group added to methane molecule (hydroxyl, halide and sulfhydryl, respectively) play an important role in their chemical reactivity. The other important factor is the strength of carbon heteroatom bond, or more generally the stability of the molecules. As a consequence, flouromethane is much less reactive than other halomethanes because of the very high strength of C-F covalent bond, being about 60 kJ/mol stronger than a C-H bond.

The use of halide and sulfur routes is far less developed than the methanol route despite inherent similarities in their catalysts and processes. This chapter deals with the former routes for the production of higher hydrocarbons from methane.

8.2 HYDROCARBONS FROM METHYL HALIDES

Olah et al. (1985) described a three-step process for the conversion of methane to higher hydrocarbons comprising monohalogenation of methane, catalytic hydrolysis of methyl halide to mixture of methanol and DME and MTO on ZSM-5 steps.

Taylor et al. (1988) proposed a two-step cyclic process for conversion of natural gas into gasoline range (mainly C_4–C_{10}) hydrocarbons via CH_3Cl intermediate. The first step is the oxychlorination of methane with HCl followed by direct methyl chloride to gasoline conversion over ZSM-5 zeolite with recycling the byproduct HCl for oxychlorination. The process is innovative and capable of commercialization (Zhang et al., 2015). Bromine is better than chlorine in both steps because of a higher

DOI: 10.1201/9781003279280-8

TABLE 8.1

Properties of Methane and Some of Its Monosubstituted C$_1$ Derivatives

Compound	Normal Boiling Point (°C)	Dipole Moment (D)	ΔH_f^0 (kJ/mol)	ΔG_f^0 (kJ/mol)
CH$_4$	−161.5	0	−74.9	−50.8
CH$_3$F	−78.4	1.81	−234	−210
CH$_3$Cl	−23.8	1.85	−83.7	−62.9
CH$_3$Br	4.0	1.81	−35.1	−28.2
CH$_3$I	42.4	1.62	−14.1	15.6
CH$_3$SH	5.95	1.52	−23.0	−9.92

monosubstituted product selectivity and easier removal from methyl halide due to weaker C–Br bond compared to C–Cl bond.

Since then, the methyl halide to hydrocarbons (MeXTH) conversion (Eq. 8.1) has been extensively studied:

$$n\,CH_3X \xrightarrow{\text{catalyst}} [CH_2]_n + n\,HX \quad X = Cl\,or\,Br \qquad (8.1)$$

The reaction is analogous to methanol to hydrocarbons. The condensation reaction can be catalyzed by acid-base bifunctional catalysts such as zeolites and WO$_3$/Al$_2$O$_3$ (Olah et al., 1984).

A comparative study between MeX (X = Cl and Br) and MeOH to hydrocarbons was reported by Svelle et al. (2006). Reaction tests were performed in a fixed-bed reactor over SAPO-34 within temperature range of 300–450°C. An induction period was observed for all reactants which was overcome by introducing small amount of propylene before the feed entered. The product selectivities were similar for different feeds but the reaction rate was one order of magnitude (about 25-fold) higher for methanol than that for methane halides. High deactivation rates were observed for all reactants at and above 350°C. The zero-time (extrapolated) conversion and ethylene selectivity greatly increased with temperature in temperature range of 300–450°C for both CH$_3$Cl and CH$_3$Br feeds over H-SAPO-34.

8.2.1 CATALYSTS

Different zeolites and zeotypes including ZSM-5, SAPO34 have been used for this reaction. The methyl halide conversion and product selectivity strongly depend on acidity and structural topology of the catalyst.

Olah et al. (1984) investigated a number of bifunctional acidic-basic-supported transition-metal oxide or oxyhalide catalysts for conversion of heterosubstituted methanes (methanol, DME, methyl mercaptan, dimethyl sulfide, methyl amines and methyl halides) into C$_2$ hydrocarbons. WO$_3$/Al$_2$O$_3$ catalyst gave 42.8% C$_2$–C$_5$ hydrocarbons at 36% conversion of methyl chloride, at 327°C and WHSV = 5 mLg⁻¹h⁻¹, much lower than the respective values of 99% and 53.8% for methanol under similar conditions. The catalyst, however, is not shape selective.

Both protonic and cationic forms of large-pore zeolites like X, Y, ETM, BEA and MOR are attractive due to their high activity but they are susceptible to large amount of coke formation due to their large canals and cages. In contrast, medium pore zeolites such as ZSM-5 shows greatly reduced coke formation with high activity and higher stability.

ZSM-5 zeolite shows high activity whereas SAPO-34 shows high selectivity in the MeXTH reaction. Research effort has been modification of the catalysts, especially, for improving stability. These include improving pore structure by addition of mesopores and modification of the acidity by incorporating different promoters or modifying initial gel composition during hydrothermal synthesis. To overcome the diffusion limitations induced by micropores, several strategies such as synthesis of nanosized zeolites, dealumination and desilication by alkalies or steam, synthesis by using silanated organic polymers or organosilane (mesopore directing agents) have been adopted.

Using HZSM-5 zeolites with different SiO_2/Al_2O_3 ratios (30, 80 and 280), Gamero et al. (2015) studied the effect SiO_2/Al_2O_3 ratio on transformation of chloromethane and compared with SAPO-n ($n = 18, 34$). The HZSM-5 sample with moderate acidity ($SiO_2/Al_2O_3 = 80$) showed good kinetic behavior at 350°C. The HZSM-5 catalysts showed higher activities and stabilities, but lower olefin selectivities, compared to the SAPO-n catalysts.

Xu et al. (2012, 2019) showed that impregnating HZSM-5 with medium concentration NH_4F followed by calcination increased propylene selectivity and stability for chloro- and bromo-methane conversion. Characterization results showed that acidity and strong Brønsted acid sites were reduced and 0.73–0.78 nm micropores were generated by fluoride treatment. Reduced Brønsted acidity decreases hydrogen transfer and cracking reactions.

Tao et al (2009) evaluated various metal oxide-modified ZSM-5 catalysts prepared by incipient wetness impregnation in methyl bromides to aromatics. The PbO-modified catalyst exhibited superior performance. The effects of PbO loading, Si/Al ratio and calcination condition of ZSM-5 on its performance in bromomethane to hydrocarbons was also studied and the values 5 wt%, 70 and 480°C for 8 h were obtained, respectively, as optimal values. The catalyst exhibited enhanced aromatization selectivity (31.6% yield at 360°C).

The effect of metals in zeolite formulation and their synergy in bimetallic catalyst was studied by comparing the performance of CuO/HZSM-5, ZnO/HZSM-5 and CuO-ZnO/HZSM-5 catalysts by Chen et al (2018). At 360°C, the aromatic yield was in the order 2%CuO-3%ZnO/HZSM-5 (27.4%) > 5%CuO/HZSM-5 (19.3%) > HZSM-5 (14.4%) > 5%ZnO/HZSM-5 (6.0%) for bromomethane feed. Zn acted as Cu promoter and increased the strong acidity as indicted by NH_3-TPD results and thus enhanced hydrogen transfer and consequently aromatic yield.

Steaming of ZSM-5 zeolite has been used for regulating acidity and structure of the catalyst in conversion of bromomethane into benzene, toluene and xylenes (BTX). It results in dealumination of the framework, and thus reduces acidity and decreases diffusion limitation by introducing mesopores in the zeolite structure. High temperature steaming (optimally at 600°C) reduced the acidity and Brønsted/Lewis acid ratio with simultaneous production of mesopore in ZSM-5 and thereby improved the catalytic performance of the zeolite in conversion of bromomethane into aromatics (27% aromatic yield with a lifetime > 50 h) (Wang et al., 2017).

Over HZSM-5 catalysts modified by Mg/Ca impregnation, NH_3-TPD and pyridine-IR techniques revealed that strong Brønsted acid sites were reduced with improved light olefins selectivity. The best catalytic performance was obtained for 5Ca2Mg-HZ with 87.2% total olefin selectivity and 99% CH_3Cl conversion over 58 h (Huang et al., 2019).

The P-modified MgZSM-5 catalyst can catalyze the conversion of CH_3Cl into C_2-C_4 olefins at 500°C (Su et al., 1993). Reaction test and spectroscopy results illustrated that the addition of P by trimethyl phosphine to Mg-impregnated ZSM-5 reduced strong Brønsted acid sites which are responsible for hydrogen transfer (HT) reactions producing alkanes and aromatics.

Wei et al. (2007) synthesized SAPO-34 catalysts with high and low silica contents. No appreciable effect was observed in XRD patterns and in ^{31}P and ^{27}Al MAS NMR spectra. The investigation of Si coordination from ^{29}Si MAS NMR spectra proposed that the difference in acidity was caused by Si incorporation into AlPO framework. In low Si samples, the coordination was Si(4Al) resulting in low coke and high propylene selectivity. In high Si samples, Si(nAl, $n = 0$–3) coordination was also observed which increased the surface Brønsted acidity. The adsorption study of bridge hydroxyls in FTIR revealed that the number of active sites varied with Si. Nevertheless, both catalysts were active and selective in chloromethane conversion.

A comparative study of SAPO-34 catalysts prepared by hydrothermal synthesis in the presence of soluble starch, sodium dodecyl sulfate (SDS) and cetyltrimethylammonium bromide (CTAB) as initial gel additives showed that compared to conventional SAPO-34, SDS- and CTAB-modified catalysts showed lower crystalline orders with higher mesopore volumes and surface areas but lower acidities unlike soluble starch which resulted to a higher acidity (Kong et al. 2015). The first two exhibited improved stabilities and less coke formation compared to the conventional one. Kong et al (2014a) also used CTAB as mesopore generating agent to synthesize hierarchical SAPO-34. Compared to conventional SAPO-34, the hierarchical sample showed lower coke amount and higher stability due to lower total acid sites. A C_2-C_3 olefin selectivity of 80% was achieved ($T = 420$°C, WHSV = 1.89 h^{-1}).

Reducing crystallite size has similar effects as mesopores. A small submicron crystallite size (700 nm) along with reduced acidity significantly increased the SAPO-34 catalyst lifetime and olefin selectivity by reducing hydrogen transfer reactions (Wang et al. 2019).

Isomorphous substitution of metals in SAPO-34 catalysts has improved the performance in MTO reaction (Varzaneh et al., 2016). Many elements (such as Be, B, Mg, Ti, Mn, Fe, Co and Ni) have been added to SAPO-34 by isomorphous substitution to improve the selectivity (Zhang et al. 2008). In SAPO-34, Si generates acidic sites while the substituted metal modifies them and even introduces redox properties. The metal replaces Al^{3+} and changes the chemical environment to P(nAl,(4-n)Me) which affect Si coordination.

Isomorphous substitution of Zn in SAPO-34, compared to parent zeotype, showed 92% CH_3Br conversion with improved $C_2 + C_3$ olefins selectivity (73.3%), especially for a Zn/Si ratio of 0.02. XRD, SEM and NH_3-TPD characterization results showed that incorporation of Zn into the framework resulted in a smaller crystallite size, with higher acidity that leads to higher conversion and higher amount of weak acid sites that improve the selectivity (Zhang et al., 2015).

Wei et al. (2006a) incorporated Mn into SAPO-34 framework by isomorphous substitution. In Mn-substituted catalyst, the TG-DSC for removing template in dilute oxygen, in addition of two endothermic peaks for the removal of water, showed two weight losses at 200–400°C and 400–600°C illustrating different environments for the template molecules. The peak temperature was higher for Mn-containing sample which showed higher interaction of the template with the framework and thus higher acidity after calcination which was confirmed by ammonia adsorption and FTIR. It was concluded that Mn changed Si coordination, template removal and acidity. The higher coordination of Si(0Al) islands in MnAPSO4 proposed generation of stronger acid sites by incorporation of Mn into the framework. Both SAPO-34 and MnAPSO4 catalysts were active and selective in chloromethane conversion.

Later, Wei et al (2008) synthesized SAPO-34 and metal-incorporated SAPO-34 by both impregnation and isomorphous substitution. An increase in unit cell parameters and crystallite size was observed by metal substitution. Small shifts in ^{31}P, ^{27}Al and ^{29}Si MAS NMR spectra were observed and showed that the island of Si is improved predicting a higher acidity. A comparison of TPR profiles showed higher reduction temperature for isomorphous substituted sample compared to corresponding impregnated one illustrating different chemical environment on the surface and stronger chemical bonds in the former. The application of MeAPSO-34s in chloromethane transformation improved light olefins yield and increased the stability of catalytic activity. Using Co- or Fe-incorporated SAPO-34 catalysts, ethylene production was favored while over Mn or Fe-containing samples, propylene generation was more predominant.

8.2.2 Mechanisms

The similarity between MeXTH and methanol to hydrocarbon (MTH) reactions especially similar catalyst types and product distributions imply a similar mechanism. The rate is, however, slower in MeXTH because of a lower adsorption of the reactant over Brønsted acid sites compared to methanol which can be adsorbed more strongly due to hydrogen bond formation leading to higher surface coverage and thus higher reaction rates. A hydrocarbon pool (HP) mechanism similar to MTH has been also proposed (Wei et al., 2006b), but the main route to the formation of the first C–C bond is ambiguous due to a large number of uncontrolled side reactions (Kong et al., 2014b).

An important difference between MTH and MeXTH is the absence of dimethyl ether (DME) as an intermediate in the later. Methoxy group is considered as intermediate in the MeXTH reaction as revealed by *in situ* FTIR investigation (Wei et al., 2006b). The mechanism of the reaction has been studied by FTIR spectroscopy (Xu et al., 2012) and the following steps has been proposed: (i) dissociation of chloromethane (ii) formation of methoxy groups on acid sites (iii) formation of alkylbenzenes as reactive intermediates (iv) and the release of olefins along with (v) parallel reactions such as hydrogen transfer and condensation. The overall reaction is limited by steps (i) and (ii) that leads to a longer induction period compared to MTH (Ibáñez et al., 2016).

An important reversible phenomenon revealed by *in situ* FTIR studies was the breakage of Al-O-P bonds of SAPO-34 by dissociative adsorption of HCl and generation of a large number of P-OH and AlCl groups. The Al-O-P groups were restored by the removal of HCl (Wei et al., 2005).

By varying the partial pressure of MeCl (0.1 vs. 1.0 bar pressure) under constant total gas flow rate, a first-order reaction rate in MeCl partial pressure was observed over SAPO-34 at 350–450°C, but a little lower at 300°C (Svelle et al 2006). This could be indicative of low MeCl surface coverage under these reaction conditions.

Fickel et al. (2016) related the slower reaction rate for chloromethane to the longer initiation period required for the formation of hydrocarbon pool in reaction onset, as indicated by *in situ* FTIR and *ex situ* ^{13}C NMR spectroscopy.

8.2.3 CATALYST DEACTIVATION

The catalysts suffer from rapid deactivation in methyl halide condensation to hydrocarbons. The main causes of catalyst deactivation are coke formation and dealumination both of which enhanced at higher reaction temperature conditions. The former is a reversible deactivation as the activity can be largely restored by coke removal. It is the main cause of deactivation at lower temperatures (350°C) (Gamero et al., 2015). Dealumination occurs by HCl product at higher temperatures (>450°C) due to removal of Al atoms from the framework and formation of AlCl$_3$ that is not active in the reaction. It is an irreversible deactivation.

The extensive and rapid coke formation compared to MTH could be due to the absence of steam in the reaction medium acting as a coke removing agent (Argyle and Bartholomew, 2016).

Over SAPO-34, the temperature-programmed oxidation (TPO) profile of the spent catalysts showed two coke types: type I which is the less condensed linear hydrogenated coke (burnt at 430°C) increasing with time on stream and the less hydrogenated type II coke with asymptotic accumulation with time to a saturation level burning at higher temperatures (510°C) (Ibáñez et al., 2016).

Gamero et al. (2018) studied the reaction pathway in conversion of chloromethane. The counter map of coke (coke amount versus temperature and space time) showed that temperature was much more effective than space time. Under severe conditions (400°C), coke formation was more rapid for short space times where chloromethane concentration was higher and HP formation was faster. They also concluded that coke was originated in micropores of the zeolite and then grew within meso- and macropores of the matrix.

8.2.4 PROCESSES

GT-G2A^{SM}, a new process offered by Sulzer GTC Technology, is a highly efficient method of converting methane by activation with a halogen, namely bromine, instead of oxygen or sulfur as a co-reactant. Bromination chemistry is used to produce petrochemical-grade BTX, which has higher value than the paraffinic or gasoline fuel products of other recently-introduced processes that convert gas into liquids. All of the bromine is fully recycled in a novel process scheme, to give the lowest CAPEX (capital expenditure) and OPEX (operational expenditure) of any competing process. The carbon utilization is the highest among any class of methane conversion processes.

8.3 HYDROCARBONS FROM SULFURATED METHANES

The sulfurated methane conversion route is of practical interest because the sour gas fields contain considerable amount of light sulfur compounds. CH_3SH is also an industrial waste gas in paper production. Carbon disulfide is formed by the reaction of methane with elemental sulfur (Eq. 8.2).

$$CH_4 + 2S_2 \rightarrow CS_2 + 2H_2S \quad \Delta H^0_{298K} = 75\,kJ/mol \tag{8.2}$$

Chang and Silvestri (1977) reported that, methanetiol (CH_3SH) was converted into hydrocarbons and H_2S over H-ZSM-5 catalyst at 482°C, but the desulfurization was only partial, with 27.2% selectivity to dimethyl sulfide (CH_3SCH_3, DMS) but only 7.0% selectivity towards C_2+C_3 olefins.

Mobil Oil researches were the first to investigate the conversion of sulfide compounds into hydrocarbons (Chang and Lang, 1984; Chang, 1985). They showed that CS_2 can be converted into gasoline range hydrocarbons over ZSM-5 zeolites (Eq. 8.3).

$$nCS_2 + 3nH_2 \xrightarrow{\text{zeolite}} [CH_2]_n + 2nH_2S\,\Delta H^0_{298K} = -151\,kJ/mol\,CS_2\,n = 3 \tag{8.3}$$

Typical reaction conditions are 480°C and 15 bar over ZSM-5 catalysts (Chang and Lang, 1984). Over ZSM-5 at 482°C, it leads to a mixture of linear alkanes (14.8%), aromatics (20.4% C_6–C_{10}, 12.4% C_{11+}) and the balance being methane (Chang and Lang, 1984).

The sulfur can be recovered and recycled by the Claus process with the overall reaction represented as Eq. (8.4):

$$2H_2S + O_2 \rightarrow S_2 + 2H_2O \tag{8.4}$$

Thus, the hydrogen atoms are wasted as water giving low-quality steam.

Most probably, partially reduced species (CH_3SH and CH_3SCH_3) are involved in CS_2 conversion reaction (Eq. 8.5):

$$nCH_3SH \xrightarrow{\text{zeolite}} [CH_2]_n + nH_2S \quad \Delta H^0_{298K} = 9.2\,kJ/mol\,CH_3SH \quad n = 3 \tag{8.5}$$

It has been shown that CS_2 is converted into these compounds over acidic catalyst (Olah et al., 2018). This has been supported by increased yield of C_{2+} hydrocarbons over Co-promoted ZSM-5 (Chang and Lang, 1984). The conversion was enhanced using Co/SiO$_2$+ZSM-5 mixture (X_{CS2} = 40.3%, S_{C4+} = 45%). The nonselective products were methane and light hydrocarbons.

Protonic zeolites (H-ZSM-5, H-Y, and H-ferrierite) can catalyze methyl mercaptan to hydrocarbons (M2TH) reaction without adding any reagent (Reina et al., 2017). At temperatures below 400°C, the main product is dimethyl sulfide (DMS) at equilibrium (Eq. 8.6) which can be, in turn, condensed to hydrocarbons (essentially light alkanes and aromatics) at higher temperatures (>500°C).

$$2CH_3SH \Leftrightarrow CH_3SCH_3 + H_2S \quad \Delta H^0_{298K} = -6.1\,kJ/mol\,CH_3SH \quad (8.6)$$

A comparative study of M2TH and MTH in a fixed-bed reactor over HZSM-5 catalyst (Si/Al = 15) showed in high conversions both for methanol (>99%) and methyl mercaptan (>97) at 550°C and WHSV = 0.30 h^{-1} (Reina et al., 2017). In the M2TH transformation test, the product was mostly C_1–C_3 alkanes (52% carbon-based selectivity) and BTX (40%) with only a minor amount of olefins (<0.7%). The respective values for MTH were 15%, 46% and 37%, illustrating remarkable difference in product distributions. Similarly, the coke amount after 10 h on stream was much higher for M2TH than MTH (7.7% and 2%, respectively).

In transformation of methyl and ethyl mercaptan mixture over ZSM-5 optimized for ethyl mercaptan to ethylene (Si/Al = 15), the former was selectively transformed into methane and H_2S with significant amounts of aromatics and coke at 550°C. For ethyl mercaptan transformation, however, ethylene and H2S were the main products, although small amounts of C_1–C_4 alkanes, C_6–C_8 aromatics, thiophenes, CS_2 and coke were also formed (Cammarano et al., 2014).

These results show that the product selectivity in CH_3SH transformation is directed towards thermodynamically most stables products CH_4 and C_6H_6 over ZSM-5, as thermodynamic consideration has implied as ultimate products in the MTH reaction (Sahebdelfar et al. 2019):

$$9CH_3SH \xrightarrow{\text{zeolite}} CH_4 + C_6H_6 + 9H_2S \quad (8.7)$$

8.3.1 CATALYSTS

It has been found that different zeolite types with different topologies such as H-ZSM-5, H-BEA, H-MOR, H-ferrierite and SAPO-34 can transform CH_3SH into hydrocarbons and H_2S and have been used for methyl mercaptan conversion (Reina et al., 2017). H-ZSM-5 is the most active and most stable catalyst. The order of activity is as follows: H-ZSM-5 > H-Y > H-MOR > H-SAPO-34 (Hulea et al., 2014). No clear correlation was observed between activity of the catalyst and number/nature or strength of its acid sites. It was concluded, at least, under the reaction conditions employed, the activity of the protonic form of zeolites is not determined by its acidity but the topology is more crucial.

With bifunctional WO_3/Al_2O_3 catalyst, a DMS conversion of 32% and C_2–C_4 hydrocarbon selectivity of 36.2 was obtained (380°C and 50 mLg^{-1}h^{-1}) (Olah et al., 1984), both of which are poorer than the corresponding values obtained by methanol and methyl halides under similar operating conditions.

8.3.2 Mechanism

CH_3SH can be viewed as activated methane being isostructural with methanol (Taifan and Baltrusaitis, 2016). However, the sulfur atom is larger and more polarizable than the oxygen atom. As a result, sulfur compounds are more powerful nucleophiles, and compounds containing –SH groups are stronger acids than their oxygen analogues.

The methyl mercaptan to hydrocarbons (M2TH) shows similarities in analogy with MTH reaction. It is sensitive to zeolite structure as is the case for MTH (Huguet et al., 2013).

M2TH is in many ways similar to MTH, e.g., in sensitivity to the topology of zeolite catalyst. The presumed similarity has led to the use of the knowledge in the MTH reaction for the former. In contrast to MTH over ZSM-5, M2TH is characterized by little olefin formation with product being mostly C_1–C_3 alkanes (90% of which is CH_4) and C_6–C_8 aromatics with short catalyst lifetimes under similar reaction conditions. This is surprising because lower olefins are considered as the precursors of aromatics via the HT reaction. This could be due to the further conversion of any C_2H_4 formed to CH_4 and coke on strong acid sites of the catalyst at the higher operating temperature of M2TH (Taifan and Bltrusaitis, 2016). The absence of olefins in M2TH implies that the hydrocarbon pool mechanism (based on methylation steps) is not sustainable for this reaction (Reina et al., 2017). Scheme 8.1 shows a simple reaction pathway for the transformation of CH_3SH over protonic zeolites at high temperatures.

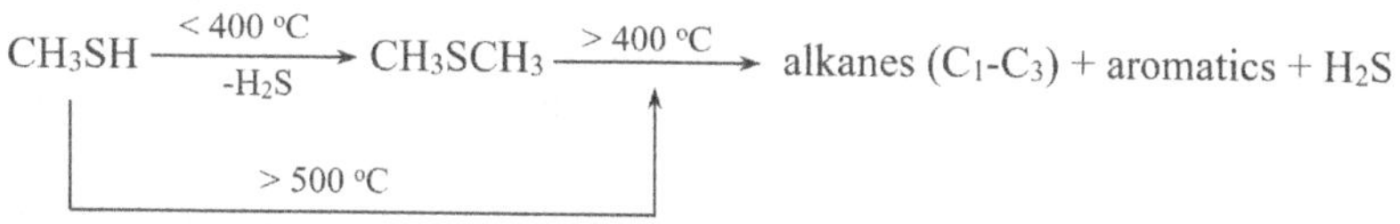

$$CH_3SH \xrightarrow[-H_2S]{< 400\ °C} CH_3SCH_3 \xrightarrow{> 400\ °C} \text{alkanes } (C_1\text{-}C_3) + \text{aromatics} + H_2S$$

$$\xrightarrow{> 500\ °C}$$

SCHEME 8.1 Methyl mercaptan conversion over H-ZSM-5 zeolite (modified from Huguet et al., 2013).

Baltrusaitis et al. (2016) conducted comparative DFT calculations between MeOH and MeSH coupling on Brønsted acid sites of chabazite as a model catalyst assuming the transformation of both reactants to ethylene has identical fundamental mechanism. With few exceptions, catalytic transformations of CH_3SH are of higher free energy when compared to those of CH_3OH. The trimethylsulfonium ion, TMS, being isostructural with the trimethyloxonium ion, TMO, has shown to be a key reactive intermediate and a thermodynamically stable species leading to ethylene formation (Baltrusaitis et al., 2016).

8.3.3 Catalyst Deactivation

The M2TH typically occurs at relatively high temperatures up to 600°C (about 200°C higher than MTH). This accelerates deactivation of HZSM-5 catalyst within 15 h on stream (He et al., 2018). The strong acid sites of pure HZSM-5 could accumulate coke. It has been found that rare-earth (RE) metals can regulate acid-base properties of HZSM-5 catalysts as they are somewhat basic. RE metals can reduce acidity and enhance basicity of the zeolite thereby increasing catalyst lifetime. In a series of RE (such as La, Ce, Pr, Ni, Sm, Y, Er) modified HZSM-5 (Si/Al = 20) catalysts prepared by incipient-wetness impregnation, the incorporation of the RE significantly improved activity and stability in CH_3SH decomposition with the best results obtained over La(13)/ZSM-5 (He et al., 2018).

The addition of 2% water to the gas feed drastically reduced coke formation and increased the catalyst lifetime (Hulea et al., 2014).

8.3.4 CHALLENGES

The conversion of methane by sulfuration needs hydrogen in conversion of CS_2 into hydrocarbons (Eq. 8.3). This is difficult from a process point of view. Unless part of H_2 produced from partial conversion of H_2S (H_2 and S recycled), independent hydrogen source is necessary which is removed and lost as water by a Claus process. These problems are eliminated by using CH_3SH, but there is no simple route for its production form methane (Anderson, 1989).

Even with metal-promoted HZSM-5 catalysts, the selectivity to methane is unsatisfactorily high. The methane should be recycled and pass the whole conversion steps.

8.4 CONCLUDING REMARKS

Despite their potential advantages over syngas routes such as milder process conditions, the sulfur and halogen substituted methane routes for natural gas conversion are far less extensively studied in academia and industry and thus much less developed than syngas-based processes.

In addition to what is presented in this chapter, a common problem to both halogen and sulfur-based routes is environmental and health concerns. The reactions involved by halogen and sulfur mediated conversions are generally considered stoichiometric rather than catalytic. Consequently, for large-scale operations, large volumes of sulfur and halogenated compounds should be handled which are mostly highly corrosive and/or poisonous. Thus, corrosion resistant materials such as high-alloys might be necessary. The contamination of the product with these impurities increases the separation costs.

Therefore, much research and development are necessary to develop the commercially viable processes.

NOMENCLATURE

BTX	benzene, toluene and xylenes
CAPEX	capital expenditure
CTAB	cetyltrimethylammonium bromide
DME	dimethyl ether
DMS	dimethyl sulfide
GTL	gas to liquid

HP	hydrocarbon pool
M2TH	methyl mercaptan to hydrocarbons
MeXTH	methyl halide to hydrocarbons
OPEX	operational expenditure
RE	rare earth
SDS	sodium dodecyl sulfate
TMO	trimethyloxonium
TMS	trimethylsulfonium

REFERENCES

Anderson, J. R., Methane to higher hydrocarbons. Appl. Catal., 47 (1989) 177–196.

Argyle, M. D., Bartholomew, C. H., Heterogeneous Catalysts Deactivation and Regeneration: A Review, In Bartholomew, C. H., Argyle, M. D. (Eds.) Advances in Catalyst Deactivation, MDPI, Basel, (2016).

Baltrusaitis, J., Bučko, T., Michaels, W., Makkee, M., Mul, G., Catalytic methyl mercaptan coupling to ethylene in chabazite: DFT study of the first C-C bond formation, Appl. Catal. B. Env., 187 (2016) 195–203.

Cammarano, C., Huguet, E, Cadours, R., Leroi, C., Coq, B., Hulea, V., Selective transformation of methyl and ethyl mercaptans mixture to hydrocarbons and H_2S on solid acid catalysts, Appl. Cat. B. Env., 156–157 (2014) 128–133.

Chang, C. D., Silvestri, A. J., The conversion of methanol and other O-compounds to hydrocarbons over zeolite catalysts. J. Catal., 47 (1977) 249–259.

Chang, C. D., Lang, W. H., Hydrocarbon synthesis, US Patent 4,480,143 (1984).

Chang, C. D., Process for producing liquid hydrocarbon fuels, US patent 4,543,43 (1985).

Chen, H., Chen, T., Chen, K., Fu, J., Lu, X., Ouyang, P., Catalytic aromatization of methyl bromide to aromatics over bimetallic CuO-ZnO/HZSM-5, Cat. Commun., 103 (2018) 38–41.

Fickel, D. W., Sabnis, K. D., Li, L. Y., Kulkarni, N., Winter, L. R., Yan, B. H., Chen, J. G. G., Chloromethane to olefins over HSAPO-34: probing the hydrocarbon pool mechanism, Appl. Catal. A. G., 527 (2016) 146–151.

Gamero, M., Aguayo, A. T., Ateka, A., Pérez-Uriarte, P., Gayubo, A. G., Bilbao, J., Role of shape selectivity and catalyst acidity in the transformation of chloromethane into light olefins, Ind. Eng. Chem. Res., 54 (2015) 2822–2832.

Gamero, M., Valle, B., Castaño, P., Aguayo, A. T., Bilbao, J., Reaction network of the chloromethane conversion into light olefins using a HZSM-5 zeolite catalyst, J. Ind. Eng. Chem., 61 (2018) 427–436.

He, D., Zhao, Y., Yang, S., Mei, Y., Yu, J., Liu, J., Chen, D., He, S., Luo, Y., Enhancement of catalytic performance and resistance to carbonaceous deposit of lanthanum (La) doped HZSM-5 catalysts for decomposition of methyl mercaptan, Chem. Eng. J., 336 (2018) 579–586.

Huang, J., Wang, W., Fei, Z., Liu, Q., Chen, X., Zhang, Z., Tang, J., Cui, M., Qiao, X., Enhanced light olefin production in chloromethane coupling over Mg/Ca modified durable HZSM-5 catalyst, Ind. Eng. Chem. Res., 58 (2019) 5131–5139.

Huguet, E., Coq, B., Durand, R., Leroi, C., Cadours, R., Hulea, V., A highly efficient process for transforming methyl mercaptan into hydrocarbons and H_2S on solid acid catalysts, Appl. Catal. B. Env., 134–135 (2013) 344–348.

Hulea, V., Huguet, E., Cammarano, C., Lacarriere, A., Durand, R., Leroi, C., Cadours, R., Coq, B., Conversion of methyl mercaptan and methanol to hydrocarbons over solid acid catalysts – a comparative study, Appl. Catal. B. Env., 144 (2014) 547- 553.

Ibáñez, M., Gamero, M., Ruiz-Martínez, J., Weckhuysen, B. M., Aguayo, A. T., Bilbao, J., Castaño, P., Simultaneous coking and dealumination of zeolite H-ZSM-5 during the transformation of chloromethane into olefins, Catal. Sci. Technol., 2 (2016) 296–306.

Kong, L., T., Shen, B. X., Jiang, Z., Zhao, J. G., Liu, J. C., Synthesis of SAPO-34 with the presence of additives and their catalytic performance in the transformation of chloromethane to olefins, Reac. Kin. Mech. Catal., 114 (2015) 697–710.

Kong, L. T., Jiang, Z., Zhao, J., Liu, J., Shen, B., The synthesis of hierarchical SAPO-34 and its enhanced catalytic performance in chloromethane conversion to light olefins, Catal. Let., 144 (2014a) 1609–1616.

Kong, L. T., Shen, B. X., Zhao, J. G., Liu, J. C., Comparative study on the chloromethane to olefins reaction over SAPO-34 and HZSM-22, Ind. Eng. Chem. Res., 53 (2014b) 16324–16331.

Olah, G. A., Doggweiler, H., Felberg, J. D., Frohlich, S., Grdina, M. J., Karpeles, R., Keumi, T., Inaba, S., Ip, W. M., Lammertsma, K., Salem, G., Tabor, D. C., Onium Ylide chemistry. 1. Bifunctional acid-base-catalyzed conversion of heterosubstituted methanes into ethylene and derived hydrocarbons. The onium ylide mechanism of the $C_1 \to C_2$ conversion J. Am. Chem. Soc., 106 (1984) 2143–2149.

Olah, G. A., Gupta, B., Farina, M., Felberg, J. D., Ip, W. M., Husain, A., Karpeles, R., Lammertsma, K., Melhotra, A. K., Trivedi, N. J., Selective monohalogenation of methane over supported acid or platinum metal catalysts and hydrolysis of methyl halides over alumina-supported metal oxide/hydroxide catalysts: a feasible path for the oxidative conversion of methane into methyl alcohol/dimethyl ether, J. Am. Chem. Soc., 107 (1985) 7097–7105.

Olah, G. A., Molnár, Á., Prakash, G. K. S., Hydrocarbon Chemistry, 3rd Ed., Wiley, New York (2018) 123–325.

Olsbye, U., Saure, O. V., Muddada, N. B., Bordiga, S., Lamberti, C., Nilsen, M. H., Lillerud, K. P., Svelle, S., Methane conversion to light olefins-How does the methyl halide route differ from the methanol to olefins (MTO) route? Cat. Today, 171 (2011) 211–220.

Reina, M., Martinez, A., Cammarano, C., Leroi, C., Hulea, V., Mineva, T., Conversion of methyl mercaptan to hydrocarbons over H-ZSM-5 zeolite: DFT/BOMD study, ACS Omega, 2 (2017) 4647–4656.

Sahebdelfar, S., Yaripour, F., Ahmadpour, S., Khorasheh, F., Methanol-to-hydrocarbons product distribution over SAPO-34 and ZSM-5 catalysts: the applicability of thermodynamic equilibrium and Anderson-Schulz-Flory distribution, Iran. J. Chem. Chem. Eng., 38 (2019) 49–59.

Su, Y., Campbell, S. M., Lunsford, J. H., Lewis, G. E., Palke, D., Tau, L. M., The catalytic conversion of methyl chloride to ethylene and propylene over phosphorous-modified ZSM-5 zeolites J. Catal., 143 (1993) 32–44.

Svelle, S., Aravinthan, S., Bjørgen, M., Lillerud, K. P., Kolboe, S., Dahl, I. M., Olsbye, U., The methyl halide to hydrocarbon reaction over H-SAPO-34, J. Catal., 241 (2006) 243–254.

Tao, L., Chen, L., Yin, S. F., Luo, S. L., Ren, Y. Q., Li, W. S., Zhou, X. P., Au, C. T., Catalytic conversion of CH_3Br to aromatics over PbO-modified HZSM-5, Appl. Catal. A. G., 367 (2009) 99–107.

Taifan, W., Baltrusaitis, J., CH_4 conversion to value added products: Potential, limitations and extensions of a single step heterogeneous catalysis, Appl. Catal. B.: Env., 198 (2016) 525–547.

Taylor, C. E., Noceti, R. P., Schehl, R. R., Direct Conversion of Methane to Liquid Hydrocarbons Through Chlorocarbon Intermediates, In Bibby, D. M., Chang, C. D., Howe, R. F., Yurchak, S. (Eds), Methane Conversion, Elsevier, Amsterdam (1988) 483–489.

Varzaneh, A. Z., Towfighi, J., Sahebdelfar, S., Carbon nanotube templated synthesis of metal containing hierarchical SAPO-34 catalysts: impact of the preparation method and metal avidities in the MTO reaction, Micro. Meso. Mater., 236 (2016) 1–12.

Wang, P., Chen, L., Xie, J., Li, H., Au, C. T., Yin, S. F., Enhanced catalytic performance in CH$_3$Br conversion to benzene, toluene, and xylene over steamed HZSM-5 zeolites, Sci. Tech., 7 (2017) 2559–2565.

Wang, P., Chen, L., Guo, J. K., Shen, S., Au, C. T., Yin, S. F., Synthesis of submicron-sized SAPO-34 as efficient catalyst for olefin generation from CH3Br, Ind. Eng. Chem. Res., 58 (2019) 18582–18589.

Wei, Y., Zhang, D., Xu, L., Liu, Z., Su, B. L., New route for light olefins production from chloromethane over HSAPO-34 molecular sieve, Catal. Today, 106 (2005) 84–89.

Wei, Y., He, Y., Zhang, D., Xu, L., Meng, S., Liu, Z., Su, B. L., Study of Mn incorporation into SAPO framework: synthesis, characterization and catalysis in chloromethane conversion to light olefins, Micro. Meso. Mater., 90 (2006a) 188–197.

Wei, Y., Zhang, D., Liu, Z., Su, B. L., Highly efficient catalytic conversion of chloromethane to light olefins over HSAPO-34 as studied by catalytic testing and in situ FTIR, J. Catal., 238 (2006b) 46–57.

Wei, Y., Zhang, D., He, Y., Xu, L., Yang, Y., Su, B. L., Liu, Z., Catalytic performance of chloromethane transformation for light olefins production over SAPO-34 with different Si content, Catal. Let., 114 (2007) 30–35.

Wei, Y., Zhang, D., Xu, L., Chang, F., He, Y., Meng, S., Su, B. L., Liu, Z., Synthesis, characterization and catalytic performance of metal-incorporated SAPO-34 for chloromethane transformation to light olefins, Catal. Today, 131 (2008) 262–269.

Xu, T., Zhang, Q., Song, H., Wang, Y., Fluoride-treated H-ZSM-5 as a highly selective and stable catalyst for the production of propylene from methyl halides. J. Catal., 295 (2012) 232–241.

Xu, T., Liu, H., Zhao, Q., Cen, S., Du, L., Tang, Q., Conversion of chloromethane to propylene over fluoride-treated H-ZSM-35 zeolite catalysts, Catal. Commun., 119 (2019) 96–100.

Zhang, D., Wei, Y., Xu, L., Chang, F., Liu, Z., Meng, S., Su, B. L., Liu, Z., MgAPSO-34 molecular sieves with various Mg stoichiometries: synthesis, characterization and catalytic behavior in the direct transformation of chloromethane into light olefins, Micro. Meso. Mater., 116 (2008) 684–692.

Zhang, Y. Y., Chen, L., Li, Y. F., Deng, Y. Q., Au, C. T., Yin, S. F., A catalytic process for the synthesis of light olefins from CH$_3$Br over Zn-modified SAPO-34 molecular sieve, Reac. Kin. Mech. Catal., 115 (2015) 691–701.

9 Outlook and Perspective

9.1 INTRODUCTION

The chemical industry has supplied the ever-increasing needs of humanity for fuels, chemicals and energy after the Industrial Revolution. It currently faces two major problems: supplying the required raw materials and energy for chemical conversions and controlling the environmental emissions (particularly carbon dioxide and heat).

The chemical industry has been mostly petroleum-based and to a lesser extent utilized natural gas and coal in the past decades. However, the fossil fuel reservoirs (oil, coal and natural gas) will be depleted in less than a hundred years (Henrich et al., 2015). Even at present, the quality of petroleum feedstock is declining as a result of inclusion of heavier oil in refinery feeds and heavier oil cuts as feedstock (Ramírez-Corredores, 2000). These cuts contain substantial amounts of heteroatoms and metal impurities making their conversion complicated and costly.

The conversion routes in petroleum-based industry are typically based on cracking of larger molecules to produce fuels and building blocks (light olefins, aromatics) for petrochemical industry. These routes are inherently with low selectivity. On the other hand, by using simple C_1 units as feedstock, the tuning of product (chain length, branching, etc.) becomes possible by using appropriate catalyst and reaction conditions. As a consequence of the preference for building a specific molecule instead of cracking, worldwide search for alternative feedstocks is underway.

The synthesis of non-petroleum-based olefins via MTO and synthetic fuels by Fischer–Tropsch synthesis are already commercialized C_1-based processes. The share of these processes in supplying the respective products should increase with availability of shale gas and hydrate-based methane in the future. Although the raw materials are still fossil fuels, the C_1 feeds produced or obtained from renewable sources would replace them in longer run.

9.2 NEW FEEDSTOCKS

The petroleum substitute should satisfy a number of criteria including economic, environmental benefits and long-term availability to ensure sustainability of the feedstock. Coal and natural gas resources are potentially much larger than that of oil, but they are fossil fuels and their resources are limited. Furthermore, their use as fuel raw materials results in a net increase in atmospheric CO_2 emissions. Therefore, there should be a transition from non-sustainable linear economy based on finite fossil fuel resources to sustainable circular economy (Figure 9.1) using renewable (solar) energy (Sheldon, 2018).

Biomass is a renewable carbon and energy resource. However, its widespread production and utilization may interfere with human food chain when crop plants are concerned. The non-edible and waste-based biomass is characterized by heterogeneity and high oxygen content. Furthermore, extensive biomass production may have

DOI: 10.1201/9781003279280-9

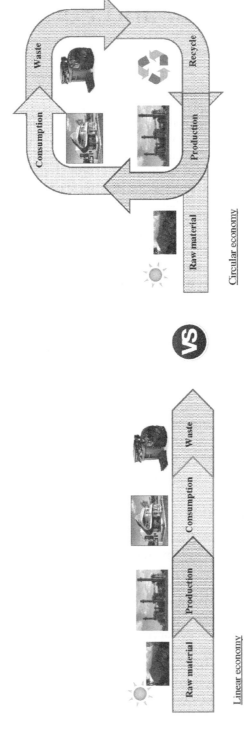

FIGURE 9.1 Linear economy versus circular economy.

ecological impacts such as deforestation. Therefore, it cannot support the total need for carbon source for production of fuels and chemicals in the future.

CO_2 is another potential carbon source with unlimited availability. However, it is highly stable and its conversion to useful chemicals needs a lot of energy that should be of a renewable source. Otherwise, with fossil fuel-based energy, the net use of fossil fuels and CO_2 emission may occur. A number of studies showed that using all potential technologies (e.g., thermocatalytic, electrochemical and renewable hydrogen), $7–9$ $GtCO_2/y$ can be avoided by 2040 which can contribute in climate change control (Aresta, 2019).

Most of the carbon sources are distributed widely and locally, only certain ones might be abundant or available.

A multisource feedstock scenario for chemical industry appears to be more realistic. However, it needs agreeing upon some platform chemical intermediates that can be easily produced from various raw materials and of sufficient activity for further transformation to more complex molecules. Simple multisource molecules such as C_1 units as building blocks are viable candidates. Among C_1 molecules, carbon monoxide and methanol satisfy these requirements.

High-energy C_1 molecules (such as CH_4 and methanol) have been used for energy storage as experienced in power-to-gas (PTG) technologies. The fluctuating renewable energies (such as solar, wind) can be stored as methane by conversion of captured CO_2 and hydrolytic hydrogen over Ni-based catalysts (Sahebdelfar and Takht Ravanchi, 2020).

9.3 ENERGY SOURCES

Along with the change of feedstock, energy sources should change from fossil fuel to renewable energies (such as solar, wind, geothermal and hydro-energy), that is, to "perennial energy" sources (Aresta, 2019).

The energy required for the conversion of high-energy carbon sources such as fossil fuels and biomass may be provided by their partial oxidation in autothermal processes. However, the production of oxygen is energy-consuming and costly and may be of a comparable or higher price than the carbon feedstock. In case of CO_2 conversion, the required energy should be provided by a high energy co-reactant such as hydrogen or by external energy sources such as heat and electricity.

9.4 METHANOL ECONOMY

Many efforts have been directed toward a non-petroleum economy such as methanol economy endorsed by Olah and coworkers (Olah, 2005). Methanol has some advantages as a platform C_1 chemical compared to CO including lower toxicity, easier transportation and a high hydrogen content. The traditional synthesis from methane via syngas is highly energy intensive due to syngas intermediate step.

Methanol economy is based on synthesis of methanol (and dimethyl ether) from captured CO_2 and renewable energy-based hydrogen which can be subsequently used as a feedstock for producing petrochemical and refining building blocks such as light olefins and aromatics.

Methanol economy is a potential route to circular economy in chemical industry. It is an anthropogenic carbon cycle that can be supplement to the nature carbon cycle via photosynthesis. However, its extent is limited to the amount of accessible renewable energy.

The first CO_2 recycling to methanol plant built by CRI (Carbon Recycling International) using geothermal energy has been in operation since 2012 in Iceland.

9.5 RESEARCH NEEDS

The reactions involving C–C bond formation from C_1 units typically show low yields due to low selectivity, low conversion (because of thermodynamic limitation) and/or unfavorable selectivity-conversion behavior. This is even the case for commercialized Fischer–Trospch synthesis by which a wide range of hydrocarbons, "syncrude", is produced. This has been a major challenge for successful commercialization of many potentially interesting C_1 conversions.

As the proceeding chapters implied, the successful supply and use of C_1 chemistry for sustainable chemical developments needs new concepts and materials for catalyst, reaction and processing among others. Thus, resource- and energy-efficient processes should be developed (Centi et al., 2013).

With the advent of new feedstocks, the traditional catalysts need to be improved or novel ones to be developed. Bifunctional (or multifunctional) catalysts could be especially effective to reduce thermodynamic constraints in equilibrium limited reactions by coupling them with another reaction as tandem reactions. This approach is most suited for coupling reactions occurring optimally under similar operation conditions (especially temperature). The components of the hybrid catalyst should be tolerant to the products of reactions under operating conditions. A well-known and potentially attractive example is direct hydrogenation of carbon oxides to dimethyl ether, which should result in higher conversion compared to methanol synthesis under similar conditions. The incorporation of zeolites in catalyst formulation can further affect the catalyst performance by introducing shape-selectivity, which is useful when a wide range of bulky molecules are potential products as in hydrogenation of carbon oxides to hydrocarbons (Zhou et al., 2019).

More efficient and cost-effective separation techniques such as membrane-based technologies should be developed for raw material purification or product separation or for *in situ* separation of one or more products in the reactor to reduce thermodynamic limitation in equilibrium-limited reactions or to protect reactive products from further transformation.

To shift from fossil fuel-based to perennial-energy driven processes, the effective use of renewable energy for production of hydrogen is another important issue. Due to large diversity of research fields, cooperative research is necessary.

Even with aforementioned improvements, which might have only marginal effects, the recycling and reprocessing of the byproducts (e.g., as in Siluria technology), as well as energy and material integration with an appropriate process, could still further improve the economy of the chemical transformation.

In addition to the conventional thermocatalytic approach, other perhaps greener alternatives are also active research topics. The direct use of solar energy in light-driven photocatalytic C_1 conversions is a green and sustainable approach, although the efficiencies are currently too low (Chen et al., 2019). Electrochemical reduction of CO_2 to oxygenates and hydrocarbons is another potential technique for CO_2 conversions. Nevertheless, the state-of-the art system exhibits low efficiency and high energy consumptions (Olah et al., 2018). Therefore, further developments are still necessary.

Finally, "as it is in nature, the concept of 'waste' must disappear from our design frameworks, such that we instead think in terms of material and energy flows" (Zimmerman et al., 2020).

REFERENCES

Aresta, M., Perspective Look on CCU Large-Scale Exploitation, Potential of large scale, Aresta, M., Karimi, I., Kawi, S. (Eds.), An Economy Based on Carbon Dioxide and Water Potential of Large-Scale Carbon Dioxide Utilization, Springer, Switzerland (2019).

Centi, G., Quadrelli, E. A., Perathoner, S., Catalysis for CO_2 conversion: a key technology for rapid introduction of renewable energy in the value chain of chemical industries, Energy Environ. Sci., 6 (2013) 1711–1731.

Chen, G., Waterhouse, G.I.N., Shi, R., Zhao, J., Li, Z., Wu, L. Z. Tung, C. H., Zhang T., From solar-to-fuels: recent advances in light-driven C_1 chemistry, Angew. Chem., 58 (2019) 17528–17551.

Henrich, E., Dahmen, N., Dinjus, E., Sauer, J., The role of biomass in a future world without fossil fuels, Chem. Ing. Tech., 87 (2015) 1667–1685.

Olah, G. A., Beyond oil and gas: the methanol economy, Angew. Chem., 44 (2005) 2636–2639.

Olah, G. A., Molnar, A., Prakash, G. K. S., Hydrocarbon Chemistry, 3rd Ed., Wiley, New Jersey (2018).

Ramírez-Corredores, M. M., Catalysis: New Concepts and New Materials, Proceeding of the 16th World Petroleum Congress, Calgary, Canada (2000).

Sahebdelfar, S., Takht Ravanchi, M., Heterogeneous Catalytic Hydrogenation of CO_2 to Basic Chemicals and Fuels, In Nadda, A., Sharma, S. (Eds.), Chemo-Biological Systems for CO_2 Utilization, CRC Press Taylor and Francis, USA (2020).

Sheldon, R. A., The road to biorenewables: carbohydrates to commodity chemicals, ACS Sust. Chem. Eng., 6 (2018) 4464–4480.

Zhou, W., Cheng, K., Kang, J., Zhou C., Subramanian, V., Zhang, Q., Wang, Y., New horizon in C_1 chemistry: breaking the selectivity limitation in transformation of syngas and hydrogenation of CO_2 into hydrocarbon chemicals and fuels, Chem. Soc. Rev., 48 (2019) 3193–3228.

Zimmerman, J. B., Anastas, P. T., Erythropel, H. C., Leitner, W., Designing for a green chemistry future, Science, 367 (2020) 397–400.

Index

A

Absorption of CO_2, 14
  chemical, 15
  physical, 15
Acetic acid, 227
  in biogas formation, 36
  by carbonylation of methanol, 227–228
Acetylene
  as a basic chemical, 90, 94
  manufacture of, 93–94
Acrylic acid, 176
Air Liquide Multi-Purpose Gasification, 32
Air separation unit (ASU), 11, 114
Aldol condensation, 147, 150, 154
Alfol process for synthesis of primary linear
        alcohols, 149
Alkali metals, in OCM, 111
Amino resins, 183–184
Ammoxidation of methane, 62
Anderson-Schulz-Flory (ASF) distribution
  in Fischer-Tropsch synthesis, 140
  in methane homologation, 95
Andrussow process for methane ammoxidation,
        63–64
ARGE process, 145
Aromatization
  of methane, 98–99
  of propane, 109
Atom economy, 153, 154, 180, 227
Autothermal
  acetylene production, 93
  gasification, 24
  reforming, 50
Aziridines, 167, 178, 181

B

BASF process for acetic acid synthesis, 229–231
Bifunctional catalysts, 3, 260
  in aromatization of methane, 103
  in CO to hydrocarbons, 147
  in HAS, 150
Biodegradable, 179–181
Biogas, production by fermentation, 35–36
Biomass, 33
  gasification of, 34–35
  sources, 33
  thermochemical conversion, 33–34
Bi-reforming of methane, 51

Block copolymer, 180
Bromomethane (methyl bromide)
  condensation to hydrocarbons, 245–246
  synthesis of, 56, 59
Brønsted acid sites, 54, 99–101, 107, 201–202,
        215, 234, 245–246, 251

C

C_1 building blocks, 3, 259
C_1 chemistry, 2
Calcor process, 47
CAMERE process, 77
Carbene mechanism in MTO, 205
Carbide mechanism, in Fischer–Tropsch
        synthesis, 138–140
Carbocation mechanism in MTO,
        204–205
Carbon capture and storage (CCS), 10
Carbon capture and utilization (CCU), 10
Carbon dioxide
  capturing, 10–19
  chemical recycling, 168, 260
  as a co-monomer, 165
  recycling into methanol or DME, 260
Carbon disulfide
  conversion to hydrocarbons, 248, 249
  synthesis of, 59–61, 249
Carbon neutral system, 2, 79
Carbon recycling, 260
Carbonylation of methanol, 227–228
Catalytic partial oxidation (CPOX), 49
Cativa process for acetic acid synthesis, 223
Cellulose, 33
Chabazite, 198, 203, 251
Chain growth probability, 137, 140
Chemical looping
  in methane to methanol, 55
  in OCM, 119
  in oxy-fuel combustion, 11
Chlorination of methane, 56–57
Circular economy, 37, 257–258
Claus process, 249, 252
Climate change, 10, 79, 259
CO_2-based polymers, 177
Coal
  chemical structure, 21–22
  liquefaction, 22
  types, 41
Coal-bed methane, 8

For Product Safety Concerns and Information please contact our EU representative GPSR@taylorandfrancis.com Taylor & Francis Verlag GmbH, Kaufingerstraße 24, 80331 München, Germany

Printed and bound by CPI Group (UK) Ltd, Croydon, CR0 4YY
01/05/2025
01858546-0004